Tax Reform in Developing Countries

WORLD BANK

REGIONAL AND

SECTORAL STUDIES

Tax Reform in Developing Countries

EDITED BY

WAYNE THIRSK

The World Bank
Washington, D.C.

nufactured in the United States of America
st printing December 1997

The World Bank Regional and Sectoral Studies series provides an outlet for rk that is relatively focused in its subject matter or geographical coverage and t contributes to the intellectual foundations of development operations and icy formulation. Some sources cited in this paper may be informal documents t are not readily available.

The findings, interpretations, and conclusions expressed in this publication those of the author and should not be attributed in any manner to the World nk, to its affiliated organizations, or to the members of its Board of Executive Directors or the countries they represent.

Wayne Thirsk is a director with the Barents Group, KPMG. Previously, he was professor of economics at the University of Waterloo, Canada, and a member of the World Bank's Public Economics Division. Currently he is working on intergovernmental fiscal reform in Kiev, Ukraine, under USAID auspices.

Cover design by Sam Ferro

Library of Congress Cataloging-in-Publication Data

Tax reform in developing countries / edited by Wayne Thirsk.
p. cm. — (World Bank regional and sectoral studies)
Includes bibliographical references.
ISBN 0-8213-3999-0
1. Taxation—Developing countries. I. Thirsk, Wayne R.
II. Series.
HJ2351.7.T38 1997
336.2'05'091724—dc21

97-29643
CIP

Contents

Tables

Box

Foreword

There is an old saying: "The only sure things in life are death and taxes." Perhaps the aphorism should be rephrased: "The only sure things in life are death and tax *reform*." Countries throughout the world have reformed or are attempting to reform their tax systems. In industrial countries, the impetus has come from the increasing complexity of the tax code, a narrowing tax base, and concerns with horizontal equity. The motivation in developing countries is similar, but the need is much more pressing. The tax base is already very narrow in most developed countries and improvements in the capacity of tax administration make it possible to shift away from inefficient and inequitable but administratively simple tax instruments (such as trade taxes).

The principles of tax reform, developed over the past quarter-century, are well known. But the practice of tax reform, especially in developing countries, has at times deviated from these principles and has certainly varied greatly across countries. This book attempts to learn from this experience by studying tax reform in eight developing countries, with a view toward sharpening our understanding of the principles and their application.

The eight case study countries provide a rich laboratory for studying tax reform in practice. The countries vary not only in per capita income but also in the context within which tax reform took place. They undertook different types of reform and the outcome differed. Yet, some important lessons emerge. For instance, although all countries are often forced to sacrifice the optimal tax structure in favor of a simpler, more uniform system, to conform to the administrative capacity of the country, the sacrifices can be especially large in developing countries. Another theme is that developing-country tax systems need not be overly concerned with equity, since redistribution in these countries is best carried out by public expenditures. Finally, throughout the book we see instances where obviously welfare-improving, even Pareto-improving, tax reforms are not undertaken. While the reason is the usual scapegoat, "politics," this situation illustrates a point that is too often ignored by economists: even reforms that benefit everyone may be opposed by the affected parties who perceive their bargaining power altered (and possibly reduced) in the longer run.

This book also illustrates how it is possible to learn from carefully managed case studies, as opposed to, say, cross-country regressions. The benefit of the case study is the richness of institutional and other details about the country; the cost is that it may be difficult to generalize beyond the case study. By fitting the case studies into a fairly tight analytical framework, this collection of essays proves that one can attain the benefits of the richness of institutional detail and depth of understanding and yet extract, with some confidence, lessons that may be applied elsewhere.

Although the book draws some lessons about the principles of tax reform, it leaves several questions unanswered and raises some new puzzles. Similarly, while the book indicates some general directions for policy, the case studies exemplify the importance of country-specific conditions in the design of tax reform. In short, this book—like other World Bank research products—should be of interest to researchers and policymakers alike.

Joseph E. Stiglitz
Chief Economist and Senior Vice President, Development Economics
World Bank

Contributors

Mukul Asher	Associate Professor of Economics and Public Policy, Department of Economics and Statistics, National University of Singapore
Roy Bahl	Dean, School of Policy Studies, Georgia State University, Atlanta, Georgia
Kenan Bulutoglu	Consultant, Barents Group LLC, Washington, D.C.
Francisco Gil-Díaz	Subgovernadour del Banco de Mexico, Mexico City, Mexico
Kwang Choi	Director, National Tax Institute, Seoul, Republic of Korea
Charles E. McLure, Jr.	Senior Fellow, Hoover Institution, Stanford University, Stanford, California
David Sewell	Fiscal Economist, Middle East and North Africa Region, Social and Economic Development, World Bank, Washington, D.C.
Wayne Thirsk	Director, Barents Group LLC, Kiev, Ukraine
George Zodrow	Professor and Chair, Department of Economics, Rice University, Houston, Texas

Affiliations as of July 1997.

1

Overview: The Substance and Process of Tax Reform in Eight Developing Countries

Wayne Thirsk

Beginning in the 1970s a number of developing countries tried to radically reform their tax systems. What can we learn from these experiences with tax reform? How can any lessons learned be made useful to other countries also on the road to tax reform? The World Bank has studied these questions for several years now, and the chapters in this volume present the fruits of this research. This overview chapter describes the genesis of this research project, outlines the common research framework adopted for each country, and summarizes the findings of the country-specific chapters that follow.

The goal of the tax reform project has been to obtain a better understanding of how developing countries can improve the performance of their tax systems. As the World Bank has become increasingly active in structural adjustment and other types of policy-based lending in developing countries, growing demands have been placed on it to suggest, and sometimes supervise, desirable tax reforms. The tax reform project carefully examines the experience of eight developing countries that have undergone, and in some instances are still undergoing, significant and comprehensive tax reform. Equal attention has been given to the process of tax reform, how it is implemented, and the substance or results of reform efforts. Throughout, the focus is on practical rather than theoretical aspects of tax reform.

It is hoped that the lessons learned provide a better appreciation of why some reform measures have worked reasonably well and others have not and thus will be helpful to other clients of the Bank considering reform. The results of this project may also contribute to the capacity of future Bank missions to analyze different tax systems by indicating the range of issues that need careful attention, the type of data required to reveal how well current systems are operating, and the reform options most likely to help them function better.

The first section of this chapter explains the criteria for selecting the eight tax reform episodes explored in the country-specific chapters that follow. The next section describes the organization and format used to investigate and interpret the tax reform experience of each country.

In the section on the intellectual foundations of tax reform, we see whether or not and to what extent the theory of optimal taxation has either inspired or been incorporated into the body of the reforms studied. Next, to better understand the impetus for reform, prereform conditions in several countries are set out, including the contribution of tax reform to macrostability, and the role of inflation both as a catalyst for reform and as a potential source of revenue. Next, the goals of tax reform and a number of important constraints on reform options are considered, since together these determine the choice of tax reform measures. Then the major elements of tax reform in each country are cataloged in order to profile a model tax reform design. Some observations follow on how governments must behave in order to successfully implement such a design. Finally, there is a summary of lessons learned that may apply to developing countries outside the sample.

Composition of the Tax Reform Sample

The developing countries that have either extensively reformed their tax systems in recent years or are currently doing so are Bolivia, Colombia, Indonesia, Jamaica, the Republic of Korea, Mexico, Morocco, and Turkey. However, within this sample there are two distinct groups. Members of the first group (Colombia, Indonesia, Jamaica, Korea, Mexico, and Turkey) have already carried out at least one comprehensive tax reform and are the major actors in the tax reform project, having consumed the majority of the project's resources. They also lend themselves to an ex post evaluation of tax reform and, in many cases, offer a sequence of reforms to examine.

Countries in the second group (Bolivia and Morocco) have either just completed a major reform or are in the midst of large reform efforts. Because their experience is so fresh, they are analyzed from a prospective viewpoint in which anticipated outcomes as well as actual accomplishments are considered. Whereas these countries comprise the minor slate of actors in the tax reform project, they nevertheless provide their own important lessons.

Other countries such as Bangladesh, Zaire, and Zambia were also considered but not included in the sample for the following reasons. At the time the study began Bangladesh had reached only a preliminary stage of tax reform, and the government was still mulling over which elements of a comprehensive reform program to adopt and implement. In Zaire and Zambia insufficient reform had been carried out to warrant a major study.

Tax reform in countries mentioned as major actors has already been investigated (see Gillis 1989; Khalilzadeh-Shirazi and Shah 1991; and Newbery and Stern 1987), making it possible to build on and extend the findings of earlier research. The authors in this volume revisit some of these earlier investigations and offer fresh insights on the tax reform experience of these countries. For example, in chapter 3 McLure and Zodrow highlight the importance of indexation as a tax reform measure in Colombia. However, for the minor group of countries, where reforms are much more recent in origin, there is a much thinner and relatively inaccessible body of literature to draw from. The approach in this volume parallels the earlier work of Gillis (1989), seeking to extract general lessons from the common core of tax reform successes and failures. In particular, Gillis focused on the broad requirements for successful tax reform, such as the importance of tax administration reform, while this volume emphasizes some of the more technical issues that arise in reforming tax systems.

As table 1.1 illustrates, there is remarkable diversity in the economic track record of the sample countries. While Indonesia and Korea enjoyed relatively high growth rates of per capita output from 1965 to 1990—rates that probably would be considered as implausible or unsustainable elsewhere—other countries, such as Bolivia and Jamaica, display a dismal record of economic growth at negative rates over the same period. Generally, there is a loose but positive correlation between rates of per capita output growth and both the relative level and the growth rate of gross domestic investment. Per capita output levels vary by almost a factor of ten between the richest country, Korea, and the poorest, Indonesia.

Uneven rates of economic growth are matched by equally sharp differences in average annual inflation rates. Bolivia and Mexico appear at the high end of the inflation scale, with Bolivia having undergone the agony of one of the world's few true hyperinflations. Indonesia, Korea, and Morocco, on the other hand, have been models of virtual price stability. In between, another group of countries, including Colombia, Jamaica, and Turkey, have recorded persistent double-digit rates of inflation over the last twenty-five years.

There is also noticeable variation in the proportionate size of the public sector and the manner in which the public sector is financed. In general, the relatively high expenditure countries of Jamaica and Morocco stand out from their lower-spending brethren in Asia and Latin America. Colombia, Bolivia, and Mexico have the lowest spending efforts followed by Korea and Indonesia in that order.

Table 1.2 charts changes in revenue effort and structure for 1972 and 1990, the period during which most of the major reform initiatives occurred. With the exception of Turkey, all of the countries in the sample experienced increases in their revenue effort, minimally in the case of Korea, and substantially for Jamaica. Comparing revenue structures across countries at a point in time, there is no obvious relationship between per capita income and the degree of reliance on direct taxes. Relatively poor Indonesia obtains more than half its total revenue from direct taxes whereas the richer economies of Korea and Mexico rely on direct taxes for only about one-third of their total revenue. However, the poorer the economy,

Table 1.1 Structural Characteristics of the Tax Reform Sample

Country	GNP per capita 1990 (US$)	GNP per capita average annual growth rate, 1965–90 (percent)	Total population in urban areas 1990 (percent)	Average annual inflation rate 1980–90 (percent)
Indonesia	570	4.5	31	8.4
Bolivia	630	–0.7	51	317.9
Morocco	950	2.3	48	7.2
Colombia[a]	1,260	2.3	70	24.8
Jamaica[b]	1,500	–1.3	52	18.3
Turkey	1,630	2.6	61	43.2
Mexico	2,490	2.8	73	70.3
Rep. of Korea	5,400	7.1	72	5.1

n.a. Not applicable.
GNP is gross national product; GDI is gross domestic investment; and GDP is gross domestic product.

the more likely it is that personal income tax yields will be dwarfed by collections from the corporate or company income tax. This feature reflects not only low levels of average household income but also the importance of the agricultural sector, which resists easy application of the personal income tax. In predominantly agrarian countries like Indonesia, personal income taxes yield less than 1 percent of gross national product (GNP). Unless a country is rich in oil, like Indonesia, or has higher-than-average tax rates, like Zimbabwe, corporate income taxes normally yield revenues amounting to between 1.5 and 2.5 percent of GNP. Some countries, notably Jamaica and Turkey, have successfully tapped revenues from personal incomes. In these two countries, personal income tax yields represent 5 percent or more of GNP.

In terms of indirect taxes, Morocco extracts an unusually large share of total revenue from trade taxes. This is because of relatively high duty and sales tax rates, combined with a relatively large import ratio in that country. Countries like Indonesia and Mexico draw relatively little of their revenue from trade taxes. Virtually every country in the sample has installed a broad-based sales tax of the value added variety. Depending on the level of the basic tax rate, the proceeds from the value added tax (VAT) amount to between 2.7 percent (Colombia) and 5.4 percent (Morocco) of GNP.

Over time, a number of the countries have made impressive strides in strengthening their revenue contributions from direct taxes. As a fraction of GNP, direct taxes rose by 5 percentage points in Jamaica, 4.4 in Indonesia, 2.01 in Turkey, 1.67 in Mexico, and 1.53 in Korea. Because both corporate and personal income tax rates were declining over this period, these higher revenue yields represent the payoff from a variety of efforts to broaden the income tax bases. Only in Bolivia did direct taxes decline in relation to GNP. In Colombia, despite several major reforms, direct taxes have remained roughly the same proportion of GNP.

As a share of GNP, indirect taxes on domestically produced goods and services also rose in a number of countries, increasing by over 5 percentage

GDI/GDP 1985 (percent)	Average annual growth of GDI, 1980–90 (percent)	Central government total revenue/GNP 1990	Central government expenditure/GNP 1972	Central government expenditure/GNP 1990
30	7.1	18.3	15.1	20.4
17	−10.7	15.7	9.3	18.8
22	2.6	23.8	22.8	n.a.
19	0.6	13.4	n.a.	15.1
23	4.1	30.7	22.7	n.a.
21	3.8	19.3	13.1	24.6
21	−3.4	14.9	11.5	18.4
30	12.5	15.7	18.0	15.7

a. Revenue data for Colombia are from chapter 3.
b. For Jamaica 1983–84 (see chapter 5).
Source: World Bank (1992).

points in Mexico and by over 2 points in Colombia and Jamaica. Proportionate yields fell in Morocco, where the decline was more than counterbalanced by a sharp rise in trade taxes, and stayed at more or less the same level in Bolivia, Korea, and Turkey. With the adoption of a broad-based VAT in most of these countries, sales tax yields were more important than the collections from excise taxes in every country. Apart from Jamaica and Morocco, reliance on trade taxes was either reduced or maintained at previous levels. Revenues from nontax sources, primarily the surpluses from state-owned enterprises, show little trend, except in Bolivia, where they rose sharply, and in Morocco, where they fell.

Changes in the relative importance of revenue from different sources display almost every conceivable pattern. Turkey, and to some extent Indonesia, have reduced their trade taxes and recouped the revenue loss from higher direct taxes. Bolivia has reduced direct taxes in favor of higher indirect taxes. Korea has moved in the opposite direction. Mexico has exchanged lower trade taxes for larger tax burdens on domestic output.

The eight countries sampled embrace a wide range of economic, geographic, and cultural conditions. The sample includes three countries from Latin America (Bolivia, Colombia, and Mexico), two from Asia (Indonesia and Korea), and one each from North Africa (Morocco), the Middle East (Turkey), and the Caribbean (Jamaica). Several of these countries have tax systems that depend heavily on the exploitation of natural resources, namely Bolivia, Indonesia, Jamaica, and Mexico. Some of these countries, such as Indonesia, are impressively large, while many others, such as Bolivia and Jamaica, are perhaps shining examples of small open economies.

Many of these countries have faced a similar set of fiscal challenges. Those plagued with recurrent fiscal deficits struggled to squeeze more revenue from existing sources and to develop new ones. Often these deficits fueled an inflationary environment that has warped and badly distorted the functioning of the tax system. All of them have grappled with the enormous administrative and economic complexity of taxing the capital

Table 1.2 Central Government Revenue Structures for the Tax Reform Sample
(percentage of GNP)

Country	Direct taxes/GNP 1972	Direct taxes/GNP 1990	Domestic indirect taxes/GNP 1972	Domestic indirect taxes/GNP 1990	Trade taxes/GNP 1972	Trade taxes/GNP 1990
Bolivia[b]	1.6	0.8	4.5	5.0	2.8	1.1
Colombia	3.9	3.7	1.6	3.7	2.1	2.4
Indonesia	6.1	10.5	3.0	4.6	2.4	1.1
Jamaica[b]	7.7	12.7	8.7	11.2	1.6	2.5
Korea, Rep. of	3.8	5.3	5.5	5.3	1.4	1.7
Mexico	3.8	5.4	3.3	8.3	1.4	0.7
Morocco[b]	3.0	6.1	8.5	6.7	2.4	9.7
Turkey	6.3	8.4	6.4	6.2	3.0	1.2
Average	4.9	7.1	5.4	7.0	2.5	3.3

n.a. Not applicable.
a. Earmarked social security taxes account for between 1 and 2 percent of GNP in Colombia, Jamaica, Mexico, and Morocco.

incomes generated in their economy. More than a few of them have suffered from the debilities of weak tax administration and a corresponding inability to effectively apply the legislative provisions of their tax systems. As a result, problems of tax evasion and tax avoidance abound and make the actual, or effective, tax system a vastly different revenue mechanism from the one formally described in the tax statutes. This volume examines the efforts that have been made to solve these and other fiscal problems.

The influence of tax designs imported from industrial countries is also clearly visible in a number of cases. Jamaica has emerged from a British colonial heritage, while Bolivia, Colombia, and Mexico have sprung from Spanish colonial origins. In every case the imprint of colonial history is clearly discernible. The Dutch, for example, have left an indelible mark on the public finance structure of Indonesia, as have the French on Morocco.

This volume contains the collaborative efforts of a number of researchers. Wayne Thirsk examined Bolivia. Charles E. McLure, Jr., and George Zodrow have analyzed the experience of tax reform in Colombia. The chapter on Indonesia was prepared by Mukul Asher. Roy Bahl has probed the Jamaican experience. Kwang Choi did the case study of Korea. Francisco Gil-Díaz and Wayne Thirsk collaborated on Mexico. The chapter on Morocco is the joint effort of David Sewell and Wayne Thirsk. Kenan Bulutoglu and Wayne Thirsk are responsible for the final chapter on Turkey.

The Project's Approach to Analyzing Tax Reform

Every country study has attempted to address a common set of basic issues. First, what has been the country's previous history of tax reform? Second, what were the pressures for tax reform? Why did it happen? Third, what was the process of tax reform? How did it happen? Fourth, what happened? How was the structure of the tax system changed? Fifth, what impact did the reforms have on the goals of revenue adequacy,

Other taxes/GNP		Nontax revenues/GNP		Current revenues/GNP[a]	
1972	1990	1972	1990	1972	1990
0.0	0.4	3.2	7.0	9.3	15.7
0.8	0.9	0.8	1.0	10.6	13.4
0.5	0.6	1.4	1.5	13.4	18.3
0.7	0.7	3.6	3.5	22.1	30.7
0.7	0.9	1.7	1.8	13.1	15.7
–0.9[c]	–2.7[c]	0.7	1.1	10.1	14.9
1.1	0.5	2.3	0.8	18.5	23.8
1.3	0.6	3.6	3.0	20.6	19.3
n.a.	n.a.	2.2	2.5	15.6	20.8

b. Initial date for Jamaica is 1980; that for Bolivia is 1975; the terminal date for Morocco is 1989.
c. Negative values indicate taxes collected on behalf of state and local governments.
Source: World Bank (1992).

economic efficiency, equity, and the capacity to effectively administer the new tax laws? Finally, what general lessons for tax reform can be distilled from this country's tax reform experience?

It is fairly easy to describe what happened and, in most cases, why. By far the hardest task is evaluating the success or merit of a particular tax reform episode. In part, the difficulty stems from the delicate balancing act required of tax systems that serve several important but often competing objectives. A large head tax, for instance, may be efficient and relatively easy to administer, but it completely neglects fairness or ability-to-pay considerations, making it politically unacceptable. Tradeoffs among competing goals are inevitable in any tax reform exercise and, for the most part, the country studies accept the tradeoffs policymakers have made and judge success or failure according to whether or not a reform has achieved its stated objectives.

A second source of difficulty in appraising a tax reform episode is the inability to measure its effect with great precision. Weak databases are the norm in most of these countries. At best, existing data sources support only partial equilibrium analysis of the effects of tax reform on efficiency and equity. In none of the countries studied was it possible to undertake a sophisticated general equilibrium analysis that might reveal how a tax reform shock interacts with the whole economy. Even if the data permitted construction of a general equilibrium model, doubts would still remain about whether the appropriate model had been used.

A final source of difficulty is conceptual, rather than data or model related, and has to do with the fact that there is more than one theoretically acceptable approach to measuring tax incidence and the efficiency benefits or costs of different tax reforms. Consider, for example, the alleged regressivity of sales taxes. If, as has traditionally been done, tax burdens are related to the distribution of annual incomes, sales taxes will appear to be regressive because higher-income households save a larger fraction of their incomes and can avoid paying sales tax by not consuming their income today. Eventually, however, income saved today will be spent tomorrow and will incur a sales tax burden when that happens. Over a longer time horizon, when tax rates are constant and there are no bequests, higher-income households only postpone paying sales taxes by saving, and the present value of taxes paid is the same whether income is spent now or later.[1] Therefore, over a household's lifetime the distribution of sales tax burdens is likely to be proportional rather than regressive.

Nowhere is this tension between the annual and lifetime approaches to measuring incidence more evident than in the contrasting policy conclusions of Kwang (chapter 6) and Gil-Díaz and Thirsk (chapter 7). Kwang adopts the annual basis for measuring indirect tax burdens in Korea and, given Korea's heavy reliance on indirect taxation, concludes that not only is the Korean tax system highly regressive but that future reforms should aim for greater progressivity. Gil-Díaz and Thirsk, on the other hand, employ a lifetime perspective in measuring the distribution of Mexico's indirect tax burdens and judge Mexico's indirect tax burdens to have had a progressive impact on income distribution.

Appraising the efficiency consequences of tax reform also raises questions about the most appropriate framework. A neutral tax system, defined as one that avoids interfering with the resource allocation decisions of the private sector, has served as a long-standing benchmark in the tax reform literature. Neutrality is generally thought to require the creation of broad tax bases and the imposition of more uniform tax rates. In the last few decades, however, this neutrality criterion has been assailed in the optimal tax literature, which generally finds that neutrality is an efficient tax rule only under exceptional circumstances and that most efficient tax regimes will involve departures from neutrality. Below, we look at this important issue more carefully.

The Intellectual Foundations of Tax Reform

Perhaps the best illustration of the modern theory of tax reform is the volume by Newbery and Stern (1987), which analyzes tax reform within the normative framework provided by the theory of optimal taxation. For any revenue objective, optimal tax reforms in this context strive to maximize an explicit social welfare function that balances vertical equity gains against tax-induced losses in the efficiency of resource allocation. Optimal tax reforms seek a pattern of tax rates on different tax bases that will minimize the efficiency cost of taxation while respecting society's aversion to extensive income inequality. An interesting and important feature of optimal tax reforms is that they almost never endorse a uniform pattern of tax rates for either direct or indirect taxes.

Nonuniform taxation can contribute to both the equity and the efficiency goals of taxation. Lower tax rates on the incomes earned by the poor and on the commodities they purchase are the hallmarks of a fair tax system. The absence of income support programs and the minor role of personal income taxes, for example, make a multiple rate VAT, with lower tax burdens on the poor, more attractive. On efficiency grounds, higher tax rates are warranted on commodities and resources that have inelastic supply or demand features just as lower tax rates are appropriate for either commodities or resources having elastic supply or demand characteristics.

Neutrality in this second-best framework is properly analyzed in terms of quantities rather than prices or tax rates. For example, most tax bases have components that differ in their substitutability with exempted goods or resources. A neutral or efficient tax reform in this context would map a system of nonuniform tax rates that induced a uniform degree of shifting between exempt and nonexempt elements of the tax base. Although such a prescription imposes information requirements that may lie well beyond the ability of many developing countries to satisfy them, Newbery and Stern (1987) argue that often the data are available and can be used to identify optimal commodity tax reforms, as in the application of the theory to India.

There is little doubt that optimal tax theory has made tax reformers more aware of the importance of improving economic efficiency in taxation. Unlike earlier pleas for broader-based taxation aimed primarily at achieving greater progressivity, more recent recommendations for broader-based taxes aim at achieving lower marginal tax rates and associated reductions in the level of tax-induced distortions. Another aim of base broadening is to reduce, if not entirely eliminate, nonneutral tax treatments that raise the efficiency cost of taxation by encouraging resource shifts to relatively low tax activities. Prior to reform many countries have applied income taxes that differentiate, for example, between capital and labor incomes, between different types of capital and labor incomes, and between different organizational forms and sectors of the economy. Broadening income tax bases and removing various kinds of discriminatory income tax treatment makes a tax system more efficient, although further departures from this norm could in principle promise even larger efficiency benefits. A tax reform achieving greater neutrality can make things better on efficiency grounds, even though it may fall some distance short of achieving the best possible result.

Neutrality is especially important for business-related taxes that affect investment decisions. One possible caveat is that this search for greater neutrality will overlook the untaxed or hard-to-tax housing sector and create a misallocation of investment between housing and other sectors of the economy. While this is true, the information needed to correct the situation—the complementarity or substitutability between housing and other sectors—is normally unavailable to policymakers. Moreover, in a simulation study for the United States, Auerbach (1989) found that the efficiency cost of uniform taxation of all types of capital income was small.

Many of the policy recommendations emerging from the optimal tax literature have been enthusiastically embraced by these countries. Among the most important is the need to avoid commodity taxes on intermediate inputs and to instead gear these taxes to final consumption. The VAT reforms have been designed to do just that. Optimal taxation considerations also urge the use of lump-sum taxes wherever possible. New methods of presumptive taxation adopted in a number of countries, particularly Turkey, have many attributes of efficient lump-sum taxes, as argued by Tanzi and Casanegra de Jantscher (1987). Also, consistent with the observation that capital is more mobile than labor, most of these countries have adopted a schedular approach to income taxation in which capital incomes are taxed more lightly than labor incomes. Finally, relatively high excise tax rates on the inelastic demand for alcoholic beverages, petroleum products, and tobacco are entirely consistent with the results of optimal commodity taxation, as well as, with the notion that excise taxes should be used to price the effects of harmful externalities.[2]

As Slemrod (1990) has noted, however, optimal tax theory assumes perfect tax administration, an assumption analogous to the physicist's simplifying assumption of a frictionless world. Once the task of implementing a tax reform and enforcing the collection of taxes is recognized, there is less appeal for a tax system having a highly differentiated structure of tax rates because it could seriously compromise the ability of tax authorities to administer taxes effectively. With imperfect administration, unintended tax rate differentials may arise that impair, rather than contribute to, the efficient functioning of the revenue system. A more complete policy framework recognizes the resource cost of implementing tax measures. The proper focus of attention for tax reform, as suggested by Slemrod (1990), is not optimal taxes but rather optimal tax systems that incorporate important constraints on administrative capacity. Burgess and Stern (1993) reach a similar conclusion, urging the blending of theoretical insights from formal models of tax reform with the results from less formal approaches to reform in developing countries that recognize the problems of tax evasion and the existence of a large informal economy.

A more recent reflection by Tanzi (1992) on the contribution of optimal tax theory in guiding the direction of tax reforms argues that optimal tax theory inappropriately neglects the informational, administrative, and political requirements needed to implement it. Because of this oversight, tax reformers have taken few of their cues from the normative analysis of optimal taxation. Thus the case for achieving a more uniform pattern of taxation that has inspired the tax reform experiences explored in this volume rests primarily on practical rather than theoretical considerations.[3]

All of the country studies adopt the rule of thumb that the cause of economic efficiency is best served by a less differentiated pattern of effective income tax rates across different sectors and activities in the economy. Most of the observed nonuniformity prior to reform has been found to be nonoptimal and highly distorting. Greater uniformity of tax rates may also enhance the horizontal equity attributes of the tax system as similar incomes will face similar tax burdens regardless of source. In some cases measurements of marginal effective tax rates (METRS) have been used as a diagnostic tool in determining either the efficiency attributes of various reforms or the pattern of prereform investment distortions. The METR tries to take account of all investment-related taxes and tax provisions in estimating how much of a marginal investment's before-tax return is absorbed by taxes.[4] According to this approach, the efficiency goal of tax reform requires a reduction in the dispersion of METRS across different sectors and investors in the economy while a smaller aggregate METR indicates success in reducing tax-induced distortions in the intertemporal allocation of resources. In their investigation of Colombia's tax reform, McLure and Zodrow have been most thorough in applying METR analysis to assess the efficiency effects of reform. Although less detailed, the studies of Morocco and Turkey also provide some METR analysis of tax reform.

In some cases tax incentives and the use of debt finance combine with inflation to generate significantly negative METRS. Such instances invariably signal inefficient resource allocation, since the after-tax rate of return exceeds the pretax rate of return and what is privately profitable is in fact socially unprofitable.[5] In other cases, however, the authors have been forced to consider qualitative evidence on the impact of tax reforms and have relied upon good judgement rather than explicit measurement to determine whether the direction of any change in taxes is likely to be efficiency enhancing.

Throughout, METR calculations also emphasize different institutional conditions in each country (for example, the importance of the curb market as a marginal source of finance in Korea), and do not lend themselves to useful intercountry comparisons. Moreover, nontax distortions are rarely considered so it is even more difficult to extract robust conclusions about the allocative impact of neutral tax measures than these exercises might indicate.

Prereform Economic Conditions

Prior to reform policymakers in most countries received two different kinds of signals indicating the need to overhaul their tax systems. Sometimes, as in Jamaica and Morocco, the combination of complex tax laws and weak tax administration has inadvertently produced a tax system characterized by high marginal tax rates applied to relatively narrow tax bases. In other countries, such as Bolivia and Turkey, the failure of the tax system to generate adequate revenues for planned expenditures has produced a climate of fiscal crisis, the most obvious symptom being persistent inflation, itself a form of implicit taxation. Below, we examine each impetus for tax reform.

Narrow Tax Bases and High Marginal Tax Rates

When separate tax regimes exist, nimble taxpayers find ways to play them against each other and minimize their tax burdens. For example, if labor income is taxed more heavily than capital income, small businesses will take their compensation in the form of capital income. Moreover, complex tax laws that make numerous fine distinctions and apply a multiplicity of tax rates to a given base invite taxpayers to engage in tax avoidance and tax evasion schemes that inevitably diminish the size of the tax base and produce pressures to raise tax rates in order to preserve revenue levels.

As the chapter on Morocco graphically illustrates, the complexity of tax laws and the incompatibility of different tax regimes severely undermine the effectiveness of the revenue system. Prior to its reform initiatives, Morocco simply had too many different taxes, too many rates, too many exemptions and fine distinctions, and too many different tax regimes. Neither taxpayer nor tax collector could easily determine the taxes owed in a particular set of circumstances. High marginal rates and narrow bases reinforced each other. Extensive initial exemptions required high marginal rates to meet revenue goals. Moreover, relatively high tax rates led to irresistible political pressures for a host of tax incentives and tax exemptions, which further eroded the size of the base.

Moroccan tax policy had neglected the myriad channels for earning and spending income in a market economy and the relative ease of diverting activity from one earning and spending channel to another in response to tax differentials. Savers, for instance, can invest either at home or abroad, depending upon the relative tax burden in each location. Many foreign firms and even some domestic ones operate at a global level and can take advantage of worldwide after-tax income opportunities. Many households—in particular, owners of small businesses and employees of large businesses—can structure their compensation in the form of either labor income or capital income according to the accompanying tax advantages. Workers may also work either on their own account or for others. Investors can invest in assets either directly or indirectly through a corporation or a financial intermediary, and they can receive their capital incomes in a variety of different forms. Firms can choose among several alternative modes of financing for their business activity as they search for the lowest-cost financial structures. Finally, within a household or a group of affiliated firms income can often be transferred from a high-tax member of the unit to a low-tax unit in an effort to reduce tax burdens.

Before they adopted a VAT, Bolivia, Indonesia, Korea, Mexico, and Turkey all relied on a raft of distorting excise taxes for indirect tax revenue. These excises were frequently tied to the purchase of productive inputs and led to significant cascading of taxes and distortion of trade and investment decisions.

In several of these countries large portions of capital income are excluded from the income tax base. Korea's personal income tax base may contain only one-third of the capital incomes received by taxpayers. That for Bolivia was less than 20 percent prior to reform in the mid-1980s. However, in both countries about three-quarters of labor income is subject to personal income taxation.

A vital issue for tax reform is whether tax policy, for equity reasons or otherwise, distinguishes among different income-earning and -spending options. If fine distinctions are made in a country's tax laws they inevitably lead to nonuniform tax treatment and a plethora of incompatible tax regimes. Households and firms in turn react to this differentiated tax treatment by shifting resources from taxed to untaxed areas, and from heavily to lightly taxed activities and spending objects, with a consequent diminution in total revenue, efficiency of resource allocation and equity. This was Morocco's situation as it stood on the brink of tax reform. The challenge facing Morocco and other countries has been to make their tax systems broader in application, simpler in administration, and make allocation less distorting.

Inflation and Fiscal Deficits: Tax Reform and Macrostability

Real resources are transferred to the public sector whenever governments spend. Taxes depress private sector spending to make this transfer possible. When tax revenues are insufficient to finance government spending, the transfer may occur through public sector borrowing, which, if it persists, often triggers a macroeconomic imbalance that must ultimately be corrected. Bolivia, Mexico, and Turkey all experienced persistent fiscal deficits during the 1980s, which heightened their awareness of the need to raise more revenue through the tax system and restore macroeconomic stability. In several countries the source of the deficit was a boom in natural resource exports (oil in Bolivia and Mexico, phosphates in Morocco) that prompted higher levels of public sector spending that could not be sustained. When the export boom faded, expenditures could not be curbed quickly enough to avoid sizable budget deficits.

The type of macroeconomic distress caused by a fiscal deficit has depended on the extent to which either domestic savers, foreign savers, or the central bank were the source of finance. During the 1970s, Mexico, along with many other developing countries, financed its fiscal deficit through external borrowing, implying a potential sacrifice of future consumption to service the foreign debt. So countries such as Mexico, which failed to spend prudently, subsequently encountered debt-servicing problems. When foreign loans dried up in the early 1980s Mexico turned to selling public bonds and imposing a severe squeeze on credit to the private sector in order to finance its deficit. The resulting high real interest rates dampened investment spending as government spending crowded out the private sector's use of investment resources.

A fiscal deficit may also be financed by an implicit inflation tax that curbs domestic consumption. In this case governments gain command over resources through access to central bank finance and their ability to monetize the fiscal deficit. The ensuing inflation forces households to reduce consumption as they struggle to maintain the real value of their monetary balances. Turkey, and particularly Bolivia, relied heavily on this method of finance in the 1980s.

In a fractional reserve banking system, where most of the monetary liabilities are matched by private sector debts, the base of the inflation tax consists of the reserves held by the banking system in addition to the cash held by households. The amount of revenue generated by the inflation tax is a function of the demand for base- or high-powered money, the economy's growth rate and the elasticity of the demand for real monetary balances with respect to inflation and income growth.

In most developing countries the high-powered money base, as a fraction of gross domestic product (GDP), seldom moves out of the 10 to 20 percent range. If this fraction were equal to 10 percent, for example, an economy that grew at a real rate of 4 percent would obtain 0.4 percent of GDP by financing the budget deficit through the noninflationary printing of money at a growth rate of 4 percent. At a faster rate of money growth, inflation would occur and, if the ratio of the monetary base to GDP remained constant, the amount of revenue collected at different inflation rates could be easily determined. For example, at an inflation rate of 20 percent the government could finance an extra 2 percent of GDP through this use of seigniorage.

However, households will attempt to avoid the inflation tax by reducing their demand for real balances as the inflation rate increases. Eventually more inflation may generate smaller real revenue yields. The historical record indicates that average rates of seigniorage are about 1 percent for industrial countries and less than 2.5 percent for developing countries.[6] However, even the higher rate for developing countries may not be sustainable unless the economy is growing rapidly. Excessive reliance on seigniorage may eventually produce a hyperinflation as occurred during 1984–85 in Bolivia, where the annual inflation rate peaked at 12,000 percent. At this lofty inflation rate, Bolivian households reverted to barter and financing their transactions in foreign currencies.

An abundance of convincing empirical evidence suggests that the efficiency cost of the inflation tax is extremely high.[7] This high cost springs from the large relative price distortions that inflation almost always imposes on the economy in general and on the foreign exchange and capital markets in particular. Negative real interest rates and chronically overvalued exchange rates are highly disruptive features of inflation that seriously interfere with the ability to achieve an efficient allocation of resources and high rates of economic growth. Inflation in an unindexed tax system also tends to create a haphazard pattern of effective tax rates in the economy that may distort the market's ability to efficiently allocate resources.

The equity dimensions of the inflation tax add to its unattractiveness, since more sophisticated and wealthier investors are better able to avoid its impact by switching their investment portfolios to a variety of inflation hedges such as real estate. The analysis of Mexican inflation indicates that it is the poor who bear most of the burden of the inflation tax. For these reasons, tax reform measures that raise total revenue and reduce reliance on the inflation tax promise to improve both the equity and efficiency performance of the public sector.

The Shifting Goals of Tax Reform

There is growing recognition of the high degree of complementarity that exists between the efficiency advantages of neutrality and the other tax reform goals of horizontal equity and greater simplicity. If tax bases are broader and tax rates are lower and more uniform, horizontal equity is easier to attain because households in similar economic circumstances before taxation will be more likely to maintain their relative economic positions after taxes have been applied. Lower tax rates also reduce the rewards for rent seeking, evasion, and tax incentive relief, thus improving the odds that equals will be treated equally under the tax system. Moreover, if differential tax treatment is ruled out and numerous distinctions among different taxpayers and different sources of income are eliminated, it is possible to simplify tax administration, improve compliance, and reduce evasion. Thus the tax reform initiatives in Colombia, Indonesia, Jamaica, Mexico, and Turkey have all stressed the basic harmony that exists among the objectives of economic neutrality, horizontal equity, and tax simplification.

Does this reordering of tax reform priorities mean that vertical equity considerations no longer matter in reforming tax systems? The correct answer to this question seems to be both yes and no. To a certain extent the goal of vertical equity has been redefined and transformed into the objective of preventing taxes from making the poor even poorer. Most of the countries studied, for example, have tried to set exemption levels high enough to remove poor households from the income tax rolls. At the same time, many countries have attempted to either remove or reduce the burden of value added taxes on the poor by either exempting or zero rating major components of food consumption by the poor.

But there also seems to be increasing acceptance of the notion that the tax system as a whole may be an extremely poor vehicle for redistributing income. In none of the countries studied has a prereform schedule of progressive tax rates successfully generated a progressive distribution of tax burdens. If developing countries are unable to make personal income taxes work effectively (because of tax evasion, accounting difficulties, long collection lags, and other enforcement problems), it may be the better part of wisdom to aim instead for the achievement of an efficient indirect tax system and use its revenues to finance targeted subsidies for the poor. Ultimately it is fiscal rather than tax incidence that affects people's welfare.

The goals of tax reform may be much more modest than they were before but they are also more realistic. Tax systems may work better if the aim is to achieve a measure of rough justice rather than some ideal, but administratively hopeless, objective. Simpler tax rules may ignore the fine distinctions that some equity considerations demand, but they also serve the broader interest of tax equity by encouraging better compliance with tax laws and by making tax evasion more difficult. For example, the policy of excluding dividends from the personal income tax, as currently practiced in Colombia, Indonesia, and Mexico, is preferable to the conceptually more accurate but administratively more complex imputation approach as a method of integrating corporate and personal income taxes. With imperfect administration, exclusion may yield better results than a more refined imputation procedure.

Similarly, it could be argued that a single rate of personal income tax combined with a generous personal exemption, as in Jamaica, operates more satisfactorily and imparts greater progressivity to the tax system than an alternative system with sharply progressive tax rates and a variety of deductions and credits. Streamlining deductions, disallowing deductions for losses against unrelated income, and imposing thin capitalization rules for domestic and foreign corporations all attempt to tailor the tax system to fit existing administrative capacities. Although these measures and others like them inevitably create some inequities by ruling out fine distinctions among taxpayers, they may also curb larger inequities arising from tax evasion and avoidance and are thus considered to be the lesser of two evils. Simpler measures may succeed where more sophisticated measures would flounder.

Constraints on Tax Reform

While tax reform debates are frequently fueled by several normative criteria, the actual reform measures that are adopted often reflect the existence of four binding constraints on the ability to choose new tax directions: political, international, technical, and institutional. Political constraints come in a variety of shapes. In most developing economies certain sectors and activities enjoy a privileged tax status from political protection that is strong enough to resist any attempt at change. In Colombia, for example, repeated efforts to include income from cattle raising in the income tax base have met with a singular lack of success. Other countries invariably have their own set of "sacred cows."

Apart from vested interests, some political constraints may be a natural outgrowth of a federal tax structure. Constitutional provisions which assign certain tax fields to subfederal levels of government may require that federal tax reform initiatives accept as given some of the tax decisions made by lower-level governments. The restrictions on choice implicit in a federal tax system raise larger issues of how to design a consistent set of tax reforms when tax changes can only be accomplished in a piecemeal fashion.

Comprehensive tax reforms, in which all or many of the components of a country's tax system are adjusted simultaneously, are relatively rare. It is more common to reform individual parts of the tax system, since the political process in most countries is better equipped to deliver marginal rather than wholesale tax reform. The ability to introduce useful reform to one part of a country's tax system in isolation from the rest of the system turns on the extent of any interactions between the two parts. It may be possible, for instance, to effect significant indirect tax reforms without reference to the direct tax system because of the absence of important repercussions on any direct tax. On the other hand, reforms occurring within the set of direct and indirect taxes will ordinarily require some coordination if consistent outcomes are to be realized.

The open economy, in particular the increasing globalization and integration of the world's capital markets, imposes severe constraints on the ability of developing economies to tax the foreign source capital income of residents and the domestic source capital income of both residents and nonresidents. A number of industrial countries, including the United States, have repealed withholding taxes on nonresident interest incomes and indirectly encouraged capital flight from developing countries that attempt to include interest income in their personal tax base. The combination of highly mobile financial capital and foreign tax havens seriously compromises the capacity of developing countries to reach residents' capital income and satisfactorily apply the residence principle—inclusion of worldwide income in residents' personal tax bases.

If interest income cannot be effectively taxed in an open economy, there is an unsettled question of whether interest expense should be allowed as a deduction from the taxable incomes of individuals and companies. Failure to disallow interest expense as a deduction creates opportunities for tax arbitrage in which investors borrow heavily to invest in either tax-exempt or tax-preferred assets, a practice referred to as "back-to-back loans" in Mexico and "round-tripping" in Indonesia. In both cases taxpayers make deposits to financial institutions conditional upon obtaining equal-valued loans from them. Such paper transactions only enrich the taxpayer at the treasury's expense and contribute nothing to the economy. However, disallowance of interest deductibility to forestall arbitrage activity may place purely domestic companies at a significant tax disadvantage in their efforts to be competitive with either foreign or domestic multinationals who can obtain tax-deductible offshore financing. Disallowance could also jeopardize the availability of a foreign tax credit. The most appropriate tax policy to pursue is not obvious under these circumstances.

Corporate tax design encounters different kinds of tradeoffs. Unless a developing country aligns its nominal corporate rate with those found in industrial countries large portions of its corporate tax base may be shifted to lower tax rate jurisdictions. At the other extreme, efforts to stimulate investment by reducing nominal (and effective) tax rates may be frustrated by foreign tax credit mechanisms which translate host country tax relief into higher home country taxes. Similarly, the attempt to introduce a more neutral cash flow business tax may incur the risk of being deemed noncreditable in those capital-exporting countries that offer a foreign tax credit to their multinationals. Finally, the introduction of creditable minimum taxes on companies may likewise run afoul of creditability restrictions and sacrifice revenue otherwise obtained without cost from foreign treasuries.

An important technical constraint on the design of tax policies is the desirability of assigning particular tax instruments to the service of specific tax goals. Clearly all of the instruments of taxation will operate to raise more revenue but some of them may have a comparative advantage over other instruments in achieving the important nonrevenue objectives of tax policy. Tariffs, for example, have the distinct nonrevenue objective of providing protection for domestic industry and therefore should not be generally used for revenue purposes as long as other instruments are available. Another common presumption is that broad-based sales taxes do a better job of raising revenue in a neutral fashion than any form of direct taxation. Direct taxes, on the other hand, usually allow greater scope for achieving some of the vertical and horizontal equity objectives of taxation than most forms of indirect tax.

Although in principle all of a country's tax instruments should be applied in some form and with some concern for the matching of instruments and objectives, these targeting considerations may have to be modified in the face of serious limitations on administrative capacity. Because of rampant tax evasion, enforcement difficulties, and a variety of accounting and measurement problems, direct income taxes are much harder for developing countries to successfully apply than indirect consumption taxes. Within the group of indirect taxes, tariffs are normally easier to administer than nontrade taxes in view of the limited number of physical locations for assessing tariffs while excise taxes are generally easier to operate than sales taxes for the same reason. An important feature of all the reforms studied is that they have invested heavily in relaxing administrative constraints and improving administrative capacity whether that be in the form of revenue collection by banks (Bolivia and Mexico), reduced filing and more extensive withholding (Colombia), or greater reliance on presumptive methods of taxation (Bolivia and Turkey).

The Pattern of Tax Reform in Individual Countries

As table 1.3 shows, most of the countries in the tax reform sample have reduced rates on both corporate and personal incomes and have succeeded in aligning their corporate rate within the top bracket personal rate. This alignment facilitates integration of the personal and corporate income taxes by means of a dividend exclusion (Bolivia, Colombia, Indonesia, Korea, and Mexico) and makes it less attractive for taxpayers to change the form in which they receive compensation or the type of organizational vehicle (partnership versus corporations) which is used to earn capital income on their behalf. Although all countries have moved in this direction, Jamaica has gone the farthest by adopting a single tax rate applicable to both personal and corporate incomes.

Broader income tax bases. A wide assortment of measures have been employed to create broader personal and corporate income tax bases. As shown in tables 1.4 and 1.5, these measures include reduced reliance on tax incentives, fewer exclusions and exemptions, greater use of presumptive methods of taxation, increased withholding, taxation of a variety of fringe benefits,

Table 1.3 Reforms of Corporate Tax and Personal Income Tax Rates
(percent)

Country	*Prereform CIT rate*[a]	*Prereform top bracket PIT*	*Postreform CIT rate*	*Postreform top bracket PIT*
Bolivia[b]	30	48.0	—	—
Colombia	40	49.0	30.0	30.0
Indonesia	45	50.0	35.0	35.0
Korea, Rep. of	38	60.0	30.0	55.0
Jamaica	45	57.5	33.5	33.5
Mexico	42	60.0	35.0	40.0
Morocco	53	76.0	45.0[c]	50.0
Turkey	50	66.0	46.0	50.0

— Not available.

CIT is corporate income tax; PIT is personal income tax.

a. Some countries have operated with multiple corporate rates and in these cases only the highest rate has been reported.

b. Bolivia has replaced personal and corporate income taxes with presumptive taxes on corporate and personal assets.

c. The personal income tax rate recorded for Morocco in the last column is the one proposed before the Moroccan parliament.

Source: See relevant chapter.

introduction of creditable minimum taxes on corporate assets and net worth, and generally tighter tax administration.

Broad-based sales taxes: The VAT. In most of the countries studied the VAT has arisen as the mainstay of the revenue system and typically accounts for a quarter or more of total tax revenue. In many cases the VAT has shifted the country's tax mix toward consumption and away from income and has become the marginal source of tax revenue. The VAT has frequently replaced either a cascaded turnover tax or a motley array of excise taxes, both of which were serious impediments in many countries to improved export and investment performance. Most of the countries have chosen the consumption type of VAT, levied on a destination basis. However, a few countries, such as Colombia and Turkey, have only partially refunded the VAT paid on capital goods and have, therefore, elements of an income type of VAT.

Table 1.6 summarizes the main characteristics of the new VAT's. Most countries have applied the VAT at the retail level rather than at the manufacturing or wholesale level. Countries operating at the retail level have put in place special tax regimes for small retailers for whom some simplified bookkeeping procedures are desirable on administrative grounds. With the exception of these regimes and the zero rate applied to exports and some items of food consumption, a number of countries, such as Indonesia and Korea, have operated with a VAT that has a single rate and allows exemptions for either basic necessities or certain hard-to-tax sectors such as finance. However, in some countries, the lag in providing VAT refunds implies that exports are subject to some unknown and presumably undesirable tax burden.

It could be argued that international considerations have also helped to propel the VAT to prominence. As more developing countries have turned to an outward-looking development strategy they have sought to make their

Table 1.4 Tax Reform Measures: Personal Income Tax Base Broadening

Country	Inclusion of fringe benefits[a]	Presumptive assessments[b]	Limited deductions[c]	Increased withholding[d]
Bolivia		X	X	X
Colombia		X	X	X
Indonesia	X[e]		X	
Jamaica	X		X	X
Korea, Rep. of		X		
Mexico		X	X	
Morocco		X		
Turkey	X	X		X

Note: X indicates that a particular measure was adopted.
a. Benefits to employees such as housing allowances in Jamaica, and heating allowances in Turkey, have been incorporated in the definition of taxable income.
b. Presumptive assessments of taxable income are based on gross turnover in Turkey, net worth in Colombia, or consumption indicators in Turkey.
c. Many countries have swept aside numerous personal deductions and replaced them with a single personal allowance.
d. Withholding of tax at source may apply to labor income or to all income sources as in Bolivia, or interest income or product sales as in Turkey.
e. In Indonesia fringe benefits are included in taxable income by denying a deduction for them at the corporate level.
Source: See relevant chapter.

indirect tax systems more consistent with this goal. The VAT is particularly well suited to removing the burden on imports and competing domestic production. To a degree not shared by other forms of broad-based sales taxes, the base of the VAT is also capable of being confined to consumption and avoiding the distorting effects of tax cascading that arise from the taxation of business inputs.

A comprehensive VAT is equivalent, under certain conditions, to a tax on labor incomes and the element of economic rent contained in various capital incomes. As such, the VAT also compensates for the weakness of many countries in applying direct taxes to labor and capital incomes.

Often the VAT has been seen as a panacea for a wide variety of fiscal ills, including the need for higher revenue. Experience in a few countries, however, tempers this unqualified enthusiasm for the VAT by pointing to the possibility of a poorly designed VAT. If preparation for the VAT is inadequate and enforcement is weak, as was the situation in Bolivia in an initial attempt to adopt the tax in 1975, or if the VAT has numerous rates and is interlaced with extensive exemptions, as in Morocco, it may operate with the same defects as, and indeed may be even worse than, the indirect taxes it replaced.

There is also substantial skepticism about the efficacy of certain so-called tax gimmicks that have been used to obtain greater compliance with the VAT. Turkey's expenditure rebate system, for example, and Bolivia's complementary tax both offer tax reductions for the collection of VAT invoices in an effort to obtain better compliance with the VAT. The jury is still out, however, on whether the revenue cost of these schemes justifies whatever gains in compliance are obtained.

Table 1.5 Tax Reform Measures: Corporate Base Broadening

Country	Minimum taxes on assets	Inclusion of public enterprises	Fewer tax incentives	Reduced debt finance incentives
Bolivia	X	X	X	X
Colombia	X[a]	X	X	X
Indonesia[b]		X	X	
Korea, Rep. of				X
Mexico	X		X	X
Morocco	X			
Turkey	X			

Note: X indicates that a particular measure was adopted.

a. Colombia imposes a presumptive income tax on farms and firms that is geared to a measure of net worth.

b. Indonesia experimented with, but ultimately abandoned, thin capitalization rules limiting corporate interest deductions.

Source: See relevant chapter.

Tariffs and the VAT. In choosing a destination-based VAT to replace an odd assortment of taxes on imports and domestic inputs, most developing countries have significantly enhanced the coordination of their trade and domestic indirect taxes. Sales taxes tend to be applied with equal force to both imports and domestic production. Nonetheless, trade and tax policies are seldom reformed simultaneously, and the danger remains that efforts to reform tariff policies may inadvertently distort a country's indirect tax system. For example, imposing tariffs on imported inputs and imports of final goods at the same rate may produce greater uniformity in effective rates of protection but at the cost of reintroducing some distortions that the VAT had previously removed.

Excise taxation. All of the countries studied collect substantial revenues from traditional excise taxes on alcoholic beverages, fuels or petroleum products, and tobacco products. Some levy excise taxes on exports, for example coffee in Colombia, in an effort to tap some of the rents generated in the agricultural sector. High taxes on alcohol and tobacco, so-called sin taxes, are justified because consumption of these products imposes social costs on the rest of the economy in the form of additional police, fire, and health care services. Taxes on automotive fuels are justified by the environmental costs of smog and congestion in urban areas and are often earmarked for the construction and maintenance of roads.

Excise taxes may also be linked to the consumption of luxury items to secure a more progressive distribution of commodity tax burdens. Some countries like Mexico have attempted to tax luxury consumption by imposing a differentially higher VAT rate on these items. This practice is not recommended because it complicates the administration of the VAT and is an example of poor targeting between tax instruments and objectives. Excise taxes can be collected with relative ease at the factory gate if the item is produced domestically (for example, watches) or at the customs gate if not. Mexico's experience in taxing fur coats suggests, however, that smuggling increases and excise revenues drop if tax rates are set too

Table 1.6 Main Features of New Value Added Taxes

Country	Adoption date	Present level of coverage	Current rate structure	Basic food (unprocessed)	Processed food
Bolivia	1975 & 1986	Retail	Single	Exempt	Taxable
Colombia	1976	Retail	Single	Exempt	Zero rated
Indonesia	1982	Wholesale	Single	Exempt	Taxable
Korea, Rep. of	1977	Retail	Single	Taxable	Taxable
Mexico	1978	Retail	Multiple	Zero rated	Zero rated
Morocco	1986	Wholesale	Multiple	Exempt	Exempt
Turkey	1984	Retail	Multiple	Zero rated	Taxable

Source: See relevant chapter.

high. The issue of whether excise rates should be established on a specific or ad valorem basis is discussed below.

Indexation. Countries susceptible to prolonged bouts of high inflation have found it necessary to index both direct and indirect taxes in order to maintain real tax revenues and avoid inflation-induced inequities and distortions in the tax treatment of commodities and both capital and labor incomes. Table 1.7 indicates the range of indexation measures recently applied by various countries in the sample. Contrary to the conventional wisdom in industrial countries that indexing is destabilizing, the absence of indexing measures in countries where indirect taxes dominate may frequently be destabilizing as inflation erodes real revenue yields. As the experience of Bolivia graphically illustrates, unindexed commodity tax bases are especially vulnerable to the effects of inflation and can contribute to explosive macroeconomic instability as inflation creates growing fiscal deficits and accelerating rates of inflation.

The experience of Turkey indicates that inflation in an unindexed tax system will often raise effective tax rates on labor incomes and systematically reduce them on capital incomes. This effect occurs as inflation lifts more workers into higher tax brackets and higher interest deductions reduce the size of the corporate income tax base. Inflationary episodes in Bolivia, Mexico, and Turkey, and also to some extent in Colombia, suggest that the failure to index the personal income tax and the consequent bracket creep stimulates the growth of untaxed fringe benefits. However, provision of these fringe benefits occurred unevenly across different groups of the labor force, with consequent damage to both vertical and horizontal equity.

Inflation also produces substantial inequities and distortions if the tax treatment of capital income is not indexed. Without indexation, debt finance is artificially encouraged, while taxes on nominal interest income are transformed into covert wealth taxes. Under the new indexing schemes used in both Colombia and Mexico, only real interest income is taxable and real interest expense is deductible. Both of these countries have also adopted inflation-proof methods for adjusting depreciation allowances, while Colombia also provides for indexed treatment of capital gains. Though some of these measures may shrink total revenue, all of them re-

Table 1.7 Tax Reform Measures: Inflation Adjustment

Country	Real interest income taxable and expense deductible	Indexed depreciation allowances	Indexed inventories	Indexed personal tax brackets	Ad valorem commodity tax rates
Bolivia					X
Colombia	X	X	X	X	
Mexico	X	X	X	X	
Turkey		X		X	

Note: X means that a particular measure was adopted.
Source: See relevant chapter.

move significant investment distortions that detract from both efficiency and equity. Indexation has a cost, however, as it always complicates the task of tax administration.

Tax administration. Tax administration improvements include creating a unique taxpayer identification numbering system; computerizing the processing of tax returns in order to detect nonfilers and stopfilers; enlisting the resources of commercial banks to receive, and to some extent process, tax payments, and creating new audit capacities focused on the country's largest taxpayers. Enhanced detection capacities have complemented stiffer enforcement rules and procedures to induce better taxpayer compliance. Less tax evasion, in turn, has efficiency and equity payoffs and contributes to higher revenue yields.

Taxation of capital income: Unfinished business. Although it has been easy to identify some of the key elements in the reform of direct and indirect taxes and the improvement of tax administration, a few important issues of tax policy remain unresolved. For example, the appropriate policy for treating interest income is not entirely clear. If a country exempts interest income from tax out of fear of capital flight and its inability to tax foreign deposits held by domestic taxpayers, such an exemption would probably create irresistible pressures to reduce tax burdens on other forms of capital income, particularly dividend income and capital gains, if for no other reason than to mitigate the debt bias in a company's financial policies that would otherwise occur. Table 1.8 shows clearly that when interest income is included in the personal tax base it is often taxed on a schedular basis through withholding at the source.

Taxing the domestic source interest income of either residents or foreigners in the context of a global capital market might cause domestic interest rates to rise as the effective tax burden is shifted forward to capital users and domestic consumers in general and to owners of immobile land and labor in particular. As emphasized by Gil-Díaz and Thirsk (see chapter 7), higher interest rates in turn will enhance the value of corporate interest deductions and erode the size of the corporate tax base while adding to government interest costs on the expenditure side of the budget. Under these circumstances it would be better to tax immobile resources directly and avoid the distortion, otherwise created, in the capital market.

Table 1.8 Tax Reform Measures: Treatment of Capital Income

Country	Inclusion of interest income[a] in the personal tax base	Schedular withholding taxes on interest and/or dividends	Schedular capital gains tax on real assets	Capital gains tax on stock market transactions
Bolivia		X		
Colombia	Real basis		X	X
Indonesia	X	X	X	
Jamaica	X			
Korea, Rep. of		X	X	
Mexico	Real basis		X	
Morocco		X	X	
Turkey		X	X	

Note: X means that a particular measure was adopted.
a. Small interest-earning deposits are exempted in Indonesia and Jamaica.
Source: See relevant chapter.

The situation is further complicated by the provision of foreign tax credits. If foreigners are able to credit host country tax liabilities against their own domestic tax obligations, interest rates in the capital-receiving country may not increase when taxes on interest income are imposed. It has also been argued that if interest income is to be excluded from the tax base, interest expense should also be made nondeductible to curb tax arbitrage. However, such an approach could easily jeopardize the availability of foreign tax credits and raise the economy's cost of capital. Moreover, large foreign firms operating in the domestic economy may turn to cheaper financing from off-shore sources that are unavailable to domestic corporations, placing them at a competitive disadvantage. There seems to be no easy answer, but one approach that would not jeopardize tax credit eligibility would be to deny an interest cost deduction up to the amount of any income received from tax-exempt debt.

Tax burdens on the poor: Are they still too high? Other areas of controversy concern the equity implications of the model blueprint and whether more effort should be made to reduce tax burdens on the poor. The results of the incidence studies cited in this volume are displayed in table 1.9 and suggest that the bottom 10 percent of taxpayers in these countries typically sacrifice 10 percent or more of their income to pay for indirect tax obligations. None of the reforms examined appear to have achieved notable success in reducing indirect tax burdens on the poor, although, as table 1.6 shows, most countries try to remove some of the VAT burden on the poor by either exempting or zero rating unprocessed food products.

Tax Reform Model for Developing Countries

Because of the multiple constraints, no country ever manages to advance in a single stride to the complete package of desirable tax reforms available. Bolivia, Colombia, Indonesia, Jamaica, Korea, Mexico, Morocco, and Turkey are all examples of developing countries that have undertaken major tax reforms in the last decade and a half. The first four countries have been

Table 1.9 Indirect Tax Burdens on the Poorest Decile of Households
(as a fraction of household income)

Country	*Average indirect tax rate*[a] *(percent)*
Bolivia	n.a.
Colombia	7–12[b]
Indonesia	n.a.
Jamaica	6–10[c]
Korea, Rep. of	15
Mexico	14[d]
Morocco	34
Turkey	13–14

n.a. Not applicable.
a. For a description of how these estimates were obtained see chapters in this volume on Korea, Mexico, Turkey, and Morocco.
b. The higher rate in the case of Colombia occurs when the corporate income tax is assumed to be borne by consumers, as explained in the chapter on Colombia.
c. In the case of Jamaica the higher rate is from Wasylenko (1990) and the lower rate is from Bird and Miller (1990).
d. The estimate for Mexico differs from the others by using a calculation of permanent rather than current household income.
Source: See relevant chapter.

engaged in continuous major tax reform since the 1950s. The others have made more recent comprehensive efforts to refashion their tax systems. A common thread running through the fabric of all of these reforms is the earnest attempt to replace narrow, distortionary tax bases with broader bases that raise revenues at lower rates and that treat most sources and uses of income in a less discriminatory manner for tax purposes. This single strategy is consistent with the simultaneous achievement of a trinity of tax reform objectives: greater simplicity and neutrality in taxation and a larger measure of horizontal equity.

While different countries have adopted somewhat different sets of reforms, it is possible to compile a list of major tax reform measures that together provide a composite profile of desirable tax reform measures for developing countries:

- Reducing marginal tax rates on both personal and corporate incomes, typically below 50 percent.
- Flattening rate schedules in the personal income tax and selecting a single rate for the corporate income tax.
- Aligning the top marginal personal tax rate with the corporate tax rate to diminish the incentive to shift income among the categories of personal income, partnership income, and corporate income.
- Eschewing selective tax incentives and the tendency of government authorities to pick winners and losers in the economy. Incentives automatically narrow tax bases, diminish revenue yields, most likely misallocate investment resources most of the time, facilitate tax evasion, and seriously compromise the goal of horizontal equity. Korea is the only country that made a serious effort to employ tax incentives in support of industrial policy but, as Kwang discusses in his chapter, Korea quickly abandoned this policy because of these concerns.

- Avoiding the imposition of indirect taxes on interindustry transactions. Taxes on productive inputs distort production decisions, discourage exports, lead to undesirable tax cascading, and create an unknown, and largely unknowable, pattern of tax incidence. A VAT to replace either turnover taxes or an unsystematic array of excise taxes has the potential to eliminate these kinds of problems in indirect taxation.
- Coordinating tariffs and domestic indirect taxes so that tariffs play only a protective, and not a revenue, role in the economy. Conversely, domestic indirect taxes should not be designed to either inadvertently or purposely enhance levels of trade protection.
- Attempting to give the same tax treatment to all components of capital income under an income tax system. This problem involves complex issues of capital gains measurement, integration of corporate and personal taxes, and the ability of tax administrations to reach their residents' interest income. With the appearance of a global capital market, increasing international tax competition, and, in particular, the U.S. decision to discontinue withholding on interest payments to foreigners, residence-based taxation of interest income may no longer be feasible. On the other hand, the prospect of capital flight may make source taxation of interest income undesirable as only a higher cost of debt capital would result.
- Indexing the tax system if inflation persists at rates of 10 percent or more. High rates of inflation play havoc with a country's tax system if it is not adequately indexed and introduces several unintended distortions. In Turkey, for example, bracket creep pushed even modest-level wage recipients into high marginal tax brackets and triggered a proliferation of untaxed fringe benefits as a way of shielding workers from bracket creep. The unsystematic supply of fringe benefits severely eroded both the vertical and horizontal equity characteristics of the tax system. Investment decisions were also seriously distorted as the tax system began to heavily subsidize debt-financed investments. Indexing provisions, such as those introduced by Colombia, require only the deduction of real interest costs and the taxation of real interest income.
- Introducing schedular tax rules and withholding to limit tax evasion opportunities. Although schedular elements are an unattractive aspect of an ideal income tax, in the real world schedular taxation of capital incomes may be necessary on administrative grounds to cope with the problem of effectively taxing capital incomes. Limiting the deduction of agricultural losses only to the amount of agricultural income obtained, for example, also takes a schedular approach in curbing tax abuse, even though it, too, violates the principles of global income taxation.
- Enlarging the personal income tax base through the inclusion of fringe benefits, presumptive taxation of hard-to-tax groups, greater reliance on withholding at source, and fewer, and less generous, deductions and exemptions (apart from the personal exemption).

- Enlarging the corporate tax base through the inclusion of state-owned enterprises in the base, the imposition of minimum taxes on private sector companies, and the elimination of many tax incentives and preferences.
- Eliminating tax-induced choice of debt rather than equity in company financing. Measures having this effect include allowing only real interest cost deductions, integration of corporate and personal income taxation to avoid double taxation of dividends, and, as in Korea, rejection of presumptive dividend taxation.
- Investing in improved administrative capacity, installing more effective audit techniques, applying stiffer penalties for noncompliance, computerizing tax records to detect evasion and to keep registration and assessments up to date, using banks as collection agents, reducing filing through more accurate withholding, and relying on reasonable presumptive methods of taxation.

Although no single country has managed to meet all of these reform targets, a surprising number of them have made significant progress in adopting several of these measures. In part, the similarity of the reforms reflects an element of growing tax competition between developing and industrial countries. Greater global competition for investment resources has made it more costly for developing countries to allow their corporate tax systems to depart significantly from those that exist in capital-exporting countries. When the United Kingdom and the United States reduced their corporate rates in the mid-1980s they exerted considerable pressure on many developing countries to follow suit. Given the desirability of aligning the top bracket personal rate with the corporate rate, corresponding pressure for lower rates was brought to bear on personal income taxes. To maintain revenues, developing countries were required to consider measures to broaden the income tax bases and to shift revenue-raising efforts to indirect taxes. This chain of events occurred in several countries. Reinforcing these tendencies was a shared intellectual environment in which widespread agreement existed on the desirability of achieving more neutral and simpler taxes.

The Process of Tax Reform

It is one thing to devise a model blueprint for tax reform; it is quite another to have it ready when the moment is ripe for reform and to apply it successfully. A number of the country studies illuminate the important steps that should guide the reform process. It is important, for example, to have the appropriate policy measures "on the shelf" before a fiscal crisis strikes in order to forestall the adoption of ad hoc and ill-advised tax reforms. Otherwise countries may be tempted to correct for their revenue shortfall by the simple expedient of raising marginal tax rates on narrow, fragmented tax bases. The time for a tax reform appears ripest either when a new government assumes power, as in Mexico or Turkey, or when political opposition is weak, as in Indonesia and Korea.

Successful tax reform efforts also require detailed knowledge of the defects of the current system, especially a sense of who pays taxes at the industry, firm, and household level, as well as a feeling for how the distribution of tax burdens would be affected by alternative tax measures that attempt to improve matters. Reforms are also more likely to be successfully adopted if local policymakers are actively involved in their design and implementation and the reform results in the creation of a cadre of local tax experts who identify with, and assume responsibility for, the success of the reform. Without this active local engagement, reforms are less likely to be woven into the country's institutional structure and important opportunities for institution building will be missed.

Moreover, all of the successful reforms mentioned in this volume have benefited from detailed and careful planning and preparation prior to the introduction of a reform, as well as close monitoring after its implementation. Along with monitoring, there should be efforts to educate the public on how a new tax system will operate and strengthen the administrative apparatus responsible for implementing the reform. In Jamaica and Mexico, for instance, considerable resources were devoted to getting the private sector "on board" and gaining their support for the tax reform proposals.

Potential improvements in simplicity and horizontal equity are strong selling points in persuading both the public and legislators about the merits of a particular reform. Gains in economic efficiency and vertical equity, on the other hand, are much harder to sell. Fairness is often perceived by the public as requiring a more even distribution of tax burdens across different firms and industries in the economy and across different households at the same income level. The efficiency benefits of less distortionary taxation may have to be cloaked, therefore, by an appeal to the apparent equity advantages of playing on a level tax field. Similarly, because complex tax rules are often perceived by voters as invitations to evade taxes, replacing them with simpler rules will receive widespread public support.

Countries are most receptive to tax reform when their backs are hard against the fiscal wall and nonsustainable fiscal deficits clearly signal the need to mobilize resources. Experience in several countries suggests that, while economic recession is not a barrier to introducing reforms, an unstable macroeconomic climate may doom reforms after they have been put in place as they will be blamed for any coinciding macroeconomic distress. The weight of this accusation may be sufficient to force either the repeal of the tax reforms or some modification of them, as happened in Colombia. Of course tax reforms that raise more tax revenue and eliminate fiscal deficits may themselves be a prerequisite for macroeconomic stability. In this sense fiscal reforms that restore price stability to the macroeconomy should logically precede trade and exchange rate reforms, which normally require an economy that is not buffeted by strong inflationary winds.

Although there may also be a logical sequence to the different components of fiscal reform, the empirical evidence on this topic can hardly be said to be overwhelming. Most of the country studies have tacitly assumed that fiscal deficits should be eliminated from the revenue side of the budget. Revenue-raising tax reforms make little sense if their proceeds are used to finance an inefficient amount of public expenditures. Thus, if expenditure reform does not accompany tax reform, it should ideally precede it. Moreover, reform of structural tax policy should ideally precede the reform of tax administration, since there is little merit in making a bad tax system work better. Improved tax administration can never compensate for bad tax design. At the same time, even good tax policies will never work properly unless they can be administered effectively. In this sense, improved tax administration is frequently the key element in a successful tax reform. Finally, the experience with tax reform in Morocco makes it clear that tax reform should logically precede tariff reform. If tariffs constitute a significant fraction of total revenues, as they did in Morocco, tariff reductions will bring fiscal deficits and macroinstability in their wake unless, or until, the tax system can be effectively revamped to replace foregone tariff revenue.

Conclusions

Given the diverse group of countries studied here, many of the lessons derived from a particular country's tax reform experience may indeed be unique to that country. Also, this is not the only attempt to generalize from the experience of these countries (Thirsk 1991). If the tax reform experience of the "group of eight" teaches us anything that can be usefully transferred to other developing countries, it is probably along the lines of the five caveats that follow.

Caveat 1. While successful tax reforms always invest in better tax administration, in the end reform measures must be compatible with existing tax administration capacities. High tax rates imposed on narrow tax bases not only violate the requirements for efficient and equitable taxation they also encourage tax evasion, trigger requests for preferential tax treatment, and undermine compliance. Broader and more simply defined tax bases, along with more uniform rate structures, describe the essence of the tax reform path taken by Bolivia, Colombia, Indonesia, Jamaica, Korea, and Mexico. The prereform experience of Morocco highlights the administrative pitfalls that await those who walk the other path of a narrowly based and highly differentiated tax system.

Caveat 2. Tax and expenditure systems design should avoid both current and future fiscal deficits. Small, open economies that are reliant on export taxes as revenue sources (such as Bolivia and Morocco) need to cultivate domestic revenue sources that will leave them less vulnerable to the international business cycle. Otherwise they run the risk of recurrent fiscal deficits financed by inflationary means. The inflation tax is arguably the worst tax invented by man. However, where high rates of inflation are for some reason unavoidable, indexing commodity tax bases and the tax treatment of both labor and capital incomes becomes an important reform objective.

Caveat 3. Broad-based indirect taxes are easier to design and implement than broad-based income taxes. At one extreme, Bolivia has given up on traditional income taxes. Although broader income tax bases are desirable in principle, in most developing countries a variety of structural and economic barriers thwart the creation of a comprehensive income tax base. Inability to reach incomes in the informal sector, measurement difficulties, vested interests, and the use of tax incentives ensure that large portions of both labor and capital income are excluded from the personal tax base. By contrast, a broad-based VAT has worked reasonably well in most countries by raising substantial amounts of revenue and removing undesirable indirect tax burdens on exports and investment spending.

Caveat 4. Tax instruments should be aligned with the objective they are intended to achieve and most, if not all, tax incentives should be eschewed. Poorly targeted tax measures always produce undesirable side effects. Revenue-raising tariffs, for example, lead to inefficiently high levels of trade protection. Most tax incentives are difficult to target and appear to either waste revenue or encourage inefficient resource allocations.

Caveat 5. Whereas tradeoffs among the goals of tax reform are almost inevitable, the three objectives of simplicity, neutrality, and horizontal equity frequently complement, rather than compete, with each other. Nevertheless, all three normally conflict with the achievement of greater vertical equity. Most of the tax reforms in this study have aimed at achieving greater simplicity, neutrality, and horizontal equity even if it meant sacrificing some vertical equity. In many of these countries the vertical equity objective has shifted from seeking heavier taxation of the rich to avoiding taxation of the poor.

The countries studied have struggled to deal with a variety of tax reform issues and produced a collective experience that leads to one major conclusion. With few exceptions, properly designed taxes have more of a passive, rather than active, role to play in the development process. However, one exception to this is a tax on pollutants that delivers the double benefit of curbing a harmful externality and generating revenues which reduce reliance on costly sources of public finance. Good tax policy more generally minimizes the damage in extracting resources for the public sector. Good expenditure policy uses these resources productively. The tax system may be unable to deliver much that is good but, harnessed to the wrong set of objectives, it can certainly cause considerable harm. Discriminatory tax policies do little to stimulate economic development and redistribute income. Any attempts to rely on such tax policies to achieve these objectives are likely to be counterproductive because they impair effective tax administration, open up opportunities for tax evasion, and create tax-induced inequities and inefficiencies.

The expenditure side of the budget is more likely to deliver improvements in both growth and equity. Tax policies will work best if they are limited in scope to avoiding fiscal deficits, regressivity in the distribution of tax burdens and large tax-induced resource misallocations. Broad-based taxes, applied at low and uniform rates, seem to offer the best means of achieving these more modest goals.

Notes

1. If the current sales tax rate is 10 percent, then a dollar spent today incurs a sales tax liability of 10 cents. A dollar saved today and spent tomorrow will be worth $1.10 if the interest rate is also 10 percent. Sales tax on that amount is $.11 or $.10 in present value terms.

2. As Pogue and Sgontz (1989) demonstrate, the optimal "sin tax" on alcoholic beverages balances the benefits of higher tax rates (fewer external costs) against the cost of the distortion in consumption patterns for beverage consumers.

3. Slemrod (1990, p. 167) reinforces the point made by Tanzi that "the ascendency of uniform taxation . . . is due to the lack of strong evidence pointing to a clear alternative and the sense that a uniform tax system is less susceptible to political pressures favoring tax changes that serve special interests and are unrelated to optimal tax considerations."

4. All of these countries have utilized a version of the King-Fullerton approach to calculating METRS. Mathematically, the METR can be defined as METR $=(p-s)/p$, when p is the pretax real rate of return on a marginal investment project and s is the post-tax rate of return paid to a saver. The size of the tax wedge $(p-s)$ is influenced by a combination of taxes levied on corporate incomes, property values, purchases of capital goods, and special provisions of the personal income tax. METRS may be calculated for different sectors, assets, and groups of savers in the economy and the analysis typically incorporates the maintained assumptions of perfect tax administration, perfect capital markets, and perfect certainty. If firms can expense their investments and are not "tax exhausted," their METR is zero. Underexpensing firms may deduct the entire cost, C, of the investment from taxable income so that their after-tax cost of the investment is $(1-t)C$ where t is the nominal corporate tax rate. If R is the present value of future returns from the investment, the firm pursues investment to the point where its after-tax cost equals its after-tax return or $(1-t)C=(1-t)R$, or $C=R$, as in the case when the METR equals zero. If firms are permitted to deduct, in present values, more than the cost of the investment, the METR will be negative and the firm will enjoy a subsidy. Unfortunately, it is impossible to compare estimates of METRS across countries in this project because of different levels of aggregation and the lack of uniformity in the choice of measurement techniques. Many of the METRS reported were made before this project began, and even those that were made afterward could not be constrained to adopt common assumptions and the same measurement basis.

5. For example, if the interest rate is 10 percent a negative METR of -66 percent implies that an investment project earning only a 6 percent rate of return before taxes would be viable. Resources would be misallocated if this occurred, since society would have to sacrifice projects yielding 10 percent in favor of ones that returned only 6 percent. A negative METR is the fiscal counterpart of a negative effective rate of protection (ERP) found in the trade literature. A negative ERP indicates that, at world prices, an activity cannot generate sufficient revenue to cover its intermediate purchases and is highly inefficient.

6. For further evidence on the size of seigniorage in developing countries, see Easterly and Schmidt-Hebbel (1993). Estimates of sustainable seigniorage are given in an earlier study by Kiguel and Liviatan (1988).

A few simple equations illustrate how the inflation tax mechanism works:

(1) $\Delta R = M/P$

The real resources transferred to the government, ΔR, are equal to the monetary emission associated with the fiscal deficit, ΔM, divided by the price level P.

Equation 1 can be rewritten as:

(1') $\Delta R = \Delta M/M \times M/P$

In a steady-state equilibrium in which prices have fully adjusted to the fiscally induced monetary impulse, prices will rise proportionally to the growth of the money supply:

(2) $\Delta M/M = \Delta P/P$

Substituting (2) into (1') gives the basic result:

(3) $\Delta R = \Delta P/P \times M/P$

where $\Delta P/P$ is the inflation "tax rate" and M/P are real balances or the inflation tax base. Inflation tends to gradually reduce the size of this base while real economic growth tends to raise it.

7. See Easterly and Schmidt-Hebbel (1993), who suggest there is a strong correlation between high deficits and low growth.

References

Auerbach, Alan. 1989. "The Deadweight Loss from Nonneutral Capital Income Taxation." *Journal of Public Economics* 40: 1–36.

Bird, Richard, and Barbara Miller. 1990. "The Incidence of Indirect Taxes on Low Income Households in Jamaica." In Roy Bahl, ed., *Tax Reform in Jamaica*. Cambridge, Mass.: Lincoln Land Institute.

Burgess, Robin, and Nicholas Stern. 1993. "Taxation and Development." *Journal of Economic Literature* 31(June): 762–826.

Easterly, William, and Klaus Schmidt-Hebbel. 1993. "Fiscal Deficits and Macroeconomic Performance in Developing Countries." *The World Bank Research Observer* 8(July): 211–37.

Gillis, Malcolm, ed. 1989. *Tax Reform in Developing Countries*. Durham, N.C.: Duke University Press.

Khalilzadeh-Shirazi, Javad, and Anwar Shah, eds. 1991. *Tax Policy in Developing Countries*. Washington, D.C.: World Bank.

Kiguel, Miguel, and Nissan Liviatan. 1988. "Inflationary Rigidities and Orthodox Stabilization Policies: Lessons from Latin America." *The World Bank Economic Review* 2(September): 273–98 .

Newbery, David, and Nicholas Stern, eds. 1987. *The Theory of Taxation for Developing Countries*. New York: Oxford University Press.

Pogue, Thomas F., and L. G. Sgontz. 1989. "Taxing to Control Social Costs: The Case of Alcohol." *American Economic Review* 79: 235–43.

Slemrod, J. Winter. 1990. "Optimal Taxation and Optimal Tax Systems." *Journal of Economic Perspectives* 80: 157–78.

Tanzi, Vito. 1992. "Theory and Policy: A Comment on Dixit and on Current Tax Theory." *International Monetary Fund Staff Papers* 39(1): 957–66.

Tanzi, Vito, and Milka Casanegra de Jantscher. 1987. "Presumptive Income Taxation: Administrative, Efficiency and Equity Aspects." International Monetary Fund Working Paper No. 54. Washington, D.C.

Thirsk, Wayne. 1991. "Lessons from Tax Reform: An Overview." In Javad Khalilzadeh-Shirazi and Anwar Shah, eds., *Tax Policy in Developing Countries*. Washington, D.C.: World Bank.

Wasylenko, Michael J. 1990. "Tax Burdens Before and After Reform." In Roy Bahl, ed., *Tax Reform in Jamaica*. Cambridge, Mass.: Lincoln Land Institute.

World Bank. 1992. *World Development Report 1992: Development and the Environment*. New York: Oxford University Press.

2

Bolivia's Tax Revolution

Wayne Thirsk

Better known for political rather than fiscal revolutions, Bolivia is one of few countries to uproot and completely replace its tax system. Bolivia tinkered with improvements to its tax system during the 1970s and then witnessed the inflationary destruction of its revenue system during the early 1980s. In response to this fiscal carnage, Bolivia introduced ambitious and far-reaching tax reform in 1986. The first part of this chapter describes and discusses the tax system prior to this major tax reform, as well as the main features of the fiscal system that provoked the fiscal crisis of the early 1980s, culminating in one of the world's few examples of hyperinflation. In the following section the new tax system is compared with its predecessor. After discussing the new tax system and the improvements in fiscal performance it is expected to deliver, a concluding section summarizes some of the major lessons for reforming taxes based on the Bolivian experience.

Bolivia is a small, open, and relatively impoverished economy with a per capita income in 1986 of US$600, the second lowest in the Western Hemisphere. For centuries its prosperity has depended upon the successful exploitation and exportation of a few natural resource products. First it was silver, then tin and hydrocarbons and, more recently, coca. It is also a country of enormous regional, ethnic, and income class diversity. This heterogeneity, and the highly unequal distribution of income associated with it, has produced a long and troubled history of economic and political instability. Since becoming independent from Spain over 160 years ago, Bolivia has seen more than 100 government administrations come and go. These long-standing and deep-seated political divisions lie at the root of many of Bolivia's fiscal failures.

The Tax System in the 1970s

During the 1970s a traditional array of direct and indirect tax instruments was employed. With small yields from a tax on individual incomes and a profits tax on enterprises, the bulk of government revenue needs were met by a wide assortment of indirect taxes, including hefty trade taxes as well as property taxes levied at the municipal level. Although the tax structure is typical of most developing countries, two features make it stand out from the crowd. As table 2.1 shows, Bolivia is, was, and still is unusually dependent upon trade taxes as the major source of revenue, with export and import taxes together accounting for almost two-thirds of total revenue prior to reform. Secondly, in the case of export levies, most of the revenue is collected from state enterprises. There is considerable overlap among exports, publicly owned enterprises, and the natural resource sectors.

At first glance, table 2.1 suggests that direct taxes are more important than indirect revenues in the fiscal system. Being a small, open economy, Bolivia faces fixed terms of trade and, as a result, any export duty is shifted backward onto agents of production in the export sector. Export taxes, therefore, directly reduce the incomes of resources employed in the export sector. One can also see from table 2.1 that the fraction of total revenue obtained from export taxes nearly matches the fraction

Table 2.1 Revenue Tax Structures before and after Reform

Source of revenue	*1975*	*1987*	*1989*
Direct taxes[a]	52.3	55.8	61.5
Domestic direct taxes (excluding exports)	18.1	10.4	11.6
Amnesty tax (regularization tax)	—	3.6	0.1
Complementary tax	—	2.2	3.3
Enterprise tax (private companies)	6.9	—	—
Personal income taxes	4.6	—	—
Presumptive tax on profits	—	0.7	2.5
Presumptive tax on property owners	—	2.9	3.1
Public enterprises (closely related to export taxes)[b]	34.2	46.1	49.9
Export taxes	(33.4)	(45.1)	(42.5)
Indirect taxes	47.7	44.2	48.1
Domestic indirect taxes (excluding imports)	18.1	26.7	33.2
Excise taxes (alcoholic beverages and tobacco products)	4.3	1.6	3.9
Mining properties	—	0.5	0.9
Simplified tax	—	0.3	0.4
Transactions tax	—	2.1	3.6
Value added tax	3.2	13.5	23.4
Import taxes	29.6	17.5	11.0

— Not available.

Note: Table includes data only for major tax categories in each year, so columns may not add up.

a. For 1975 direct taxes are considered to fall on income and include tax on exports.

b. According to Mann (1990), the state petroleum company accounts for most of this revenue through payments of royalties and more recently the VAT and the transaction tax.

Source: For 1975, Musgrave (1981); for 1987 and 1989, World Bank (1989).

derived from publicly owned enterprises. Domestic direct taxes, on the other hand, omitting any export tax revenue, were responsible for only about 19 percent of total revenue. Below we look at the various components of the revenue system as it existed in 1975.

The Personal Income Tax

Prior to 1975 personal income was taxed on a schedular basis. However, a tax reform in 1975 converted the personal income tax into something resembling a modern globalized tax. Despite the attempt to tax income from all sources on a uniform basis, the personal income tax contributed only 4.6 percent of total tax revenues in 1975. In part, this poor revenue performance reflected the relatively low level of average income in the economy and hence the absence of significant taxable capacity. For the most part, however, the insignificance of the personal income tax was attributable to its uneven coverage and application. In 1975, for instance, out of a total labor force of 2.3 million workers, 80 percent were self-employed and very few of those filed an income tax return. Only about half of the employed labor force (approximately 200,000 workers or 9 percent of all income earners) filed a return.

The personal income tax was effectively a tax on labor incomes earned in the modern sector of the economy. In 1975, 62 percent of personal income tax revenues was obtained from the withholding of tax on wage payments made in the modern sector. Musgrave (1981) estimated that in 1975 actual tax collections as a fraction of potential revenue yield were 75, 33, and 20 percent, respectively, for recipients of labor income, self-employment income, and capital income. Virtually all agricultural incomes, including those earned by profitable large farmers, were outside the scope of the personal income tax. Urban professionals also managed to escape the personal income tax.

A number of factors explain the failure to adequately tax capital incomes. Most forms of capital income were excluded from the personal income tax base, and even those that were not could not be adequately captured under a system of self-assessment. The exclusion of interest income earned on domestic deposits from the personal income tax base was justified on the grounds that interest income received from lending in the informal and foreign credit market went largely untaxed for administrative reasons. Like many other Latin American countries, Bolivia applied the territorial principle in defining the jurisdictional limits of personal income tax, making it relatively easy for wealthy depositors, at least, to earn interest income from holding foreign bank accounts.[1] Dividend income was theoretically taxable but there was no withholding unless bearer shares were involved.

In principle, capital gains were fully taxable. In the case of real property, the tax base was defined as the sale price less the greater of the acquisition price or the cadastral value (both were indexed). In the case of shareholders' capital gains, the absence of a formal stock exchange led the tax authorities to define the base as the difference between book value (initial capital subscription plus capital reserves plus accumulated profits) and the cost of acquisition. Once again, because of administrative constraints, the tax on real estate capital gains was suspended in 1985 and very little in the way of additional revenue was collected from shareholder transactions. Because of these exclusions and omissions, the Musgrave Report obtained a potential yield for the personal income tax of 625 million bolivianos in 1975 compared to the actual yield of only 291 million bolivianos. About two-thirds of this gap between performance and potential was attributed to the ineffective taxation of capital income.

The nominal rate structure for personal income taxes ranged between an initial rate of 4 percent and a top marginal rate of 48 percent. Given large exclusions from the tax base and a liberal set of allowable deductions, effective income tax rates, however, were much lower than this nominal rate progression would suggest. An upper limit on tax payments equal to 35 percent of net income also served to mitigate the impact of a progressive rate schedule.

Interestingly, in an economy that had been prone to earlier bouts of rapid inflation, there were virtually no explicit features of the personal income tax to make it immune from the effects of inflation. Also the Musgrave Report rejected the option of indexing personal tax on the mistaken conviction that inflation in an unindexed tax system would raise real revenue yields and act as a built-in stabilizer. Instead of allowing for explicit inflation adjustments, the government introduced a series of ad hoc measures to cope with the tendency of ongoing inflation to shift taxpayers into higher tax brackets. For example, fringe benefits and bonuses for workers were excluded from the personal income tax base as a means of providing some protection against inflation. Only when inflation accelerated markedly in 1982 did the government attempt to adjust rates in order to offset inflationary bracket creep.

Even though the personal income tax made a modest contribution toward achieving a progressive pattern of tax burdens, it was less able to satisfy the requirement of horizontal equity because of the rather narrow definition of the tax base. Failure to reach most elements of capital and self-employment incomes meant that taxpayers with the same total incomes faced widely varying tax burdens.

As outlined in the Musgrave Report, about 60,000 taxpayers of all types were enrolled with the tax administration in 1975, including 15,000 or so small taxpayers who were assessed under a special regime. This list excluded taxpayers whose sole source of income was employment and from whom income tax was withheld. By comparison, in 1974, 17,000 income tax payers filed a return, of which about 10,000 were individual rather than company taxpayers. In contrast to withholding on wage income, self-employed taxpayers were only required to pay one-half of the previous year's tax bill by the middle of each year. The balance was not due until

March of the subsequent year. This, and other collection lags, became serious problems when inflation ratcheted upwards during the early 1980s.

Weak collection capacity was not the only administrative shortcoming noted in the Musgrave Report. In general, the tax administration received low marks for its inability to carry out effective audits, collect new data, and process and make use of existing information sources.

The Enterprise Tax

Unlike the tax treatment accorded to businesses in most other countries, the enterprise tax is applied to all forms of business organization—corporations and partnerships as well as sole proprietorships. All of these organization types were subject to a uniform tax on net business income of 30 percent. While the enterprise tax was neutral with respect to the choice of organizational form, it excluded all mining, electrical power, and hydrocarbon enterprises; these firms were subject to separate tax regimes. All state-owned enterprises were taxed at a lower rate of 20 percent, which they frequently neglected to pay. There was also a special tax regime for small businesses employing fewer than three workers. In 1976 these firms were required to pay a tax that was the simple average of the amounts that had been paid in the previous two years. In future years, the tax was subject to a 5 percent annual rate increase.

Since the 1960s, state enterprises have become the bulwark of the Bolivian revenue system. In 1975 taxes and royalties on hydrocarbons comprised 57 percent of all natural resource taxes and nearly 20 percent of total central government revenue. Taxes obtained from natural resource production made up 48 percent of total central government revenue in 1975, compared to only 8.4 percent in 1968.

Two state-owned natural resource companies, the tin company (Comibol) and the petroleum company (YPFB), together accounted for 44 percent of central government revenue in 1975. Also, both negotiate an annual transfer of any surplus income to the central government. In the case of the state petroleum company, this surplus accrues after paying a 30 percent royalty on crude oil at the wellhead, 20 percent of all export earnings and a consumption tax of 12 percent imposed on the pump price, plus 30 percent on any net increase in pump price over the year.

Offsetting these taxes in 1975 was the pricing of domestic petroleum at 25 percent of the world price, which provided a subsidy to domestic consumers worth more than three-quarters of the value of all three of the petroleum taxes mentioned above. This subsidy, really an implicit tax imposed on the state enterprise, was the equivalent of nearly 20 percent of the central government's total revenue and, according to the Musgrave Report, inflicted a welfare cost on the economy that was worth between 0.4 and 0.6 percent of gross domestic product (GDP).

As table 2.1 shows, the enterprise tax generated only about 7 percent of total central government revenue in 1975. It also proved to be a difficult tax to administer. Large companies were rarely audited and tax administrators faced numerous problems in preventing business owners from converting business income into more lightly taxed labor income and distinguishing between personal and business expenses. These difficulties, combined with generous depreciation rates for buildings and equipment, produced effective tax rates that were significantly lower than the statutory rate.

Indirect Taxes

While indirect taxes accounted for almost 48 percent of total central government revenue, the bulk of indirect tax revenue came from the imposition of import duties (70 percent), whereas the general sales tax and five excise taxes accounted, respectively, for 10 and 20 percent of total indirect tax collections. Nominal tariff rates ranged between 1 and 100 percent, with higher rates on finished goods and lower rates on imports and raw materials. The highest tariff rates were reserved for luxury items. Duty exemptions were provided for imports used in the mining and oil sectors, goods used in the northwest region, firms operating under the investment incentive laws, and goods supplied from the Andean group trade partners (LAFTA). On the whole, the tariff system was geared to provide revenue rather than protection.

The general sales tax, of the value added type (VAT), was established in 1972–73, levying a 5 percent tax on both imports and domestic output and various stages of large retailer production. Because producers were denied credit for the purchase of capital goods, the Bolivian VAT included both consumption goods and investment goods in its base. If the VAT had been applied on a universal basis, its tax base would have consisted of gross national product (GNP). As it was, the coverage of the VAT was relatively broad and exemptions were available only for purchases of basic foods and grains, certain consumer necessities such as soap and candles, petroleum products, live animals, all printed matter, and all items subject to excise taxation. The inclusion of capital goods purchases in the tax base was argued to be a desirable feature of the VAT, working to compensate for the alleged overvaluation of labor and foreign exchange.

Although the tax authorities were empowered to close down businesses that failed to comply with the VAT, tax evasion was alleged to be widespread. Many firms were slow to pay the tax or paid less than they should have, while others managed to avoid registering for the tax altogether. Moreover, there was no effective audit program to aid in tax enforcement.

Alongside the VAT, a separate services tax was collected using a multitiered rate structure. Services supplied by hotels, clubs, meatpacking plants, interior decorators, and telephone and telegraph operators were subject to a 10 percent tax. A lower rate of 3 percent was applied to the services of international transport companies, barber shops, beauty parlors, electric utilities, and funeral parlors. The lowest rate of tax, 2 percent, was for domestic transportation services. Banking, medical, dental, and hospital services were exempt from tax, while a special 25 percent rate was applied to admission tickets for public entertainment.

Because the tax applied to a wide range of business services, it had undesirable cascading features and became embedded in the structure of production costs throughout the economy. By the same token, the tax encouraged vertical integration as a method of avoiding the impact of the tax. Moreover, even though small firms were subject to the tax they managed to successfully evade it.

As table 2.1 indicates, the excise taxes on alcoholic beverages and tobacco products brought in more revenue than the sales tax. Although limited to domestic products, the excise taxes were coordinated with import duties imposed separately on competing products from abroad. In November 1973 selective consumption taxes were levied on a number of so-called nonessential goods at rates of 10, 20, and 30 percent. Although the tax base for these luxury taxes consisted of both imports and the manufacturers' sales price of competing domestic products, in practice almost all of the revenues generated came from imports. Besides the indirect taxes already noted, there was a host of relatively insignificant revenue contributors like a stamp tax, a tax on soft drinks, an agricultural land tax, and a tax on real estate transfers.

An Evaluation of the Tax System Prior to Reform

In 1976 the Musgrave Report critically assessed the performance of the revenue system. It discovered numerous instances in which the tax system departed substantially from neutral or uniform tax treatment. Examples included the inability to tax commercial agriculture and small businesses in the informal sector, preferential treatment for state-owned enterprises, and significant undertaxation of urban professionals. Further distortions were attributable to a tariff structure that created large disparities in effective rates of protection across sectors and commodities, and the cascading effects of the sales and service taxes on business inputs.

On the basis of skimpy data, the Musgrave Report concluded that the burden of indirect taxes was more or less proportional to household income. Within the indirect tax structure, the tax levied on automobile purchases stood out as being highly progressive, while the taxes imposed on the consumption of beer and cigarettes appeared to be highly regressive. Personal and corporate income taxes also tended to be significantly progressive in their impact. However, because reliance on these taxes was minimal the distribution of the total tax burden across income groups was at best only mildly progressive. Overall, the tax system did little to alter the distribution of income in the economy. The report also pointed out that income level disparities were reduced more by nationalization of the tin and petroleum industries than by changing the tax system.

The Musgrave Report underlined the extremely fragile foundations on which the tax system appeared to rest. With rising oil prices and increased oil production during the early 1970s, Bolivia had become dependent on export taxes as its major source of revenue. As table 2.2 shows, total tax revenues as a fraction of GDP increased from 9.3 percent in 1972 to nearly 13 percent in 1974. The Musgrave Report noted with alarm that the growth in export-related taxes fueled higher levels of current government spending. It therefore recommended that a stabilization fund be established in order to prevent temporary export bonanzas from creating irreversibly higher levels of government expenditure. In a situation in which domestic tax revenues accounted for only one-third of total revenues, the Musgrave Report expressed concern that the revenue system was unusually susceptible to foreign shocks. Because of this critical vulnerability, the most important policy aim was to augment the revenue yield of the personal and enterprise income taxes, as well as the sales tax and domestic excises.

The Musgrave Report urged the government to "prepare itself for future contingencies before any pressure arises and while the adjustment is still easy to make. Bolivia is fortunate in that it can undertake a broadening of its domestic tax base at a time when the pressures on domestic tax revenue are low. This time may be running out, however, and the opportunity should not be lost by undue delay" (Musgrave 1981, p. 271). Looking back now, these admonitions have a prophetic ring.

Table 2.2 Central Government Tax and Spending Efforts
(percent)

Year	*Tax revenue/GDP*	*Expenditure/GDP*
1970	10.5	10.6
1971	9.2	10.8
1972	9.3	10.5
1973	11.5	11.2
1974	12.8	10.5
1975	11.8	14.2
1976	12.1	14.2
1977	11.7	13.9
1978	11.2	13.8
1979	9.2	14.3
1980	9.6	16.0
1981	9.4	15.1
1982	5.0	26.9
1983	2.9	20.1
1984	2.6	33.2
1985	1.3	6.1
1986	10.3	7.7
1987	12.8	14.5

Source: Tax revenue/GDP for 1970 to 1975, Musgrave (1981); for 1976 to 1984, UDAPE (1985); fo 1985 to 1987, World Bank (1989). Expenditure/GDP for 1970 and 1975, Musgrave (1981); for 1979 to 1986, Sachs (1986); for 1976 to 1979, UDAPE (1985); for 1987, World Bank (1989).

The Fiscal Collapse of the Early 1980s

In a sense, the temporary prosperity enjoyed during the 1970s set the stage for the fiscal calamity to follow. Bolivia did not adjust its revenue base to prepare for future contingencies when it had the opportunity to do so. The sound counsel of the Musgrave Report was ignored, and time indeed did run out.[2]

Throughout the 1970s Bolivia experienced an era of sustained prosperity. As table 2.3 shows, annual growth rates were ordinarily in the 6 to 7 percent range. Beginning in 1975, however, a significant increase in the level of government spending was not matched by any corresponding growth in total tax revenues. The divergence in expenditure and revenue growth can be clearly seen in table 2.2.

A temporary export boom in the later 1970s elicited higher levels of government spending and confirmed the worst fears of the Musgrave Report; the growing fiscal deficit was financed by resorting to easily obtained foreign loans.[3] The weak fiscal foundations for Bolivia's prosperity were quickly exposed with the onset of the global recession in the early 1980s. The recession was felt initially through a fall in export revenues and an adverse shift in the terms of trade.

Coupled with the appearance of higher real interest rates in the world economy, declining export revenues soon created doubts about the capacity

Table 2.3 Macroperformance
(*percent*)

Year	*Annual inflation rate*[a]	*Annual growth rate*[b]	*Gross domestic investment/GDP*	*Gross domestic saving/GDP*	*Fiscal surplus (+) or deficit (–)/GDP*[c]
1971	3.6	6.00	24.6	21.5	—
1972	6.5	5.40	29.4	26.4	—
1973	31.6	6.50	18.5	28.0	—
1974	62.7	5.00	21.7	17.2	—
1975	8.0	7.10	17.8	31.2	—
1976	4.5	6.10	29.1	27.2	3.6
1977	8.1	3.20	26.6	25.0	4.0
1978	10.4	2.90	24.0	16.0	1.8
1979	19.6	–1.40	20.2	16.8	1.5
1980	47.2	–2.70	14.1	19.3	–2.2
1981	32.2	0.80	11.9	11.5	–0.8
1982	123.5	–6.90	9.7	14.1	–1.8
1983	275.6	–6.90	3.4	6.3	–18.3
1984	1,284.4	–0.01	5.7	8.6	–29.9
1985	11,857.1	–1.70	–0.8	0.1	–12.7
1986	66.0[d]	–1.90	8.0	3.0	4.0
1987	10.7[d]	3.40	9.5	3.9	10.0

— Not available.
a. Consumer price index, year-to-year percentage change.
b. GNP, year-to-year percentage change.
c. Overall public sector deficit or surplus on current account.
d. January to December change.
Source: Bolivian National Accounts data; for 1986 and 1987 fiscal deficit/surplus data, Mann (1990).

of the government to service its foreign debt, and foreign lending shut down in 1982. Although Bolivia tried to trim both its current and capital expenditure programs between 1982 and 1983, the interest cost component on external debt virtually doubled. Moreover, the decline in export prices produced a sharp falloff in total central government revenue, with the inevitable consequence of a large and rapidly growing fiscal deficit. This deficit soon began to fan a spectacular inflationary fire.[4] Caught between rising expenditures and falling revenues, the government substituted central bank finance for foreign lending, thus unleashing an inflationary expansion of the money supply.[5]

Inflation got off to a running start in 1982 and 1983, when annual rates of economic growth turned sharply negative. Inflation produced a corrosive effect on the domestic components of the revenue system, as is evident from table 2.4. Real tax yields for both direct and indirect taxes declined precipitously between 1979 and 1984. During that period, the real value of tax collections obtained from both the personal income and the enterprise income taxes fell by a factor of four. Lags in the collection process, or the so-called Tanzi effect, combined with inadequate penalties for late payment or even no payment of taxes, were responsible for this outcome. For the most part, real indirect tax collections fell by an even larger factor. The evaporation of real value from a wide variety of indirect taxes was caused by the specific nature of these levies. Tax rates were defined in nominal terms as so many bolivianos per unit output. Because inflation proceeded so rapidly, the tax authorities were unable to adjust nominal rates to keep pace with the accelerating rate of inflation. The only exception to this pattern was the excise tax on soft drinks. Unlike the other excise taxes, it was levied on an ad valorem basis, making it possible to maintain the real value of the tax yield.

As was the case with domestic taxes, trade taxes were likewise eroded by the high inflation. Adjustments in the exchange rate were unable to keep pace with the inflation rate and the resulting overvaluation of the exchange rate diminished real collections from import and export taxes.

Table 2.4 Inflationary Erosion of Tax Bases

	Real tax yields (1980 Bolvianos)	
Tax base	*1979*	*1984*
Alcoholic beverages	42,739	6,487
Beer consumption	651,040[a]	158,452
Cigarettes and tobacco	369,105	62,448
Enterprise profits	942,367	258,947
Personal income	1,121,330	260,910
Sales (value added)	740,198	194,101
Services tax	234,408	67,894
Soft drinks tax	11,712	10,775
Stamp taxes	488,110	84,578

a. Because of an unexplained drop in beer production in 1979, the real yield for 1980 is used.
Source: UDAPE (1985).

On the import side, not only did the real value of nominal tariff revenues fall but the overvalued exchange rate became an incentive for increased smuggling, an activity that generated no tariff revenue. Smuggling also occurred with exports. Neighboring Peru, for instance, which has no tin mines, became a tin exporter during 1983–85. Petroleum was likewise smuggled to Peru in response to controlled domestic prices, which were only about one-seventh of the world market price. Faced with an overvalued exchange rate and unprofitable domestic sales, the state petroleum company paid neither taxes nor transfers to the central government.

A combination of collection lags, an overvalued exchange rate, and per unit commodity taxes was sufficient to doom the performance of the revenue system under conditions approaching hyperinflation. Declining real revenue yields contributed to ever larger fiscal deficits, mushrooming inflation rates, and a continuing decline in real tax yields. During 1982–83 current spending as a fraction of GNP doubled due to the rising interest cost on the public debt and prompted a rescheduling of repayments on that debt. The consequences of this vicious cycle of inflation feeding upon itself is seen in table 2.2. By 1985, real tax revenues had shrunk to only 1.3 percent of GNP, inflation was soaring at almost 12,000 percent annually, and the government faced a dismal economic situation. Between 1980 and 1985 per capita income had declined by almost a third and per capita consumption by nearly 17 percent.[6]

Confronted with an economy in growing disarray, the newly elected government of Paz Estenssoro introduced a bold package of economic reforms under the banner of the New Economic Policy (NEP). The NEP embraced a radically different philosophy, which discarded the previous model of state capitalism and made new commitments to a deregulated and market-oriented economy. Credit, foreign trade, exchange rate, and labor markets were simultaneously deregulated and liberalized during 1985–86. The comprehensive tax reform of May 1986 was one significant component of the new economic policy.

Remarkably, these measures stopped inflation in its tracks. As table 2.3 shows, by 1987 the annual inflation rate was down to 10.7 percent. After a large devaluation of the exchange rate in August of 1985 and an associated hike in domestic price levels, inflation was promptly brought to heel.

Two factors explain this rapid return to conditions of overall price stability. The first was the success of the authorities in stabilizing the exchange rate. Because the economy was effectively "dollarized" and the dollar, therefore, served as the economy's unit of account, although not its medium of exchange, the creation of a stable and realistic exchange rate limited the increase in domestic prices to the world rate of price inflation. The inflation-induced absence of any long-term credit or labor market contracts assisted the authorities in maintaining a stable exchange rate. The second important factor at work was the government's pledge not to incur new fiscal deficits. A new Office of Tax Collections was established under the NEP as an independent fiscal entity. Government expenditure levels on current accounts were constrained to the amount of revenue taken in by this office. By this time, the rise in domestic oil prices alone had raised the tax collection rate to about 7 percent of GDP.

The restoration of a noninflationary environment in 1987 was unaccompanied by any contraction of total output in the economy. The higher taxes that came in the wake of the tax reform represented a shift from reliance on the inflation tax to other, less distorting sources of taxation. Moreover, the drastic depreciation of the boliviano also raised public sector revenues substantially. However, while total output in the economy did not contract during this stabilization exercise, economic growth has remained in the doldrums.[7] The challenging task facing Bolivia today is to undertake measures that will revive the engine of economic growth. What started out as the debt crisis of the early 1980s became the growth crisis of the late 1980s.

The Tax Reform of 1986

The tax reform of 1986 was more of a revolution than a revamping or modification of the existing tax system. With the exception of export taxes and import duties, which remained intact, all of the taxes in operation prior to the reform in 1986 were swept aside, replaced with a battery of new taxes on domestic activity. As a transition device, all taxpayers were subject to an extraordinary, one-time, regularization tax in 1986. For companies not in the resource sector, this tax was set at 3 percent of the current value of fixed assets and inventories. In the case of individuals and trusts owning real estate, vehicles, boats, or airplanes, the tax was set at 50 percent more than the amount normally paid under the presumptive income tax to be levied on owners of property.

As table 2.5 shows, the personal income tax and the enterprise income tax disappeared, along with the inheritance tax and all social security taxes. The earlier excise taxes on cigarettes, beer, and alcoholic beverages, along with some components of the selective consumption tax, were rolled into the new tax on items of specific consumption. A host of minor revenue sources, including the tax on soft drinks, the stamp tax, small business tax, tax on real estate transfers, and the agricultural land tax, also vanished from the revenue landscape. In total, seven new taxes on domestic activity replaced about fifteen previously existing taxes.

What were the goals of the 1986 tax reform? One of the primary aims of the reform was to enhance the tax system's revenue capacity and to make it less vulnerable to both inflation and economic shocks. This was to be accomplished by reducing reliance on trade taxes and, within the group of domestic taxes, by increasing reliance on domestic indirect taxes. Another objective was to introduce a simpler tax system that could be more easily administered, one that would significantly reduce the gap between revenue potential and revenue collections. Equity concerns were given short shrift.

Table 2.5 Central Government Taxes before and after Reform

Taxes before 1986	*Taxes after 1986*
Agricultural land tax	Complementary tax to the VAT
Alcoholic beverages tax	Export taxes
Cigarette excise tax	Import duties
Enterprise income tax	Simplified tax for small business
Export taxes	Tax on presumed corporation profits
Import duties	Tax on presumed income of property holders
Inheritance tax	Tax on specific consumption
Personal income tax	Transaction tax
Sales tax (VAT)	Value added tax
Selective consumption tax	
Services tax	
Social security taxes	
Soft drinks tax	
Small business tax	
Stamp tax	
Tax on real estate transfers	

Source: Author.

In short, the goals of the tax reform were to improve revenue adequacy, to simplify the overall tax system by reducing the number of taxes that were imposed and also by adopting new taxes that would be easier to administer, and, finally, to obtain greater efficiency in the administrative techniques for collecting taxes. To reach these objectives, tax reform was guided by the principles of taxing what households spend and own instead of what they earn, and by enlisting the willingness of taxpayers to become tax agents in securing better compliance with the VAT. Consumption and wealth, not income and payrolls, were to become the preeminent bases for greater domestic taxation.

The key component of the new revenue system is the new VAT, which is levied at a 10 percent rate and allows almost no exemptions. This reconstructed VAT generated about half of all of the revenue from domestic taxes in 1987. Only housing and financial services, along with many purchases made in informal markets, particularly food, remain outside the scope of the VAT. Exclusion of the informal sector lessens the VAT's regressivity. The VAT relies upon the credit-invoice method of tax collection and uses the destination as a basis for the treatment of tradable goods. Thus exporters pay no tax on their sales and receive a rebate for any tax imposed on the purchase of their inputs, while imports are taxed at the same rate as competing domestic products. Since the VAT rate is applied to the price of a good or service, which includes a new 1 percent transaction tax on the same purchase, the effective VAT rate is somewhat larger than the nominal 10 percent rate. For example, if the producer's price is $100, the consumer price becomes $101 with the transactions tax and $110.10 with the VAT, for an effective rate of 10.1 percent. Tax is payable on a monthly basis. Small taxpayers pay a special lump-sum tax.

To encourage the issuing of invoices and better compliance with the tax, the administration of the VAT was reinforced by a complementary tax. Under this complementary tax, 10 percent of source labor income and most capital income is withheld at source. The tax applies to salaries, rents, royalties, dividends, and professional income but not interest income. However, from this tax, the individual may deduct all of the VAT paid on previous purchases verified by invoices. In addition, there is a lump-sum exemption equal to twice the minimum monthly wage.

The complementary tax is, in effect, a tax on household savings. Since the VAT is a tax on household consumption, the two taxes together are equivalent to a flat rate personal income tax on incomes earned in the formal sector. By giving households an incentive to acquire receipts for their consumption outlays, it is hoped that retailers will remit any taxes they have collected and that they, in turn, will be given an incentive to request receipts from their own suppliers and so forth down the chain of production and distribution. While this effort to create a paper trail is laudable on the grounds of improving tax administration, it is not obvious that the scheme will work. If the choice for consumers is either to pay the VAT and deduct it from their income tax, or to avoid paying the VAT at purchase but pay the full amount of the complementary tax, it is not at all clear that the incentive to demand documentation is very strong.[8]

To support the operation of the VAT and generate a verifiable paper trail, a 1 percent tax is imposed on gross monthly income. The tax applies to virtually all transactions of individuals and companies and is really a turnover tax applied to most sales transactions in the economy. Only publishers, exporters, private sector educators, and government agencies are exempt. Wage and interest payments are also exempt. Because of its turnover nature, activities may be exposed to multiple taxation. It is also not deductible from the complementary tax to the VAT. Although exports are exempt from tax, the inputs of production are not. As table 2.1 shows, only a small fraction of total revenue, 3.6 percent in 1989, was accounted for by this tax.

Sumptuary taxes are also imposed on a number of specific consumer items. These items fall into two categories. Included in the first category are beer, brandy, liqueurs and other alcoholic beverages, and perfumes and cosmetics, all of which are taxed at 30 percent. In the second category are tobacco products and jewelry, which are taxed at 50 percent.[9] For domestic products, the tax base is the retail price, while in the case of imports, the base is the cost, insurance, and freight value (CIF). Neither base includes the VAT, although the VAT base includes the excises.

Taxes on assets are of two types. For individuals there is a new tax on the presumed profits from property ownership. Assets subject to tax include urban and rural real estate and vehicles (airplanes, cars, and motorboats). Rates of tax on urban real estate vary between 1.5 and 3 percent, depending on the type of property and increasing with its value.[10] Municipalities receive 0.35 percentage points of the revenue from this source; the central government receives the balance. The tax on rural real estate was implemented in August 1989 (Mann 1990). According to the legislation, this tax was to be determined on the basis of potential land yield in different areas of the country and also according to the farming techniques used. As adopted, there are separate classifications for arable land, structures, and grazing land. A 1 percent tax applies to landholdings above a certain size and three-quarters of the revenue is earmarked for rural development purposes. Vehicular taxes range between 1 and 5 percent of value and increase according to the value of the vehicle. Municipalities are entitled to the initial 1 percent of revenue from this source; the central government obtains the balance.

Taxes on the presumed profits of enterprises operating in either the private or public sectors are calculated as 2 percent of net worth. Previously, public enterprises were exempt from the profits tax. Cooperatives and resource-based companies are excluded from the effect of this tax. Assets are defined comprehensively to include both real and financial kinds. Fixed assets are valued on the basis of either replacement cost or current market value and annual asset revaluation is to occur according to the rate of inflation. Any taxes previously paid from the ownership of real estate or vehicles are creditable under this tax. A presumptive tax base was chosen in order to overcome some of the problems previously encountered in trying to adequately measure enterprise profits.

Another form of presumptive taxation is the simplified tax, which was designed to apply to very small enterprises operating in the informal sector and to be paid in lieu of all other taxes. To qualify for this tax treatment, a small business is subject to certain capital and sales value limitations. This tax replaces an earlier presumptive levy on small business, which was based on declared capital value. The purpose of the new tax is to eliminate bookkeeping requirements for small business. Under the new tax small businesses are not required to issue invoices or account for their purchases. The only requirement is designating gross revenue to one of three broad income groups. Once this declaration is made, capital is presumed to be 20 percent of gross earnings. Profits, in turn, are presumed to equal 20 percent of earnings, and consumption is presumed to equal 95 percent of profits. Given these presumptions, the VAT is assessed at 10 percent of earnings, the transactions tax is calculated as 1 percent of estimated capital, and the tax on the presumed income of enterprises is determined as 2 percent of presumed capital value. On the basis of the foregoing assumptions, the complementary tax to the VAT is assessed as 0.1 percent of gross earnings. The taxpayer, however, does not have to carry out this complicated computational exercise. Instead, special tax tables indicate the taxpayer's total liability according to the estimate of gross business earnings. This approach to the presumptive taxation of groups that are hard to tax in the economy appears to have borrowed generously from the recommendations of the Musgrave Report. The structure of Bolivia's postreform revenue system is shown in table 2.6.

Table 2.6 Tax Structure after Reform, 1988

Source of revenue	*Revenue as percentage of total current revenue*	*Tax revenue as percentage of total current revenue*
Complementary tax	3.4	7.1
Hydrocarbons (direct taxes)	48.3	—
Import duties	10.5	6.5
Presumed income tax	3.1	6.5
Presumed profits tax	2.5	5.2
Simplified tax	0.4	0.8
Specific consumption tax	4.0	8.3
Transaction tax	3.7	7.7
Value added tax	20.4	42.5

— Not available.

Note: Tax revenue excludes the operating surpluses of the hydrocarbons sector.

Source: Author.

Accompanying the tax reform of 1986, but separate from it, was a more restrictive stance on the provision for investment incentives. Guided by the investment law of December 1981, a generous array of tax incentives had been offered on new investments in most sectors of the economy, including agriculture, cattle, mining, construction, and tourism. The law was applied on a discretionary basis and much of the new investment from 1981 onward was exempt from most, if not all, forms of taxation. Under the new economic policy no new tax incentives will be provided. However, incentives extended between 1981 and 1985 will continue to be honored. Thus, for example, companies that were previously exempted from paying the enterprise income tax would now be exempt from paying the tax on the presumed profits of enterprises.

The tax reform of 1986 also ushered in a new set of fiscal arrangements between the central government and subordinate fiscal entities. Prior to tax reform, certain revenue sources had been earmarked to finance the activities of particular fiscal entities. For example, all of the revenues collected from the excise tax on beer consumption were allocated to cover the cost of operating the regional development corporations. Under the new scheme, referred to as coparticipation, all of the new domestic tax revenues will be collected by the central government and then shared between the central government and other fiscal entities according to a formula under which the central government receives a three-quarter share. Municipalities and the regional development corporations each receive 10 percent, and the universities get 5 percent. The principle of derivation has been respected in the new law so that municipalities and regions receive the amounts collected from their particular jurisdiction. It also means that the principle of regional equalization has been ignored. Certain revenue guarantees were also provided to the subordinate fiscal units. For example, if the 10 percent share allocated to the regional development corporations turned out to be less than the amount of beer tax revenues collected by the department that previously provided the sole source of finance for these corporations, the central government would be required to make up the difference.

Tariff reform was another facet of the new economic policy that had fiscal implications. Before 1985 there was wide dispersion in both nominal and effective tariff rates. Bolivia had developed a tariff structure, typical of many developing countries, in which imported final goods were subject to considerably higher duties than those which were imposed on intermediate inputs and capital goods. As part of the new economic policy, the government adopted a uniform tariff to reduce disparities in the extent of protection across different sectors, control corruption in the application of tariff rates to particular goods, and avoid giving undue encouragement to smuggling. In August of 1986, the government imposed a uniform 20 percent tariff on the CIF value of all imports. In March of 1988 the government announced a staged reduction in the uniform tariff from 20 percent to 10 percent effective in 1991.

Between 1977 and 1988 a number of different techniques for removing taxes on inputs used in the production of exports were tried. From 1977 until March 1984, a tax credit was provided for exporters equal to 10 percent of the value of agricultural exports and artisanal goods and 5 to 25 percent for manufactured goods. However, during the fiscal crisis from 1984 to 1986, the provision for removing the tax on inputs used to produce exports was repealed. An attempt to reintroduce this export tax credit in 1987 met with only partial success. Facing political pressure from exporters, the government passed new legislation in 1988 in support of the input tax credit. Under this legislation, the tax credit was set equal to 5 percent of the value of traditional exports and 10 percent of the value of nontraditional exports. Denominated in U.S. dollars, the credits could be used for payment of any tax or tariff and as collateral for credit or sold to another taxpayer if unused.

A Preliminary Assessment of the Tax Reform

There are two conceivable approaches for determining the success of the 1986 tax reform. One is to ask how well the reforms have achieved the goals that were set out for them. The other is to evaluate the reforms against the traditional triad of normative objectives in public finance: the ability of the reforms to enhance economic efficiency, equity, and ease of tax administration. Although both approaches are briefly explored below, more experience with the new tax system will be required before any definitive assessment can be made.

As mentioned earlier, the primary aims of the 1986 tax reform were to restore the overall tax effort to historical levels and to simplify the system of domestic taxation by reducing the number of revenue sources, and adopt a new set of taxes which would be easier to administer. Improved ability to collect taxes and reduced reliance on trade taxes were other important objectives. Table 2.5 offers a comparison of the tax structure before and after the 1986 reform. As table 2.1 shows, the proportion of total revenue received from direct and indirect forces remained about the same between

1975 and 1989, but there were important changes in the composition of both taxes. Although the relative importance of export taxes changed very little, the diminished importance of import duties reduced overall reliance on trade taxes. Within the higher share of domestic taxes in total revenue, the larger weight of domestic indirect taxes, the VAT in particular, is evident from table 2.1, as is the reduced role assigned to domestic direct taxes. Direct taxes on personal and enterprise incomes have been replaced by presumptive taxes on company profits and on the presumed capital income of asset owners. It is expected that these presumptive tax bases will now be easier for tax administrators to measure than corporate and personal incomes were in the past. From table 2.3 we see that revenue yields have been restored to their previous levels in the early 1970s.

Bolivia also appears to have scored some important gains in tax administration by conserving scarce administrative resources in a number of ways. In addition to reducing the number of taxes and relying more heavily on a presumptive tax basis, the commercial banks now play a major role as collection agent. All of the new taxes may be paid at a commercial bank branch (no effort is made to verify the returns they receive), which also receives the returns filed by individual taxpayers. The branches then make revenue allocations prescribed under the tax-sharing formula for coparticipation and send the balance to their main office in La Paz. Main offices have the dual responsibility of transferring tax revenues to the treasury and transmitting data received from the branches to the automatic data processing center in Internal Revenue. For their services, the commercial banks are allowed to withhold 0.8 percent of the tax revenues they collect as compensation. This procedure also minimizes contact and dialogue between government workers and taxpayers and removes several opportunities for corruption that had plagued the earlier tax regime.

The government has also conducted several registration campaigns for companies and individuals, in which forms were circulated to taxpayers soliciting requests for basic information. In response to these campaigns, twice as many taxpayers are now registered under the current tax system. A single tax registry has also been created assigning a unique identification number to each taxpayer no matter how many taxes he or she might pay. Before, each tax had a separate registration.

Several steps have been taken to secure better taxpayer compliance. One measure has been the creation of an office for large taxpayers in La Paz. This office has established its own database on large taxpayers and carefully checks their tax declarations. There is a special audit program for them as well which provides a quick comparison of the taxpayers' financial accounts with the information kept on file. In addition, it serves as a pilot program for a nationwide system of audits. Enforcement has also been assisted by the threat of jail sentences for tax evasion, the indexation of tax arrears to the value of the U.S. dollar, and the imposition of high interest rates on late payments. New fines established at 50 to 100 percent of the amount of tax evaded, accompanied by the loss of a business license or, in some cases, the closing down of the business for as long as six months, have increased the cost of tax dishonesty. Tax inspectors have already closed a number of retail establishments in La Paz for failing to issue invoices with the amount of VAT indicated. VAT collections originating from small businesses have apparently been stimulated by each experience with closure.[11] Computers used to track tax receipts and monitor payments facilitate the administration of the new tax system.

When the neutrality of the new tax system is compared with that of the earlier regime, it seems likely that the reforms, on balance, have reduced the extent to which the tax system distorts resource allocation. The current VAT is more uniform and avoids the cascading effects inherent in the earlier VAT's taxation of capital goods. The complementary tax to the VAT provides a degree of self-enforcement to the VAT, but is itself something of a mixed blessing. By itself it is a tax on household savings and may, therefore, distort the intertemporal allocation of resources. However, to the extent that high-income groups are able to purchase VAT invoices in the black market, perhaps from low-income taxpayers who do not pay this tax, the effective tax rate on household savings and, therefore, the size of the intertemporal distortion, may be reduced.

Because it is a tax on turnover, the transactions tax is of questionable merit. Its chief virtue is that it is relatively easy to administer and may also aid in the enforcement of the VAT. Neither of these benefits, however, may compensate for the cascading effects of this tax on resource allocation. On the other hand, it is possible that the now defunct taxes on services produced even worse cascading effects. With regard to excise taxes, the sumptuary levies on tobacco products and alcoholic beverages are nonneutral from a narrow perspective, but from a wider perspective, may offset important externalities associated with their consumption.

The tax on the presumed profits of enterprises has many of the features of a lump-sum levy. As such its marginal effective tax rate on investment spending is zero and it will not distort investment decisions within firms.[12] On the other hand, the tax base can be reduced by resorting to debt finance. The tax may thus encourage the use of debt. A more satisfactory base for this tax might be corporate assets, especially in light of the fact that interest income is not highly taxed elsewhere in the system. In its present form, the tax also has the disadvantage of deterring investment in the capital-intensive sectors and failing to meet the criteria for credibility under foreign tax treaties.

The tax on the presumed profits of property owners makes no allowance for how the asset was acquired. Thus, within the group of asset holders it discriminates against those who used debt to acquire their assets. Its relatively narrow base also makes it an imperfect proxy for taxation of capital income. Wealth held in forms other than real estate and vehicles escapes the tax entirely. This tax is, therefore, nonneutral in its treatment of different assets.

Finally, the simplified tax on small businesses is, in principle, a neutral tax because of its presumptive nature. However, because the declarations businesses make in order to qualify for this tax treatment are not carefully monitored, it is possible that many large companies in different industries may be tempted to register under the tax. To the extent that behavior of this type occurs, the simplified tax will not be neutral, either within or among different sectors of the economy.

Conspicuously absent from the debate about the attractions of the new tax system is any discussion of equity. The absence of data makes it impossible to compare different income groups' tax burdens under the old and new tax systems. Some concern has been voiced regarding the equity implications of greater reliance upon the VAT. Because this tax allows for no exemptions whatsoever, it is feared that it may have a regressive impact on the distribution of tax burdens. Detailed analysis will be required to determine if there are any grounds for this concern.

One's perspective on this equity issue depends a great deal on which part of the revenue system is replaced by the VAT. As a substitute for the personal income tax, it is not at all clear that there has been any increase in regressivity. As mentioned earlier, the personal income tax functioned as little more than a tax on modern sector labor income, while the VAT may largely skip over most of the activities of low-income households in the informal sectors of the economy. If, on the other hand, the VAT is perceived as having raised total government revenues and reduced reliance on the inflation tax, a less regressive distribution of tax burdens is even more likely. Most empirical studies indicate that the inflation tax is highly regressive and fully deserves the label of being the "cruelest tax of all."[13] Finally, even if the VAT and its companion, the transaction tax, are judged to be regressive, they must be weighed against the progressive effects one would expect from the excise taxes imposed on luxury items, the presumptive tax on enterprise profits, and taxes on the assets held by property owners, as well as any progressive effects which may be associated with the complementary tax to the VAT.

As a final accomplishment of the tax reform, it can also be argued that the tax system is now less susceptible to the ravages of inflation. All of the indirect taxes now in place are assessed on an ad valorem basis and are, therefore, automatically indexed. In addition, the bases of the presumptive taxes have been indexed to the value of the U.S. dollar and collection lags have been shortened for both direct and indirect taxes. As a result of these institutional changes, the tax system is less likely to collapse under the weight of future inflation than it was prior to the reform.

Lessons

Although it is still too early to draw strong conclusions from the Bolivian experience with tax reform, it is tempting nonetheless to extract some broad generalizations.

1. *Small, open economies that make trade taxes the centerpiece of their revenue systems* are extremely vulnerable to the vicissitudes of the world economy.

2. *Countries that rely heavily on trade taxes and other unstable revenue sources* should avoid making expenditure commitments that cannot be easily reversed when total revenue falls unexpectedly.

3. *Countries seldom undertake tax reform exercises until unpleasant economic events force them to do so.* In part, this is because the deficiencies in a country's tax system do not become readily apparent until an economic crisis occurs to reveal them. In the case of Bolivia, the fiscal crisis of the early to mid-1980s catalyzed reform of the whole economy.

4. *Poor countries, weak in tax administration, are better able to apply indirect than direct taxes.* It may be easier, and perhaps more equitable and neutral, to tax consumption, and possibly wealth, rather than income.

5. *Presumptive taxes on both companies and individuals may compensate for intractable deficiencies in tax administration* and conform more closely to the administrative capabilities of low-income countries than more conventional forms of taxation.

6. *The personal income tax is ill equipped to include many forms of capital income in its base unless withholding taxes are applied to this source of income.* The role of the corporation tax, or a more widely assessed company tax, is to tax rents and other forms of capital income that escape taxation at the personal level. Poor countries, however, may have great difficulty in administering a corporate income tax, especially if transfer pricing practices are widespread.

7. *A proliferation of tax incentives may severely undermine* the ability of the corporate income tax to collect a portion of the economy's economic rents.

8. *Unindexed tax systems may be highly volatile in the face of inflation and apt to unravel in a way that makes it more difficult to bring inflation under control.* A key concern of any tax reform is to reduce the sensitivity of the revenue system to inflation. Thus unindexed tax systems may contribute much more to inflationary pressures than indexed tax systems.

9. *Persistent fiscal deficits and a lack of fiscal discipline may be indicative of firmly entrenched political divisions within a country.* The inflation tax may be considered the ultimate fiscal compromise, since it allows populist governments to spend more and conservative legislatures to resist higher levels of explicit taxation.

10. *Tax reforms that eliminate fiscal deficits and restore macroeconomic stability may be essential* for promoting higher rates of economic growth.

11. *Simpler tax systems are easier to administer* and allow tax officials to concentrate on obtaining better compliance with the tax laws.

Notes

1. The elimination of U.S. withholding taxes on interest income paid to foreigners in 1973 meant that this source of income went completely untaxed. This added to the difficulties of any developing country attempting to tax interest income from domestic sources.

2. The Musgrave Report also pointed to the relatively low elasticity of the revenue system. While the elasticity of total revenue with respect to GDP was slightly larger than one, the elasticity diminished to less than one when hydrocarbon revenues were removed and both income taxes (personal and enterprise) exhibited elasticities much smaller than one over the period 1970–75. This spelled doom for the revenue system during the inflationary era that followed.

3. A World Bank (1989) study argues that the proceeds from foreign borrowing were poorly invested at home because of a badly distorted price structure in the economy, and helped to underwrite capital flight abroad.

4. From a wider historical perspective Morales and Sachs (1988) contend that fiscal deficits symbolize an ongoing income distribution struggle between alternating populist governments, which strive to increase spending, and conservative regimes, which resist higher taxes and often hand out business-related subsidies. In their view political pressures to spend vastly outweigh the power to tax, leading to endemic fiscal deficits and periodic bouts of inflation. In light of this deep-seated political disunity, Bolivia is viewed as a victim of its own internal contradictions in which the state is expected to do more than its resources permit. U.S. aid in the 1960s, like foreign lending in the 1970s, masked the true dimensions of Bolivia's fiscal weakness.

5. According to Sachs (1986), by 1985, the inflation tax, defined as the growth in the real reserves of the banking system, had reached 12 percent of GDP, nearly the same percentage value as the shift in net resource transfers (new foreign loans less interest costs on foreign debt) from Bolivia to the rest of the world. As described in the text, this inflation tax replaced other revenue sources which were undermined by the virulent inflation of the early 1980s.

6. Compounding the misery was a sharp drop in agricultural output of nearly 27 percent during 1982–83 as a result of drought in the highlands and floods in the lowlands. World tin prices also reached unprecedented low levels in 1985.

7. For more details of this stabilization program, see Sachs (1986).

8. Since the effective VAT rate is slightly more than 10 percent, households may prefer instead to pay the 10 percent complementary tax and not bother to request receipts from vendors.

9. The possibility of evading tax through smuggling limits the level to which these taxes can be raised. Recently, a porous border has forced the government to scale back some of these excises.

10. More precisely, the 3 percent rate applies to planes while the 1.5 to 3 percent range is applicable to the value of cars and motorboats. Because of the registration requirements for these assets, the taxes imposed on them should be relatively easy to administer. The recently implemented tax on rural real estate has the interesting feature that three-quarters of its yield will be used for investment in rural areas. The benefit feature of this levy should promote better compliance with the tax.

11. There is some concern, however, that stricter enforcement of the VAT may drive some firms into the informal sector, where they cannot be reached by the tax.

12. It could be questioned whether it is truly beyond Bolivia's administrative capacity to develop a satisfactory, and higher-yielding, income tax on enterprises and argued that the appropriate role of the presumptive levy is that of a minimum tax which should be fully creditable against any form of income tax.

13. In Colombia, for instance, Berry (1980) concludes that inflation has had a significantly regressive impact on the distribution of income, both before and after the imposition of taxes. Gil-Díaz (1987) reached the same conclusion in the case of Mexico.

References

Berry, R. Albert. 1980. "The Effects of Inflation on Income Distribution in Colombia: Some Hypotheses and a Framework for Analysis." In R. Albert Berry and Ronald Soligo, eds., *Economic Policy and Income Distribution in Colombia*. Boulder, Colo.: Westview Press.

Gil-Díaz, Francisco. 1987. "Some Lessons from Mexico's Tax Reform." In Nicholas Stern and David Newbery, eds. *The Theory of Taxation for Developing Countries*. New York: Oxford University Press.

Mann, Arthur J. 1990. "Bolivia: Tax Reform, 1988–89." *Bulletin for International Fiscal Documentation* 44(January): 32–36.

Morales, Juan Antonio, and Jeffrey Sachs. 1988. "Bolivia's Economic Crisis." In Jeffrey D. Sachs, ed., *Foreign Debt and Economic Performance: Summary Volume*. Chicago: University of Chicago Press.

Musgrave, Richard A. 1981. "Fiscal Reform in Bolivia: Final Report of the Bolivian Mission on Tax Reform." Harvard Law School, International Tax Program, Cambridge, Mass.

Sachs, Jeffrey. 1986. "The Bolivian Hyperinflation and Stabilization." NBER Working Paper No. 2073: 1–51. National Bureau of Economic Research, Cambridge, Mass.

UDAPE (Unidad de Análisis de Politicas Economicas). 1985. "El Sistema Tributario en Bolivia." La Paz, Bolivia.

World Bank. 1989. "Bolivia: Country Economic Memorandum." Report No. 7645-BO (restricted access). Latin America and Caribbean Region, Country Department III, Washington, D.C.

3

Thirty Years of Tax Reform in Colombia

Charles E. McLure, Jr.
George Zodrow

The tax system of Colombia has been in almost constant flux during the last thirty years. This paper examines the process and results of tax reform over this period. It begins with an essentially chronological summary of the many episodes of tax reform that have occurred since the 1960s. This discussion focuses on the major features of the various episodes, but also considers those details that illustrate best the nature of the evolution of tax reform. Particular emphasis is placed on the details of changes that reduced the complexity of the individual income tax. In addition, the chapter considers the extent to which changes in the tax system over the past thirty years have reflected changes in the professional "conventional wisdom" of the day. The most important changes in conventional wisdom include (1) a loss of faith in economic planning and targeted investment incentives; (2) decreased emphasis on income distribution and vertical equity; (3) increased concern for administrative simplification, economic neutrality (including neutrality toward investment decisions), and horizontal equity; and (4) the growth of attention to the problems created by the interaction of inflation with a tax system based on nominal (unindexed) historical values.

Colombia experienced significant structural changes in its tax systems in the early 1950s and 1960s, and, as a result of comprehensive reform initiatives, in 1974, 1986, and 1988. Dramatic policy reversals occurred during this span. For example, dividends were made taxable in 1953 and nontaxable in 1986. Widespread tax incentives were embraced in the early 1960s and discarded a decade later. Inflation adjustment was rejected completely in 1974, but it has gradually been enacted, and the 1986 and 1988 reforms provided full indexation of the measurement of income.

Normally sparked by recurrent fiscal crisis, the episodes of tax reforms have accompanied a fluctuating ratio of total revenue to gross domestic product (GDP) and a gradual shift from trade taxes to domestic revenue sources. A value added tax (VAT) was introduced in 1966 and has become a revenue mainstay.

The reform of 1974 added large amounts of capital income to the personal income base, established a single corporate rate (but with a preferential rate for economically similar limited partnerships), eliminated many tax incentives, extended the presumptive income tax based on net wealth, and broadened the sales tax base to include many services while increasing sales tax rate differentials. It is estimated that progressivity increased significantly as a result of these reforms: perhaps as much as 1.5 percent of GDP was extracted from the top quintile of taxpayers.

More recent reforms in 1986 and 1988 were driven by the different goals of achieving greater simplicity, neutrality, and horizontal equity and reversing the observed "decapitalization" of firms. In 1986 dividends were excluded from personal taxation, corporate and personal rates were reduced significantly, the corporate and top personal rates were equalized, more labor income was included in the personal tax base, and the inflation component of interest income was made nontaxable while the inflation element in interest expense became nondeductible. The personal tax threshold was also raised, while alterations to withholding and filing requirements reduced the number of taxpayers required to submit returns. The 1988 reform completed the indexing initiative begun in 1986 by extending inflation adjustment to depreciation and moving gradually towards an "integrated" or balance sheet approach. In 1989 the net wealth tax was scrapped and the rate of presumptive income taxation was reduced; these changes have dramatically reduced progressivity at the top of the income scale.

In this chapter we calculate the marginal effective tax rates (METRS) on capital income and provide quantitative estimates of the net impact of the tax structure on marginal decisions to invest for various combinations of asset type, finance method, and type of saver. Calculations are performed for the pre-1986 tax structure, the tax system under the 1986 law (when fully implemented), and the tax structure of the 1988 law (when fully implemented). The METR rates under the reforms proposed in a 1988 report on the taxation of business and capital income are also sometimes reported to serve as a nearly neutral point of reference. These calculations demonstrate that the tax system has improved in terms of economic neutrality with respect to the allocation of investment across asset types and business sectors, the choice of financial policy (whether to finance new investment with debt, the issuance of new shares, or the retention of earnings), and the choice of business organization.

Emphasis is given to the progress of efforts to simplify the individual income tax system, including the implications of tax reform for simplification. It focuses on the inevitable tension between accurate income measurement and administrative simplicity, especially in the areas of personal exemptions and the ceding of income, deductions for personal expenses, the tax treatment of owner-occupied and rental housing, filing requirements, and the double taxation of dividends.

Finally, we draw some conclusions about the process and results of tax reform in Colombia. The Colombian experience is described as one of gradual groping toward an improved tax system. This process has been heavily influenced by foreign advisers. Reform has recently emphasized the goals of simplicity and neutrality in an effort to make the tax system conform more closely to administrative realities. The study makes a plea to supplement tax reform with administrative reform and warns against the dangers of carelessly importing fiscal fads from other countries.

Background

The tax system has been in almost constant flux during the past thirty years (OAS and IDB 1965; Bird 1970; Musgrave and Gillis 1971; Gillis and McLure 1977, 1978; McLure 1982, 1989a, 1990a; Perry and Cárdenas 1986; Thirsk 1988; Perry and Orozco de Triana 1990; and McLure, Mutti, Thuronyi, and Zodrow 1990). Reforms have often been dramatic, and sometimes involved reversals of earlier fundamental decisions. For example, dividends were made taxable in 1953 and then exempted in 1986. A complex system of personal exemptions, deductions, and income splitting was provided in 1960, modified substantially during the 1970s, and then largely eliminated in 1986. A wide-ranging system of tax incentives for investment was introduced in 1960, in response to the recommendations of the United Nations (1957) Economic Commission for Latin America (ECLA), and then allowed to expire ten years later. Inflation adjustment in the measurement of income from business and capital was first rejected but subsequently enacted. In 1989 the net wealth tax, an important part of the fiscal landscape since 1935, was repealed, effective in 1992.

Two major studies of tax reform, headed by the North American experts, were conducted in the 1960s (OAS and IDB 1965; Musgrave and Gillis 1971). These greatly influenced the content of massive tax reforms in 1974 and provided support for a number of subsequent smaller improvements.[1]

Fundamental reforms enacted in 1986 greatly simplified the tax system, dramatically lowered tax rates, eliminated taxation of dividends, and provided for inflation adjustment of interest income and expense. The 1986 act also provided the president with discretionary power to change inflation adjustment during the following two years. He exercised this power at the end of 1988, introducing changes based on a third major study by foreign experts (hereafter referred to as "the 1988 report"), this one on the taxation of income from business and capital (McLure, Mutti, Thuronyi, and Zodrow 1990, chapter 11).

Changes in the tax system have often been made in response to the conventional wisdom of the day, as well as to economic and political conditions in the country. Since the conventional wisdom has often been a moving target, the apparent goals of tax policy have also changed over time, sometimes quite markedly (Thirsk 1988; McLure 1989a, 1990a). Among the most important changes have been the following: the loss of faith in economic planning and targeted investment incentives; a decline in attention to issues of income distribution and vertical equity; increased concern for simplification, economic neutrality, and horizontal equity; and increased attention to problems created by inflation interacting with an unindexed tax system.

These shifts in emphasis can be detected in the academic research on Colombia. In the 1960s and 1970s numerous books and articles were published on distributional equity and the incidence of tax and expenditure policies; by comparison, that field was relatively dormant during the 1980s.[2]

The early disenchantment with investment incentives can be traced, in part, to the devastating critique of their effectiveness provided by foreign advisers working in the National Planning Department and highlighted by the Musgrave Mission.[3] Colombia was among the first developing countries to reject the interventionist ECLA strategy for economic development based on fiscal incentives.

The 1986 reforms showed more concern with economic neutrality than any previous tax reform. One explanation may be the development of the analytical technique of calculating METRS on capital income, an important tool in quantifying distortions in investment decisions caused by differential taxation. Economists may simply be paying more attention to what they can measure—or think they can measure—and convincing policymakers to do the same.

Probably much more important was the attention devoted to neutrality in the 1986 U.S. income tax reform. Whereas neutrality had previously been largely a theoretical objective espoused primarily by academic economists, the publication of the U.S. Treasury Department's 1984 report to President Reagan, entitled *Tax Reform for Fairness, Simplicity, and Economic Growth*, and the subsequent passage of the Tax Reform Act of 1986 in the United States gave tax neutrality greater respectability and urgency. The rationale for the 1986 tax reforms, as described in "Exposición de Motivos," resembles explanations of the need for U.S. tax reform in the Treasury Department report to President Reagan.

A similar influence may have contributed to the introduction of inflation adjustment for business and capital income. Historically, such inflation adjustment has been employed by countries with much higher inflation rates—countries such as Argentina, Brazil, Chile, and Israel. No industrial country employs such adjustments on a large scale, and prior to 1984 few governments of industrial nations had seriously considered introducing them.[4] Colombia might have been expected to reject inflation adjustments, both to avoid being lumped together with the high-inflation countries and because of failure to appreciate the costs that even low rates of inflation can create in an unindexed system. Provisions for inflation adjustment in the U.S. Treasury Department's 1984 tax reform proposals may have helped overcome both these objections by giving the idea respectability and by pointing out the costs of an unindexed system, even in a country with low inflation rates.

The 1986 reforms were motivated by the belief that tax policy contributed to the "decapitalization" of the economy, especially in the industrial sector (Carrizosa 1986; Chica 1984, 1985; and Perry and Cárdenas 1986, pp. 178–81). This concern helps to explain both the elimination of the double taxation of dividends and the introduction of inflation adjustment for interest expense in the 1986 reforms.

The road to simplification has been torturous, especially for individual income tax, which from its inception onward was fairly complicated. At various times during the period surveyed, further complications were added. But the predominant trend over the period has been toward simplification. This reflects the realization that an administrable system that achieves rough justice is preferable to a conceptually superior system that cannot be administered.

The shifting emphasis of tax policy can be seen by examining three periods (Perry and Cárdenas 1986). Three developments during the first period, from the introduction of the modern income tax in 1935 until the mid-1960s, set the stage for much of the discussion that follows.

In 1953 taxes on corporations were increased and dividends received by individuals were made taxable. In 1960 the government adopted a system of investment incentives; ECLA had advocated these policies in the name of economic planning and development. These measures inefficiently redirected the allocation of economic resources, conflicted with the achievement of both horizontal and vertical equity, and eventually were allowed to expire. In 1960 the government also adopted income splitting and a system of personal exemptions and deductions intended to adjust for household circumstances, which again added considerable complexity.

Concern for distributional equity dominated the proposals of the Taylor and Musgrave Missions in the 1960s. Both missions recommended raising the top individual income tax rates from 51 to 62 percent (Taylor) or 55 percent (Musgrave). Both missions advocated broadening the tax base as a means to reduce vertical inequity,[5] and paid little attention to the objectives of simplification and economic neutrality.[6] Whereas the Taylor Mission was critical of then-existing investment incentives, its condemnation was lukewarm (perhaps because they were already being allowed to expire), and it even gave a favorable appraisal of using well-designed incentives to encourage investment in selected activities (OAS and IDB 1965, pp. 87–91).

Neither of these reports recommended inflation adjustment. By comparison, the 1974 reforms, responding to economic reality, started adjusting for inflation. Explicit inflation adjustment for nominal values and some interest income became widespread. It was extended to virtually all interest income and expenses in the 1986 law and to depreciable assets and inventories in the 1988 reforms.[7] Inflation adjustments introduced between 1974 and 1986 were almost wholly "homegrown," with little influence from foreign advisers.

Finally, much greater emphasis on simplification and economic neutrality characterized tax policy in the 1980s. The reduction in the top individual income tax from 49 to 30 percent in the 1986 reforms indicates just how much views on tax equity and the distortionary effects of high tax rates had changed in the twenty years since the Taylor Mission, which proposed a top rate of over 60 percent. Further weakening of concern for equity is evidenced by the 1989 repeal of the net wealth tax, effective in 1992.

Revenue needs have conditioned reforms to varying degrees throughout the period. Reforms intended to raise revenue, such as introduction of the sales tax, have been enacted during times of fiscal crisis. Yet other reforms intended to improve the system have occasionally been enacted when there was no budgetary pressure. Finally, in times of relatively plentiful revenues "counterreforms" frequently reversed earlier progress.

Given these shifts, it may appear difficult to appraise the success of tax reform in Colombia. Judging the various episodes of tax reform according to the criteria prevalent at the time would yield a favorable judgment of all the reform episodes (excluding counterreforms or "deforms"). The early efforts to tax dividends and provide targeted investment incentives were applauded for reasons that appeared sound at the time, but both were subsequently repealed, again for reasons that appeared sound. Those who endorsed the high progressive rate schedules of the 1960s were naturally appalled at the reduction of the top individual and corporate rates to 30 percent. Most likely they are also chagrined by repeal of the net wealth tax.

There can be little doubt that the present income tax system is structurally superior to that of the late 1950s. It contains fewer loopholes created by exemptions, special incentives, and inappropriate deductions. Comprehensive inflation adjustment helps prevent both overtaxation and undertaxation of income from business and capital. As a result, the system is more neutral and has greater horizontal equity. In many respects it was also made simpler with the cutback on efforts to fine-tune tax burdens to family circumstances, which were just too difficult to administer. On the other hand, inflation adjustment adds to complexity.

The system is probably less progressive than it was thirty years ago because of the reduction of corporate and individual tax rates, the elimination of income tax on dividends, and the repeal of the net wealth tax. Whether vertical equity has been unduly compromised is a matter on which reasonable persons can and will disagree.

Indirect taxation is also much improved. A hodgepodge of archaic and defective levies has been replaced by a VAT that is far more appropriate.

It has been said that the tax system falls short in one important dimension; it yields the same amount of revenue, as a percent of GDP, obtained during the 1960s and 1970s. Urrutia (1988) has argued that tax reform throughout Latin America should focus on raising more revenues to finance badly needed public services.

This study examines the process and results of tax reform beginning in the mid-1950s. The next section provides a chronological summary of tax reforms over the three decades that follow, highlighting proposals and changes in the law that illustrate the evolution of tax reform. The chapter then examines the evolution of tax policy in terms of revenues, distributional effects, economic neutrality, and simplification.

Virtually nothing is said about taxing international trade. To some extent this reflects the way trade taxes have been considered in earlier appraisals of tax policy. Although the Taylor Mission devoted two full chapters to this issue, the Musgrave Report contained only about three pages on it, dealing primarily with the coordination of tariffs and internal indirect taxes. In part because tariff reform had just been studied by the National Planning Department, the topic was outside the terms of reference of the Musgrave Mission (OAS and IDB 1965, chapters 8 and 9; Musgrave and Gillis 1971, pp. 12, 112–14; and for the coordination issue Gillis and McLure 1971). Although Gillis and McLure (1977) contains a short chapter on taxes and subsidies on foreign trade, Perry and Cárdenas (1986) have very little to say about the taxation of international trade. Because of its limited focus on taxation of income from business and capital—a focus that was, however, wider than the issue of inflation adjustment that motivated it—the 1988 report did not address trade taxes.[8]

Episodes of Tax Reform

Colombian tax reform has been a constant process over the past thirty years.[9] For convenience, the major episodes of tax reform are defined as (1) the incentive provisions enacted on the advice of ECLA; (2) the Taylor and Musgrave Reports and the base-broadening reforms that resulted from them, especially in 1974; and (3) the movement toward simplification, economic neutrality, and generalized inflation adjustment in the measurement of income begun in 1986 and continued in 1988.

In addition, there were many minor episodes of reform and counterreform, such as those that followed the 1974 reforms. These are also considered briefly, since they reveal important characteristics of the process of reform. Separate subsections will describe the incentive legislation resulting from the ECLA proposals, the tax system that prevailed after the 1960 reforms, the recommendations of the Taylor and Musgrave Missions, the reforms enacted in 1974, the 1974–86 changes in the tax law, the 1986 reforms, the 1988 report on inflation adjustment, the ensuing 1988 reforms, and the repeal of the net wealth tax.

The Economic Commission for Latin America Proposals and the 1960 Incentives

The history of tax incentives, which goes back almost as far as the Republic, is similar to other less-developed countries (Bilsborrow and Porter 1972, pp. 398–99; OAS and IDB 1965, p. 33; Perry and Cárdenas 1986, pp. 195–200). Before 1940 income tax exemptions were commonly granted on an ad hoc basis to specific firms. More recently they have, at least in principle, been made generally available to all firms meeting certain criteria. (Because criteria have often been vague and applied with discretion, this difference may be more apparent than real.) The government attempted to provide incentives for industries of "basic importance" to the development of the country, which often turned out simply to mean something that the country was not then producing. Iron and steel were favored, a pattern institutionalized in the 1960 tax reform. Incentives were also provided for investments in agriculture, accruals of an economic development reserve (up to 5 percent of profits each year), and reserves for replacement of industrial machinery and equipment.

The Tax System after the 1960 Reforms

In 1960 there had been a tripartite system of direct taxes (the so-called income and complementary taxes): the individual and company income taxes, a net wealth tax, and an excess profits tax (OAS and IDB 1965, chapter 1 to 4, and 10). The 1960 reforms changed the income and net wealth taxes in important ways, but did not significantly alter many of the nonincentive provisions that constituted the basic structure of the law.[10]

Income taxes on individuals. Several types of labor income were exempt from individual income tax, including interest on government securities and limited amounts of other interest and dividends. Tax on interest and dividends paid to holders of bearer securities could easily be evaded, though there was a 12 percent withholding tax on such payments. Capital gains were made taxable in 1960, but with a 10 percent exemption for each year the asset had been held. Income from owner-occupied housing, in excess of a small exclusion, was subject to tax.

The law allowed personal exemptions for the taxpayer and dependents, as well as deductions for expenditures on medical, educational, and professional services; the latter depended on the income and the number of children of the taxpayer and were thus complicated. Deductions were also allowed for interest expenses, real estate taxes, social security taxes and pension contributions, and a limited amount of residential rent. Separate returns were required of married taxpayers, but a limited amount of earned income could be ceded to the spouse for tax purposes. There were fifty-six marginal tax rates, ranging from 0.5 to 51 percent.

Net wealth tax. A net wealth tax was imposed in 1935 and repealed in June 1989, effective January 1, 1992. A substantial proportion of the net wealth tax base consists of real estate. The rationales supporting this tax were to increase the progressivity of the tax system, especially on income from capital; induce more productive use of land; offset the effects of evasion by high-income individuals; and help prevent such evasion.[11] The tax and the presumptive income tax based on it may have been useful in taxing wealth amassed from illegal activities, especially the drug trade.

Taxation of business income. Under the 1935 income tax law, corporations and individuals were taxed under the same rate schedules, and dividends were exempt when received by individuals. Partnerships were exempt, but their income was taxed to their owners. In 1953 dividends were made taxable.[12] Business taxation distinguished between corporations (*sociedades anónimas*), limited liability companies (*sociedades de responsabilidad limitada*, hereafter *limitadas*), and ordinary partnerships. Corporate rates were 12, 24, and 36 percent; these also applied to the income of all foreign-based entities. By comparison, rates of only 4, 8, and 12 percent were applied to the income of *limitadas*, the organizational form used for many important businesses. The partnership rates were 3 and 6 percent. The entire net income of *limitadas* and partnerships was imputed to the owners of the entity for tax purposes, whether distributed or not. By comparison, corporate shareholders were liable for tax only on income distributed as dividends.

Capital consumption allowances were based on the straightline method, with useful lives of five years for motor vehicles and airplanes, ten years for other personal property, and twenty years for real property. Depreciation allowances were not indexed for inflation. There was, however, a provision for deductions to add to a replacement reserve intended to compensate for the effects of a large devaluation of the peso in 1957.

Under the 1960 reforms, all foreign companies doing business in Colombia were taxed as corporations. In addition, the withholding tax on dividends paid to foreigners was raised from 5 to 12 percent. The withholding tax on profits of Colombian branches of foreign corporations was 6 percent (raised to 12 percent in a 1963 decree), and collected on income not reinvested domestically.

Administrative shortcomings. Some of the administrative problems encountered by the Taylor Mission were amenable to direct or simple legislative remedies. These included absence of withholding, even on wages and salaries, total reliance on official assessments, rather than self-assessments, and a statute of limitations so short that taxpayers could file false returns knowing the errors would not be detected within the two-year period.

One of the most important problems resulting from the lags in collection of the income tax was the shortfall in revenues produced by inflation. When prices rose, government expenditures also rose, but income tax collections lagged. This was emphasized especially by Bird (1970, p. 33). More recently it has been called the "Tanzi effect," after Tanzi (1977).

Indirect taxation. In the early 1960s internal indirect taxation had a distinctly colonial (or even medieval) flavor, concentrating more on predictability of revenues, administrative ease, and certainty of collection than on such principles of taxation as fairness, economic neutrality, and ease of compliance. There was no broad-based tax on consumption such as a VAT or a retail sales tax. More than half of the revenues from indirect taxation came from stamp taxes and the sale of stamped paper. Indirect taxes accounted for less than 10 percent of total tax revenues of the central government (OAS and IDB 1965, pp. 9, 210).

The Taylor and Musgrave Missions

The Taylor Mission was sponsored jointly by the Organization of American States (OAS) and the Inter-American Development Bank (IDB). The Musgrave Mission was organized at the request of President Carlos Lleras Restrepo and financed with Colombian funds. Although both missions produced impressive reports and similar recommendations, the Musgrave Report is remembered, both inside and outside the country, as having had greater direct impact on policy.

Since the Musgrave Mission accepted and repeated many of the recommendations of the Taylor Mission, it is difficult to differentiate between the separate and indirect impacts of the two reports. Similarly, it is difficult to identify the separate impact of Bird (1970) because its author had the benefit of the Taylor Mission's report and, after his manuscript was completed, was intimately involved in preparing the terms of reference for the Musgrave Mission.

Both the Taylor and Musgrave Missions agreed that the combination of income and net wealth taxes was good, but they disagreed fundamentally on whether the excess profits tax should be retained. The Taylor Report said, "The trinity of an income tax, excess profits tax, and net wealth tax represents a development that is essentially ingenious, progressive, and enlightened—both in terms of goals of tax policy and administration" (OAS and IDB 1965, p. 54). The Musgrave Report said that "the Colombian income tax is a relatively well-developed and sophisticated statute in comparison with others in Latin America," (Musgrave and Gillis 1971, p. 35) and that "the net wealth tax fulfills a valuable role in the Colombian tax structure" (Musgrave and Gillis 1971, p. 98). It recommended in the strongest terms that the excess profits tax should be abolished.[13] The excess profits tax was abolished in 1974 with little controversy.

Individual income taxes. The Taylor and Musgrave Missions found glaring omissions from the bases of both income and net wealth taxes. Both missions advocated the taxation of most exempt income and the elimination or sharp curtailment of most nonbusiness deductions available to individuals. Whereas the Taylor Mission would have eliminated the partial exclusion for income from owner-occupied housing, the Musgrave Mission offered an additional alternative, the complete exemption of such income, combined with disallowance of deductions for mortgage interest and property taxes. The Musgrave Mission would have made many exemptions and deductions subject to vanishing provisions, under which tax benefits would gradually be reduced (and ultimately eliminated) as income rose above certain levels. It also favored introduction of a standard deduction to simplify compliance and administration. The Taylor Mission would have eliminated the ceding of income from one spouse to the other, whereas the Musgrave Mission endorsed this provision.

The Taylor Mission recommended taxation of nominal capital gains as ordinary income. By comparison, the Musgrave Mission offered two alternatives, taxation of real (inflation-adjusted) capital gains as ordinary income or taxation of nominal gains at a rate of 5 percentage points below the rate otherwise applicable. This was one of the few instances in which the Musgrave Mission considered inflation adjustment. A two-year holding period would be required for capital gains treatment, and the 10 percent per year reduction in taxable gains would be eliminated. Gains on assets transferred at death would be constructively realized.

Neither mission favored inflation adjustment of amounts stated in nominal terms such as personal exemptions and the bracket limits of the rate structure. The Taylor Mission thought that personal exemptions were too high and favored allowing inflation to erode them to more appropriate real levels. The Musgrave Mission argued, on the other hand, that "provision for automatic adjustment tends to remove resistance to inflation and to institutionalize a high inflation rate. These effects are detrimental to sound economic development" (Musgrave and Gillis 1971, p. 51).

Taxation of wealth. Whereas the Taylor Mission made no major recommendations for reform of the net wealth tax, it did propose adding a presumptive income tax based on agricultural wealth to the tax system in order to deal with evasion of income tax in that sector. The Musgrave Mission endorsed this measure and also recommended the elimination of many existing exemptions and that debts be deductible only if related to or secured by taxable assets.

Taxation of business. Both missions explicitly recognized the inequities and distortions that can result from a classical system in which business income is taxed at both the company and individual levels (OAS and IDB 1965, pp. 54–58; Musgrave and Gillis 1971, pp. 80–82). Nevertheless, both rejected the idea of "integrating" the taxes by either taxing all business income as if earned by partnerships or providing relief from the double taxation of dividends. The Taylor Mission favored some upward adjustment in the tax rates applied to the income of limited liability companies, whereas the Musgrave Mission proposed a more far-reaching reform in which graduated business tax rates would be abolished and both corporations and *limitadas* would be subject to the tax regime applied to corporations.[14] Both missions favored more generous treatment of losses.

Both missions considered and rejected indexing of depreciation allowances. The Taylor Mission thought the inflation rate was not high enough to justify explicit inflation adjustment, but favored more generous depreciation allowances as an ad hoc means of offsetting inflation and encouraging investment. The Musgrave Mission recognized that accurate inflation accounting requires adjustment of the entire balance sheet (that is, if adjustment is made for debt or interest expense it must also be made for depreciable assets). It rejected both ad hoc inflation adjustment of depreciation allowances and complete revaluation. Instead, it favored accelerated depreciation in the form of double declining balance depreciation for assets with useful lives of at least five years and increased depreciation for assets used for more than one shift. Neither mission noted the potentially adverse effects on corporate financial policy, resource allocation, and the perception of equity that could result from the combination of accelerated depreciation and full deduction of nominal interest expense.[15]

The Taylor Mission generally did not favor incentives for investment in particular activities such as those introduced as a result of the ECLA report. Rather it preferred general incentives such as accelerated depreciation and the liberalization of loss carry-forward.[16] The mission appears not to have been opposed in principle to incentives based on a careful appraisal of their potential for achieving particular objectives. The Musgrave Mission took a similar line, opposition to the existing incentives and a preference for direct grants. It also offered guidelines for incentives in three areas: general investment, targeted investment, and investment in particular regions.[17]

The Taylor and Musgrave Missions had little to say about taxing the income of foreign investors, aside from endorsing the changes enacted in 1960 and described above. The Taylor Mission recommended that service expenses incurred abroad but related to the earning of Colombian source income be allowed as a deduction. It also favored eliminating limitations on the deductibility of salaries, which were likely to impinge heavily on Colombian branches and subsidiaries of foreign firms, and replacing them with the authority to identify dividends disguised as salary payments (OAS and IDB 1965, p. 82). The Musgrave Mission favored a rebuttable presumption that home office and other foreign expenses in excess of 10 percent of taxable income are not deductible and a fairly complicated sliding scale to eliminate deductions on salaries above a certain amount paid to those owning at least 10 percent of the stock of the employing company (Musgrave and Gillis 1971, pp. 86–87; Slitor 1971, p. 508).

Administrative improvements. The Taylor Mission proposed withholding on wages, salaries, interest, and dividends; self-assessment by taxpayers; advanced payments based on estimated income; extension of the statute of limitations; and stiffer application of penalties. The Musgrave Mission suggested improved taxation of sectors that are hard to tax (for example, small business, independent professionals, agriculture), and that agricultural losses not be allowed to offset income from other sources.

Indirect taxation. Besides advocating the rationalization of indirect taxes through the elimination of the most outmoded forms of taxation, the Taylor Mission made the following recommendations in the area of indirect taxation: elimination of stamp taxes levied only for revenue, rationalization of other stamp taxes, and introduction of "a broad system of excises on semiluxury and luxury goods," which it favored as being less regressive than a general sales tax.[18]

In 1965, during the interval between the two missions, the government introduced a broad-based sales tax levied at rates ranging from 3 to 10 percent.[19] Rather than using the invoice/credit technique found in standard value added taxes, the authorities opted for a manufacturers' tax on finished goods and imports. The system was converted to an invoice/credit-type VAT in mid-1966. Like many VATS imposed by developing countries, the Colombian tax initially extended only through the manufacturers' level; contrary to common practice, no credit was allowed for tax paid on capital goods.[20] The Musgrave Mission recommended relatively little in the area of sales taxation, but did note that it might eventually be desirable to move to a retail level sales tax and exempt capital goods from the tax (Musgrave and Gillis 1971, chapter 8).

The 1974 Reforms

The 1974 reforms were quite far-reaching (Perry and Cárdenas 1986, chapters 2 and 13). They repealed the excess profits tax, rationalized many aspects of the income and net wealth taxes, added a presumptive calculation of income based on net wealth, and improved the sales tax.

The process of reform. Shortly after his election in 1974, President Alfonso Lopez Michelson decreed several fundamental reforms.[21] In doing so, he used emergency powers provided by the constitution. However, the Council of State (Consejo de Estado) almost immediately ruled that the first 49 articles of the reform exceeded the authority that could be exercised under the emergency powers, and were thus unconstitutional. This was a serious setback to tax reform, since these articles contained administrative reforms without which the structural reforms were largely ineffective.[22]

The 1974 reforms built on the work of both missions and created a cadre of trained tax professionals with a personal interest in improving the tax system. The Taylor and Musgrave proposals were not simply taken from the shelf, dusted off, and enacted without further consideration. Some were rejected for various reasons, and others were retained, modified, extended, or improved (or weakened, in a few cases) before enactment.

Individual income tax. The 1974 reforms eliminated many of the exemptions for nonlabor income. Among these were exemptions for interest on government debt and income from investments previously subject to incentive legislation. Exemptions for labor income were left unchanged because of a constitutional prohibition against using emergency powers to override labor laws.

Based on the advice of the Musgrave Mission, several personal deductions had already vanished with increases in income. To eliminate the resulting complexity, the 1974 reform converted these vanishing deductions to tax credits; a standard credit analogous to the more common standard deduction was also introduced.

"Occasional gains," a concept that included gambling receipts, 80 percent of gifts and inheritances, and nominal interest in excess of 8 percent per annum on certain indexed bonds, as well as gains on assets held more than two years, were taxed at a marginal rate determined by deducting 10 percentage points from the marginal rate that would apply if 20 percent of occasional gains were added to other income.[23] All taxable gains were made subject to this rule; the 10 percent per year reduction of taxable gains was retained only for owner-occupied housing.

Personal allowances, bracket limits in the rate schedule, and other amounts fixed in monetary terms were indexed for inflation, up to 8 percent per year. Complete adjustment was not provided because of the desire to preserve the countercyclical features of the income tax and because it was thought that there could be a quick return to the lower rates of inflation of earlier years (Perry and Cárdenas 1986, p. 24).

The cost of capital assets could be adjusted for inflation annually, again up to 8 percent per year. In addition, assets could be revalued to 1974 levels. Both adjustments were optional; but adjustments had to be used for purposes of the net wealth tax and the presumptive income tax, and revaluations not made currently could not be claimed upon subsequent disposition of the asset. Such revaluations could not be used for the purposes of calculating depreciation allowances.

Net wealth tax. Many of the reforms of the net wealth tax proposed by the Musgrave Mission were adopted in 1974. In particular, exemptions ended for government debt securities, assets yielding exempt income, and assets not capable of yielding income.

Business income tax. Graduated tax rates applied to the income of companies were replaced by flat rate taxes in the 1974 reform, as proposed by the Musgrave Mission. On the other hand, the taxation of corporations and *limitadas* was not unified; instead the corporate rate was set at 40 percent and the rate applied to income of *limitadas* and partnerships set at 20 percent. As before, owners of the last two types of businesses were taxed on their pro rata share of the entire (retained and distributed) net income of companies.

Rules for depreciation followed the Musgrave Mission proposals of double declining balance for assets with lives of more than five years and 25 percent extra depreciation for each extra shift an asset is used.

Inflation adjustment was not provided for depreciation allowances or for interest income and expense with some exceptions, the most important being partial indexing in the case of constant purchasing power obligations issued by financial institutions to fund mortgage lending.

Administrative safeguards. As noted earlier, the procedural "teeth" of the 1974 law were extracted by the judicial ruling declaring the relevant provisions unconstitutional. Perry and Cárdenas (1986, p. 269) note that although the congress passed a replacement for the provisions that had been declared unconstitutional in 1975, the new procedural law lacked many of the most important provisions of the prior effort. However, several substantive provisions did survive, including the prohibition on the use of agricultural losses by individuals to offset income from other sources.

Presumptive income tax. The proposal made earlier by both the Taylor and Musgrave Missions for a presumptive income tax on the agricultural sector had been adopted in 1973. In 1974 this concept was extended to the entire economy. Under the 1974 law, income was presumed to be at least 8 percent of the value of net wealth. In addition, it was presumed that any unexplained increase in net wealth constituted income for tax purposes.

Indirect taxation. The 1974 reforms made several important changes to the sales tax. First, businesses could no longer buy on a tax-exempt basis; they could use only the credit/invoice method. Second, the tax base was expanded to include many services, and the spread between the various rates was increased, by adding a preferential rate of 6 percent on nonexempt wage goods and raising the rate on luxuries to 35 percent. Virtually all food and agricultural machinery were made exempt. Finally, imported capital goods destined for basic industries were made tax exempt, even though tax would be charged on comparable domestically produced capital goods (Perry and Cárdenas 1986, pp. 221–22).

Public enterprises. The 1974 reform subjected the income of state enterprises (except for those offering public services) to tax. It also eliminated the enterprises' exemptions from customs duties; though these exemptions were quickly restored, the principle had been established that exemption was a matter of policy, not an automatic right (Perry and Cárdenas 1986, pp. 32, 267; Gillis and McLure 1977).

Counterreform and Further Reform, 1974–86

From 1974 to 1986 the tax system underwent more or less continuous change.[24] For the most part these changes were concentrated in certain areas, most notably capital gains; the definitions of exempt income and the basic structure of credits for personal exemptions and tax-preferred personal expenditures remained more or less unchanged. Many of the most important changes were clear retrogressions or "counterreforms" undertaken in response to previous reforms, presumably at the behest of powerful special interest groups. Though some reform did occur, many of the provisions identified earlier as inappropriate survived; this is especially true of exemptions for labor income. Perhaps worse, administrative problems continued, so that a system that was quite admirable on paper was much less satisfactory in practice.[25]

Individual income tax. There were serious reversals of previous reforms in the taxation of capital gains, especially in 1979. The rate of tax on capital gains was set at no more than one-half that on ordinary income, subject to a minimum of 10 percent. Compared to the complicated formula for the taxation of occasional gains contained in the 1974 reform, this conferred a substantial benefit on high-income individuals with capital gains. In addition, gains were exempt from tax as long as at least 80 percent of the gain (and the original inflation-adjusted investment) was invested in specified assets. Since the length of time for holding qualified investments was not specified, this was essentially an optional exemption for capital gains and any other income that could be recharacterized as capital gains.

Inflation adjustment of amounts specified in nominal terms was progressively allowed until, by 1979, all such values were fully indexed. Inflation adjustment was also increasingly extended to interest income. By 1983, 60 percent of the monetary correction credited to accounts held by individuals in certain fixed-value deposits (40 percent in the case of certain other debt) was exempt.[26]

Measures were taken to reduce the number of taxpayers filing returns. Returns were not required for those with at least 80 percent of their income from labor subject to withholding, provided that the remaining nonlabor income was also subject to withholding, the taxpayer was not a shareholder in a *limitada,* net wealth did not exceed a certain figure, and the taxpayer was not liable for sales tax. For such taxpayers, no income or net wealth tax was due beyond income tax withheld at source.

Presumptive income. In 1983 the value of real estate for the purpose of presumptive income tax was limited to 75 percent of its cadastral value as a concession to agricultural interests (Perry and Cárdenas 1986, p. 45). In addition, a measure of presumptive income equal to 2 percent of receipts was added to the system based on net wealth. Supposedly, this was to prevent evasion by merchants who overstated the cost of goods sold.

Business income tax. The law was further liberalized in 1976 by allowing the taxpayer to use any depreciation rate desired for personal property (that is, property other than real estate), as long as the deduction taken in any one year did not exceed 40 percent of the original cost of an asset. Thus eligible assets could be written off over a three-year period. In combination with the 25 percent premium for extra shifts, this implied that as much as 60 percent of the cost of an asset could be written off in the first year, and the entire value in two years.

A credit equal to 10 percent of corporate dividends received was provided in 1983. More generous credits were allowed for dividends received from "open" corporations (up to 34.1 percent of dividends) and for taxpayers receiving only small dividends (20 percent of the first $200,000). The tax rate applied to the income of *limitadas* was also reduced to 18 percent.

In late 1982, using economic emergency powers, a government decree gradually introduced inflation adjustment for interest income and expense. Since the country was in a serious economic recession, the timing of such a reform, which would have increased taxes for many companies, was inopportune, and the new law was repealed (Perry and Cárdenas 1986, pp. 178–80).

Indirect taxation. The sales tax was extended to the retail level in 1983. A simplified system was made available to small retailers, in order to ease compliance and administration. More services were made taxable, and the 6 and 15 percent rates were unified into a single rate of 10 percent. The second set of changes became necessary when taxes were extended to the retail level. In 1984 exemptions were eliminated for certain goods, including agricultural machinery and transportation equipment (Perry and Cárdenas 1986, pp. 222–23, 230–31; Perry and Orozco de Triana 1990).

The 1986 Reforms

The 1986 reforms followed closely on the heels of the fundamental tax reform in the United States and had much the same goals: horizontal equity, simplification, and economic neutrality.[27] Yet they differed in one crucial way; they provided for the inflation adjustment of interest income and expense, a recommendation that had been considered and rejected in the United States.

Horizontal equity. Though the 1986 reforms still favored some economic sectors, they did make further progress toward eliminating tax preferences. Rules intended to prevent evasion and fraud by independent professionals and businesses establishing related nonprofit organizations were tightened significantly. Business expenses claimed by professionals could not exceed 50 percent of income (90 percent in the case of those in construction, provided they kept registered books of account). Nonprofit organizations were subject to a 20 percent tax on income not used for specified exempt purposes. Payments to nonprofit organizations economically related to the taxpayer were not deductible.

Exemptions were eliminated for severance pay, Christmas and "thirteenth month" bonuses, vacation pay, pensions in excess of a generous floor, and representation allowances. But exemptions were retained for such politically sensitive items such as representation allowances for high government officials, teachers, and judges, and for the income of the military in excess of the basic amount, as well as for pensions below the ceiling. Preferential treatment was retained for cattle raising, forestry, commercial airlines, and navigation.

Simplification. Several changes were made to reduce filing requirements. First, the minimum limit on income and net wealth for filing were raised; this was a major simplification. Second, the increase in the zero bracket amount from $180,000 to $1 million eliminated taxes for many Colombians.[28] Third, credits for personal exemption and for the expenses of residential rent, health care, and education were eliminated. Imputed income from owner-occupied housing was no longer taxed. Mortgage interest remained deductible, but subject to annual limits. Finally, the benefits of income splitting resulting from the ability to cede a limited amount of labor income to the spouse were abolished. Tax liability is now calculated entirely on an individual basis. These changes made withholding on wages and salaries simpler, less subject to fraudulent claims for excessive numbers of dependents, and much more accurate. Filing a return is neither required nor allowed for many taxpayers.

Colombia now recognizes that the achievement of "rough justice" is preferable to attempts at a more refined standard of equity if such efforts are rendered ineffectual by administrative deficiencies. In general, simplicity facilitates administration, as well as compliance. Administrative problems include difficulty verifying the number of dependents claimed, deductions for health care, etc. Moreover, with respect to the repeal of the provision for limited income splitting, there is substantial professional support for taxation of individual, rather than family, income.[29] Whereas the traditional view has been that the consuming unit should also be the taxpaying unit, more recent analyses stress the simplicity and incentive benefits of separate (individual) filing.

Economic neutrality. The 1986 reforms unified the taxation of corporations and limited liability companies, thus eliminating an important source of distortions and inequities. Similarly, the taxation of the income of decentralized agencies, mixed enterprises, and business and financial activities of nonprofit organizations was increased to create greater parity with the for-profit private sector.

The 1986 reforms exempted dividends received by individuals from taxation.[30] Such a rough-and-ready approach to integration is not conceptually attractive, since the aggregate (company and individual) tax burden on equity investments in companies equals the company rate, rather than the marginal rate of the individual shareholder. The 1986 change can be seen as a retreat to the pre-1953 policy of exempting dividends, making it questionable on equity grounds given the concentration of share ownership. Moreover, one can question the allocative benefits of providing any relief from double taxation of dividends; under one view, such benefits can be expected only to the extent that new share issues are involved, but revenue is lost on dividends on all shares.[31] Yet, assuming one wants to provide relief from double taxation of dividends, the simplicity of this approach suggests that it is probably appropriate for Colombia.

This change, along with interest indexing, to be described below, was felt to be important to prevent or reverse the decapitalization of the economy.[32] Debt finance is attractive if either (1) the return to equity is subject to double taxation, or (2) investors anticipate an increase in inflation and lenders either do not pay tax or do not also anticipate the acceleration of inflation. Chica (1984/85) and Carrizosa (1986) presented persuasive evidence that the distortions of financial policy in favor of debt finance one

Table 3.1 Indebtedness of Companies
(debt as percentage of debt plus equity)

Year	All companies	Manufacturing companies
1950	24.0	—
1951	27.0	—
1952	27.0	—
1953	32.0	—
1954	31.0	—
1955	29.0	—
1956	32.0	—
1957	32.0	—
1958	34.0	—
1959	35.0	—
1960	37.0	34.2
1961	37.0	35.9
1962	40.0	37.9
1963	40.0	40.0
1964	43.0	42.5
1965	43.0	42.8
1966	45.0	45.7
1967	44.0	44.1
1968	44.0	44.5
1969	46.0	—
1970	43.9	42.5
1971	47.4	46.8
1972	49.8	49.0
1972	49.8	49.0
1973	53.7	52.7
1974	57.0	57.0
1975	60.5	59.9
1976	62.1	61.5
1977	61.6	60.4
1978	61.1	59.4
1979	65.1	63.5
1980	73.0	—
1981	71.2	—
1982	71.7	—
1983	71.2	—

— Not available.
Source: For all companies, Carrizosa (1986, p. 32); for manufacturing companies, Chica (1984/85, pp. 232–33).

would expect to result from the combination of an unindexed tax system and the double taxation of dividends have indeed occurred (table 3.1). Starting from a level of just under 25 percent in 1950, indebtedness as a percentage of the sum of debt and equity for all corporations rose to around 45 percent for the late 1960s and by 1980 stood at over 70 percent (Carrizosa 1986, p. 32). In the manufacturing portion of the corporate sector this ratio rose from below 35 percent in 1960 to more than 60 percent by 1979 (Chica 1984/85, pp. 232–33). Decapitalization of the corporate sector was clearly occurring.

Moreover, the term structure of debt issued by manufacturing corporations has also become shorter over time. Short-term debt (less than two years),

Table 3.2 Term Structure of Debt of Manufacturing Companies
(percentage of total debt)

Year	Short term	Medium term	Long term
1970	30.95	28.19	40.86
1971	27.21	28.03	44.76
1972	21.00	24.84	54.16
1973	27.51	21.87	50.62
1974	31.89	20.70	47.41
1975	32.26	21.24	42.50
1976	43.74	21.40	34.86
1977	45.83	26.06	28.11
1978	46.01	27.40	26.59
1979	47.59	25.53	26.88
1980	49.94	26.98	23.08

Source: Chica (1984/85, p. 220).

which had fallen from 31 percent of total debt in 1970 to 21 percent by 1972, had risen to 50 percent by 1980 (table 3.2). By comparison, long-term debt (more than five years) fell from 54 percent of total debt in 1972 (after rising from 41 percent in 1970) to only 23 percent in 1980 (Chica 1984/85, p. 220).

The composition of asset holdings of households reflects these changes. Only about 14 percent of household assets consisted of deposits in credit institutions in the early 1960s; almost 70 percent were in "investments," including shares of companies. By 1975 the former figure exceeded 27 percent, and the latter had fallen to just over 50 percent (Carrizosa 1986, p. 34).

It is no simple matter to explain these developments in the financial structure of Colombian business. It seems almost certain, however, that the incentives provided by tax policy, combined with inflation, played an important role (Carrizosa 1986, p. 19; Urdinola 1987, pp. 58–59).

It is as yet impossible to know the long-term effect of these changes in tax policy, but it may be worthwhile to report one fragmentary piece of evidence that changes are having the desired effect. In the first two years affected by the 1986 reforms, primary share issues by companies listed on the Bogotá stock exchange increased by 53 percent and 73 percent over the previous year.[33] Taken at face value, this seems to be strong testimony to the intended power of the tax changes.

However, several caveats are in order. First, though the 1986 reforms were effective retroactively for the entire year, they were only proposed in October and passed in late December. Thus any effect they would have had in 1986 could only be due to the expectation of legislation. Second, if the increases in convertible bonds (including a 56 percent drop in 1987) are included in the calculation, the rates of increase of share issues in the two years fell to 38 percent. Finally, the shares of companies listed on the Bogotá exchange comprise only a small fraction of total corporate equity; the much more important question is what happened in the unlisted companies.

Rate reduction. Perhaps the most dramatic single element of the 1986 reform was the reduction in marginal tax rates. The top individual rate was reduced from 49 to 30 percent and the corporate rate was reduced from 40 to 30 percent and applied to limited liability companies, a questionable change on vertical equity grounds. Occasional gains are now taxed at the same rate as other income—but through separate "schedular" application of the regular rate structure to such gains, rather than by adding the gains to other income.

The 1986 act raised the withholding rate on dividends paid to foreigners to 30 percent. In combination with the corporate rate of 30 percent, this left the net return to a foreigner (49 percent of the before-tax return) almost unchanged from its previous level (48 percent of the before-tax return). Since the 1986 reduction in U.S. rates meant Colombian taxes could not be claimed as a credit against U.S. taxes, the withholding rate was reduced to 20 percent in 1988 to avoid penalizing American investment in Colombia.

Inflation adjustment. The final fundamental reform contained in the 1986 package was the inflation adjustment of interest income and expense. (Adjustment was effective immediately for interest income received by individuals, but was phased in over ten years for interest expense and interest income received by companies.) Inflation adjustment was not extended to depreciable assets and inventories, but inflation-adjusted values were to be used in calculating capital gains upon the disposition of assets.[34] The 1986 act provided presidential discretion to change the provisions relating to inflation adjustment during 1987–88.

The 1988 Report

In order to further appropriate reforms under the extraordinary faculties granted by the 1986 law, the government commissioned a third major study of tax policy by foreign experts (McLure, Mutti, Thuronyi, and Zodrow 1990). The resulting report, released in late 1988 and entitled *The Taxation of Income from Business and Capital in Colombia* (hereafter the 1988 report), addressed issues in the inflation adjustment of the measurement of income, but also discussed such related matters as timing issues, net wealth tax, presumptive income tax, and the possible substitution of a system of direct taxation based on consumption, rather than income. In each case the report treated international issues, where relevant.

The report's most important feature was the provision of a three-way choice between (1) switching to a consumption-based direct tax termed the simplified alternative tax (SAT), (2) extending the ad hoc approach to inflation adjustment begun(for interest income and expense) in the 1986 reforms to depreciation and the cost of goods sold from inventories, or (3) adopting an integrated approach to inflation adjustment of the type employed in Chile.[35]

The consumption-based direct tax. The SAT contains three basic features: (1) expensing is allowed for all business purchases, whether of current services, additions to inventories (for example, raw materials), or depreciable assets; (2) no deduction is allowed for the payment of interest or dividends, and no tax is levied on interest income or on dividends received; and (3) there are separate taxes for labor and business income with the former levied at graduated rates and the latter at a flat rate equal to the top individual rate. The primary advantage of the SAT, as suggested by its name, is its potential for simplifying tax accounting. Timing issues and the need for inflation adjustment largely disappear under the SAT due to its reliance on cash flow accounting techniques.

The SAT produces a METR on capital income of zero; thus it is equivalent to granting a tax exemption to capital income.[36] This has the advantage of being conducive to saving, investment, and the repatriation of Colombian capital that has left the country in response to more generous tax treatment elsewhere (Perry and Cárdenas 1986, p. 45).

On the other hand, the SAT can be questioned on grounds of vertical equity, especially given the concentration of ownership and wealth, and several other troublesome issues. First, it is not clear whether the United States and other capital-exporting countries would allow a foreign tax credit for the SAT and withholding taxes levied in conjunction with it. Second, under the SAT the link between the income tax and the net wealth tax would be broken making it impossible to employ a presumptive SAT, as under the income tax. Third, the link would be broken between financial accounting and accounting for tax purposes. This is a potentially powerful argument against the SAT, given the widely recognized need to improve financial accounting in the country. Fourth, some of the same issues that complicate administration of the income tax also characterize calculation of the base of the net wealth tax, which the 1988 report advocated retaining. Finally, there is the daunting prospect of being the first country to introduce this novel fiscal instrument. India and Sri Lanka (then Ceylon) had a brief and disastrous experience with a different form of direct consumption-based tax in the 1960s (Goode 1960). Although that system was sufficiently different from the SAT that it provides little useful precedent, policymakers are naturally chary about being the first in recent times to go down the direct consumption tax road.

Ad hoc inflation adjustment. Extending the ad hoc measures for inflation adjustment would be relatively simple. Under the 1986 law (when fully phased in) deductions would not be allowed for the inflationary component of interest expense, and the inflationary component of interest income would not be taxed. The value of depreciable assets and inventories would be adjusted for inflation, and depreciation allowances and the cost of goods sold would be based on the inflation-adjusted values. Depreciation schedules would be based on the best available estimates of economic depreciation, and inflation-adjusted first in, first out (FIFO) inventory accounting would be required, that is, last in, first out (LIFO) inventory accounting would no longer be acceptable (McLure, Mutti, Thuronyi, and Zodrow 1990, appendix 7A).

The integrated Chilean system. A more sophisticated alternative involves using figures from inflation-adjusted balance sheets in the calculation of taxable income. The beginning balance of depreciable assets (and other "real" assets and liabilities, including indexed debt) are adjusted for inflation; depreciation allowances are based on this inflation-adjusted figure. A similar inflation adjustment is made to net wealth. When the adjustments to depreciable assets (and other real assets and liabilities) and net wealth are included in the calculation of taxable income, the result is an accurate measure of real taxable income.[37] As with the ad hoc approach, depreciation schedules must be realistic and inventory accounting would be limited to FIFO.

Other issues. The 1988 report contained proposals to deal with a number of timing issues, including depreciation, advance payments, original issue discounts, amortization of intangibles, construction period interest, long-term contracts, and installment sales. In addition, it discussed the treatment of operating losses; taxation of capital gains and losses, and below-market loans; and the preferential treatment of cattle, timber, and several other industries. Net wealth tax issues covered include limitation of deductions for debt to the reported value of associated assets (McLure, Mutti, Thuronyi, and Zodrow 1990, chapters 5 and 6).

Particular attention was given to international aspects of the taxation of income from business and capital. Note that the zero METR under the SAT eliminating the incentive for capital flight was created by the fact that Colombia taxes income from capital earned in the United States, but does not tax interest paid to foreigners. The exemption of dividends and the inflation adjustment of interest income contained in the 1986 reform reduced incentives for capital outflow, though to a lesser degree. Because of the U.S. Tax Reform Act of 1986, the fraction of American firms with excess credits has increased dramatically, implying that the potential for SAT credits against U.S. income tax liability is less important than sometimes thought.[38]

The 1988 Legislation

At the end of 1988 the government exercised the extraordinary powers provided by the 1986 legislation to extend inflation adjustment in the measurement of income from business and capital. Two legislative decrees defined a two-stage process of reform, with transition rules extending into the next century. During the first stage, from 1989 to 1991, the ad hoc approach, extended to depreciation, would give the business community and individual taxpayers time to become accustomed to comprehensive inflation adjustment. Then in 1992 there would be a switch to the conceptually more satisfactory integrated system used in Chile.[39]

Stage I changes in the tax law. The changes to be implemented during 1989–91 were relatively modest, especially when compared with the wide-ranging recommendations of the 1988 report.[40] These fell into three broad categories: inflation adjustment, presumptive taxation of income, and other provisions. The descriptions that follow focus on the most important changes in the taxation of income from business and capital; they are not intended to be exhaustive. Changes that affected the base of the net wealth tax are described, although that tax was eliminated in 1992.

The basis of depreciable assets acquired after 1988 could be adjusted for inflation. Net wealth tax and depreciation allowances on such assets would be based on inflation adjusted values. Such adjustments could be made only when also used to calculate profits and losses for financial purposes. Inflation adjustment was to be based on the change in the consumer price index (CPI) over the period ending October 1 of the year prior to the taxable year in question. The accelerated depreciation schedules of prior law were not modified. The use of LIFO inventory accounting would not be allowed after 1998. (Under the second stage of the 1988 reforms, replacement cost accounting would be provided for companies and for some individuals.) Until that time LIFO would continue to serve as a form of ad hoc adjustment for inflation in calculating taxable income. The difference in the value of inventories under LIFO and FIFO would be gradually reflected in net wealth beginning in 1992 . Such differences would not be subject to income tax. Beginning in 1989, owners of shares in publicly held companies who have a net wealth of less than $20 million could use the market value of such shares in calculating capital gains on the disposal of shares. In essence, this would eliminate the tax on capital gains from holding such shares. The portion of nominal income of companies representing inflation adjustment could either be distributed to shareholders as stock dividends (or their analog in the case of *limitadas*) or capitalized without being taxed. The phase-in of the disallowance of deductions for the inflationary component of interest expense would be frozen temporarily, with the fraction disallowed kept at its 1988 level of 30 percent through 1991. After that, the phase-in fraction would continue its previous pattern (with the fraction disallowed increasing by 10 percentage points per year). Finally, beginning in 1989, leasing companies would be subject to the general rules for inflation adjustment.

The measure of presumptive income based on gross income would be phased out by 1990. The tax rate applied to net wealth in the calculation of presumptive income would be reduced from 8 to 7 percent. The value of real estate for purposes of the net wealth tax and the presumptive taxation of income would be 100 percent of its cadastral value, beginning in 1990. In addition, for the purpose of the presumptive income tax the exemption of 60 percent of the value of beef and dairy cattle was repealed. To the extent that presumptive income exceeds income as calculated under the ordinary income tax system, the excess could be carried forward to the next two years as a deduction from gross income.

Law 84 of 1988 provided the president with emergency powers to reduce or eliminate the net wealth tax. This law also applied a 20 percent tax to the income of nonprofit organizations to the extent such income is not employed for the social purpose of the organization. For these organizations there would be no inflation adjustment in the measurement of income, including interest income and capital gains. Immediate deductions would be allowed for nonfinancial investments. In addition to the provisions described here, the decrees dealt with the treatment of business deductions for future pension payments, imposed a tax of 2 percent on payments to foreigners for the rental of certain equipment, and eliminated the income tax on dividends received by Colombian branches of foreign corporations.

Stage II changes in the tax law. The second stage of the 1988 reforms dealt almost exclusively with inflation adjustment. In particular, these reforms defined the technique by which the integrated system would be implemented.

The integrated system would be mandatory for taxpayers who are required by law to keep books of account, except for individuals qualified to use the simplified regime under the sales tax and nonprofit organizations. Most profit-seeking companies would be required to use the system, leaving most individuals under the ad hoc system.[41] Those not required to use the integrated system could do so voluntarily. Those using the integrated system must do so for purposes of both tax and commercial accounting.

The cost of fixed assets (including improvements made during the year and assets acquired before 1992) would be adjusted for inflation using the change in the CPI. If the cadastral value of real estate exceeds the inflation-adjusted cost, the former value would be employed as the value of the asset. Depreciation allowances and gains and losses on the disposition of such assets would be based on such inflation-adjusted values, except that depreciation would be disallowed for land. No change would be made in the depreciation schedules provided under prior law. Inventories would be valued at replacement cost.[42]

Corporate shares traded on a stock exchange would be valued at market prices. Untradable *limitadas* and corporate shares would be valued on the basis of book value certified by the company once it had revalued its assets.

Inflation adjustment would be applied to indexed monetary or financial assets. It would not be applied to ordinary monetary assets (those fixed in nominal terms). In the case of assets and liabilities denominated in foreign currencies, inflation adjustment would be based on the change in the exchange rate for the relevant currency. Interest and similar expenses incurred before an asset is placed in service (for example, construction period interest and the cost of acquiring inventories) would be capitalized; only interest occurring after such time would be subject to inflation adjustment and deducted in the year incurred.

In addition to specifying procedures for implementing the integrated system of inflation adjustment, the 1988 reforms included the following provisions. First, a loss carry-forward of five years was specified. Second, the distinction between ordinary income and capital gains and losses was eliminated for taxpayers subject to the integrated system.

Appraisal of the 1988 Changes

The 1988 reforms are consistent with the recommendation of the 1988 report in most, but not all, significant respects (McLure, Mutti, Thuronyi, and Zodrow 1990, chapters 5 and 7). It is inconsistent to introduce inflation adjustment for depreciable assets without modifying the schedule of accelerated depreciation allowances to reflect economic depreciation more accurately. This can easily be seen in the changes in the METRs that resulted from the 1988 reforms. Under prior law, at an inflation rate of 20 percent the acceleration of depreciation allowances roughly offset the failure to allow inflation adjustment, producing a METR on income from investments in depreciable assets (equipment and structure) by tax-exempt organizations approximately equal to the statutory tax rate. By comparison, the uncompensated introduction of inflation adjustment reduces the two METRs to well below the statutory rate, creating a tax distortion favoring investment in such assets. At an inflation rate of 20 percent, the two METRs under 1985 law were 35.4 percent and 42.2 percent, compared to a statutory rate of 40 percent. By comparison, the 1988 law reduced these METRs to 9.3 percent and 20.7 percent, respectively, compared to the statutory rate of 30 percent. These are the METRs for income used for tax-exempt purposes.

The 1988 law required that during the 1989–91 transition period, inflation adjustment of depreciation allowances had to be used for financial accounting if it was also used for tax accounting and that companies had to use the integrated system for both purposes beginning in 1992. These "conformity requirements" should make financial accounting more accurate and prevent abuses based on questionable bookkeeping practices.

The decision to base inflation adjustment on the change in the CPI during the year ending October 1 of the year before the tax year in question is questionable on policy grounds, even though it may provide administrative simplicity. At the very least, it means that real income will not be measured accurately during periods of changing inflation rates. A potentially more serious problem is that this approach may encourage "game playing" by taxpayers.

Basing the valuation of inventories on replacement costs and basing the calculation of exchange rate gains and losses on changes in the values of various currencies produces a more accurate measure of net wealth than indexed FIFO and the use of the internal rate of inflation for calculating gains and losses on foreign exchange. But they are also more complicated and do not provide a satisfactory measure of taxable income. Indexed FIFO provides a conceptually better measure of income, since it captures the effects of movements in relative prices that have occurred since acquisition of inventories. The same line of reasoning applies to the indexation of assets and liabilities denominated in foreign currencies.[43]

The politics of the matter. Although it is difficult to fully understand the political forces reflected in the 1988 reforms, a few useful comments can be made on the latest tax reform episode. The SAT was rejected for very practical reasons. First, it would have involved substantial departures from prior law, including the provision of expensing, the disallowance of interest deductions, and the exemption of interest income. Also, it might have been ruled unconstitutional and beyond the purview of the 1986 emergency powers, which provided only for inflation adjustment. Second, the Spanish version of the 1988 report was only released in mid-November, less than two months before expiration of emergency powers.[44] It would have been politically unwise, and perhaps irresponsible, to attempt to introduce by decree, for implementation in 1989, a reform as far-reaching as the SAT without adequate time for public discussion. It is hardly surprising that the SAT was rejected in favor of inflation adjustment.

The two-stage approach to the introduction of inflation adjustment was a political decision. As with the SAT, it might have been unwise and irresponsible to propose immediate implementation of the integrated approach used in Chile. At the very least, it would be desirable to gain experience under the ad hoc approach before switching to the integrated approach. Moreover, there was some risk that the introduction of reform by emergency powers would have been found unconstitutional. Using two separate decrees to extend the ad hoc system and then replace it with the Chilean system had two benefits: (1) it posted implementation of the new system until 1991, giving time for public discussion of the scheme and allowing fiscal authorities the opportunity to fine-tune the plan, and (2) it guaranteed a fall-back position (continuation of the ad hoc system) if the Chilean system was found unacceptable.

Repeal of the net wealth tax. In June 1989, acting under emergency powers contained in the 1988 legislation, the minister of finance eliminated the tax on net wealth, effective in 1992.[45] The reason given for this radical move was the need to keep the tax law up to date, considering that cadastral values were being updated and that the integrated system of inflation adjustment would keep calculation of net wealth and presumptive income tax up to date.[46]

This reasoning is not persuasive. As noted at the beginning of this section, the net wealth tax has been advocated as a means of increasing progressivity, including more productive use of land, preventing tax evasion, and offsetting the effects of evasion. Although it may be true that the introduction of inflation adjustment will increase the progressivity of the tax system and the potential for the presumptive income tax to affect wealth from illegal activities, such as improved valuation of real estate, it will not eliminate tax evasion. It is worth noting that the 1988 report explicitly favored retaining the net wealth tax, perhaps at a reduced rate.

Revenue Performance

When considering the lessons gained from examining episodes of tax reform it is useful to understand the fiscal context underlying such reforms. This section provides information on the revenue performance of the tax system during the period in which the reforms took place and on the tax burden (ratio of tax to GDP) relative to other countries.

Tax Effort and Composition

Perry and Cárdenas (1986, chapters 4 and 5) report that from 1925 until 1949 the ratio of taxes to GDP fluctuated within a narrow range, from just below 5 percent to just above 6 percent.[47] During this period reliance on indirect taxation fell from almost 96 percent of the total from 1925 to 1929 to just over 62 percent from 1945 to 1952. This reflected a reduction in taxes on imports and exports from 46 percent of the total to less than 24 percent. The modernization of the income tax in 1935 also explains the relative decline of indirect taxation. Yet taxes on exports, particularly coffee, and on imports remained extremely important.[48]

In 1950 the ratio of taxes to GDP stood at 6.5 percent, and by 1955 it had reached 8.6 percent. This rapid growth reflected the increased taxes on exports and imports made possible by the strong coffee prices of the early 1950s. In addition, the 1953 tax reform introduced taxes on dividends (Perry and Cárdenas 1986, p. 79).

During the remainder of the 1950s the tax to GDP ratio stayed level despite declines in revenues from coffee exports and from imports. During the latter part of the decade revenue was sustained through higher taxes on imports and exports and a surcharge on the income tax, but by 1962 the tax to GDP ratio had returned to its earlier level of 6.3 percent. The precipitate decline in this ratio from 1960 to 1962 can be explained by the 1960 tax reform, which included income splitting and expensive investment incentives, lags in collections during the period of inflation, and the implementation of import controls (Perry and Cárdenas 1986, p. 80).[49]

The switch from a system dominated by taxes on international trade to one in which direct taxes played a major role had largely been completed by the early 1960s. This is seen from a comparison of the composition of tax revenues at all levels of government from 1963 to 1965 and from 1979 to 1981 (table 3.3). The most important feature is the apparent drop in the relative importance of taxes on property and wealth and the increase in social security taxes.

During the remainder of the 1960s the tax to GDP ratio recovered, reaching the previous high of the coffee boom years by 1966–67. By 1971 it stood at 11 percent. This can be explained in part by the continuing tax reforms that occurred, including the introduction of withholding and estimated tax payments, the repeal of income tax exemptions, and the introduction of the sales tax, as well as improved economic performance (Perry and Cárdenas 1986, p. 79).

Following a brief dip from 1971 to 1974, revenues as a percentage of GDP recovered in 1975, reflecting the effects of the 1974 reforms. During the remainder of the 1970s, revenues as a percentage of GDP declined once again, due in part to the counterreforms of the period, and by 1982 the

Table 3.3 Composition of Total Tax Revenues, All Levels of Government
(percent)

Taxes	*1963–65*	*1979–81*
Income taxes	26.95	23.97
Individual	14.78	10.95
Corporate	12.17	12.80
Property and wealth	16.52	2.45
Exports and imports	19.13	18.84
Other individual taxes	34.79	32.85
Social security	2.61	15.53
Other	—	6.36
Total	100.00	100.00

— Not available.
Note: Definitions underlying calculations of columns may differ.
Source: For 1963–65, Musgrave and Gillis (1971, p. 27), for 1979–81, Tanzi (1987, p. 214).

Table 3.4 Buoyancy of Central Government Revenues
(percentage increases)

Year	*Taxes*	*GDP*	*Buoyancy*
1979	23.9	30.7	77.9
1980	32.6	32.8	99.4
1981	18.1	25.6	70.7
1982	24.8	25.9	95.8
1983	11.1	22.3	49.8
1984	30.2	26.3	114.8
1985	50.7	28.8	176.0
1986	43.3	34.9	124.1
1987	38.9	30.8	123.1

Source: IMF (1988, p. 281).

tax to GDP ratio stood again at about 8 percent, where it had been during the coffee boom of the 1950s (Perry and Cárdenas 1986, pp. 80–81; IMF 1988, p. 281). Reflecting a familiar pattern of fluctuations, revenues recovered during the late 1980s to a level of about 11 percent of GDP.

Table 3.4 shows the buoyancy of central government revenues over the period 1979 to 1987. Buoyancy relates the percentage growth in tax revenues to the percentage growth in GDP. It thus takes account of changes in tax law and administration, as well as the inherent elasticity of the tax system. In each year from 1979 to 1983 tax revenues grew less rapidly than GDP; in 1983 revenues grew only half as fast as GDP. By comparison, from 1984 to 1987 revenues grew considerably more rapidly than GDP.

The 1986 reforms were intended to be revenue-neutral. Whether this has been realized is difficult to determine. Like the 1986 reforms, the 1988 changes included provisions that should increase revenues (revaluation of inventories, real estate, and depreciable assets) as well as those that would lose revenues (inflation adjustment of depreciation allowances).

This quick historical review indicates just how dependent tax collections have been on external events, as well as on tax reform. When exports have been strong, imports have generally been allowed to rise and the government has taken revenues from both sides of the international flows of funds. When exports and imports fell, a fiscal crisis often developed, spurring most tax reforms to increase revenues.[50] A prime example is the introduction of the sales tax in the 1960s. By comparison, when revenues have been plentiful, tax changes that both worsen the system and lose revenue have often been enacted (as in 1977 and 1979). In a few cases, revenue-neutral changes were made for principled reasons and arguably improved the tax system (1986 and 1988 reforms).

There have, of course, been tax reforms that were not motivated solely—or even primarily—by a fiscal crisis (1953, 1974, and 1986–88). In each case reforms have occurred for reasons that might best be characterized as philosophical or even ideological. Thus in 1953 the leftist government of Rojas Pinilla introduced taxation of dividends received by corporate shareholders. The 1974 reforms followed many of the broad outlines of the Musgrave Mission's report, their chief objectives being base broadening and equity. The 1986 and 1988 reforms were intended to make the system simpler and more neutral.

Relative Tax Burdens

The Musgrave Mission noted that from 1963 to 1965 the ratio of taxes to GDP (11.5 percent) was only 69 percent of the average ratio for a group of seven Latin American countries.[51] It was higher than the corresponding ratio for Mexico (9.9 percent), but lower than those for Argentina (20.1 percent), Brazil (21.4 percent), Chile (20.9 percent), Ecuador (16.7 percent), and Peru (16.0 percent). The ratio for the individual income tax was actually the highest of any of the seven countries; by comparison, the tax on business income was well below the average. Colombia's ratio for all direct taxes was also higher than average, presumably reflecting the net wealth tax as well as the individual income tax. Its reliance on indirect taxes was some 50 percent below the average, reflecting its lack of sales tax. This ratio rose subsequent to implementation of the sales tax in 1965.

In the aggregate the comparable picture for the period 1979–81 had not changed much.[52] Colombia's ratio of tax to GDP stood at 12.2 percent, compared to an average of 17.6 percent for all developing countries in the Western Hemisphere and an average of 18.2 percent for developing countries with comparable levels of income per capita (table 3.5). (Income per capita in Colombia was US$1,380, placing it near the top of the group of countries with incomes of US$850–1,699.) The ratios for the other Latin American countries covered in the comparisons made by the Musgrave Mission were Argentina (19.9 percent), Brazil (23.5 percent), Chile (24.8 percent), Ecuador (12.2 percent), Mexico (16.6 percent), and Peru (17.0 percent). Thus, Colombia is in a virtual tie with Ecuador for last place in this list; its ratio of tax revenues to GDP, as a fraction of the average of 18.0 percent for the seven countries, at 68 percent, was virtually unchanged

Table 3.5 Ratios of Tax to GDP, Colombia and Selected Groups of Countries
(percent)

Type of tax	*Colombia*[a] *(1979–81)*	*Western Hemisphere developing countries*	*Upper-middle-income developing countries*[a]
Total taxes	12.21	18.16	17.62
Total income taxes	2.93	5.75	4.86
Individual income	1.34	2.15	1.42
Corporate	1.57	3.25	2.93
Other	0.03	0.84	0.62
Domestic goods services	4.01	4.73	5.45
Sales, VAT, etc.	1.96	1.89	2.14
Excises	0.78	1.91	2.20
Other	1.27	0.92	1.11
Foreign trade	2.30	5.31	3.96
Import duties	1.50	4.35	3.03
Export duties	0.77	0.81	0.76
Other	0.02	0.15	0.17
Social security	1.90	1.12	2.02
Wealth and property	0.30	0.53	0.53
Other	0.78	0.66	0.81

a. Income per capita in Colombia was US$1,380; the income range in upper-middle-income countries was US$850–1,699.
Source: Tanzi (1987, pp. 210–11, 216, 239).

from fifteen years earlier. The tax to GDP ratios for all major categories of tax were below the average ratios for the Western Hemisphere.

Income Distribution, Tax Incidence, and Tax Reform

The distribution of income is quite unequal. Many of the studies of income distribution in Colombia have been compiled and compared in a study by the United Nations (1988). That study is careful to point out the weaknesses of the data and methodologies used. Given these weaknesses and the focus of the present discussion—tax incidence, rather than income distribution, per se—it seems best to concentrate on the basic distributional patterns found in these studies, which appear to be common to all and beyond doubt, rather than on details, which are subject to much greater uncertainty.[53]

Using what it considers to be the most reliable data from surveys of households by the national statistical office (Departamento Administrativo Nacional de Estadística), the U.N. study finds that in 1971 the 20 percent of households with the lowest incomes living in the seven largest cities received a mere 3.1 percent of all cash income and that the 10 percent with the highest incomes received 41.3 percent. For 1979 it reports corresponding figures of 2.3 percent and 42.3 percent, suggesting a worsening of the urban income distribution (United Nations 1988, p. 81).[54] It appears that the distribution of income may be somewhat less unequal in the rural sector, but that the distribution for the nation as a whole is more unequal than that for either sector because of the difference in urban and rural income levels.[55]

Tax Incidence

There have been several studies of the incidence of the entire tax system (OAS and IDB 1965, pp. 221–28; McLure 1971, 1975).[56] All have taken essentially the same two-step approach to calculating the percentage of income taken by taxes at various income levels: (1) specification of the likely incidence of each tax in terms of its burden on particular groups, and (2) attribution of the tax burdens on particular groups to the income levels in which the members of the various groups of consumers and income recipients fall. In some cases an effort is also made to split burdens between urban and rural sectors.[57] A basic problem common to most studies of tax incidence is the difficulty of dealing with issues raised by the "permanent income hypothesis" or by inaccurate data. It is quite common to find that consumption exceeds income in the lowest income classes. Of course, this is not a sustainable pattern in the long run. One assumes it occurs because households base consumption decisions on income prospects over a period longer than that covered by the survey, experience temporary reductions in income, or receive income not covered by the survey.[58] The result is an estimated pattern of tax incidence that probably overstates the burden of indirect taxes on genuinely low income households (those with permanently low incomes) and thus the regressivity of the tax system in the lowest-income classes. This problem can be ameliorated by ranking households according to consumption rather than by income or by constraining consumption to be no greater than income in any income class. No study of tax incidence in Colombia makes such corrections in a systematic manner, though at least one, Berry and Soligo (1980a), attempts to take account of the problem in a qualitative way.

It is commonly assumed that individual income taxes are borne by the recipients of the income. Since such taxes are paid only by a small minority of the population, they contribute powerfully to the progressivity of taxation. They are also likely to be borne primarily by urban households.

The incidence of the corporate income tax is subject to the greatest uncertainty. The traditional assumption would be incidence on shareholders, but few analysts would now accept such a view, especially given the internationalization of capital markets. Competing assumptions include incidence on owners of all capital in the country, on consumers and/or labor, especially consumers of goods produced in the advanced sector or labor employed in that sector, or on land. Because of the existence of foreign tax credits in many capital-exporting countries, taxes paid initially by foreign multinationals may reduce revenues of foreign governments.[59]

These different assumptions have important implications for the distributional impact of company taxes. Note that these assumptions also have important implications for the estimation of the distribution of income. If the tax is borne by shareholders, it should be imputed to shareholders in estimating the distribution of income, as well as for the purpose of estimating tax incidence (McLure 1975). A tax borne by shareholders would be extremely progressive, given the concentration of wealth in Colombia. A burden shared by owners of all capital would be only slightly less progressive, and attribution of the tax to landowners might be even more progressive. In all these cases, the burden of taxation would fall primarily on high-income urban households.

By comparison, if the tax is shifted to consumers or to labor, it would be substantially less progressive; it would probably not be regressive, because consumers and workers in the rural sector might be affected relatively little. It is readily apparent that whether and how a tax affects those in various sectors depend crucially on the isolation or interconnectedness of various parts of the economy (McLure 1979). Finally, to the extent that a tax is borne by foreigners, including foreign treasuries, it is best omitted from incidence calculations.

The incidence of the property tax raises many of the same problems as the company tax except that such taxes generally would not be creditable in capital-exporting countries. Such incidence is particularly dependent on administrative procedures and capabilities, an aspect of reality that can only be incorporated into studies of incidence analysis with considerable painstaking research. Being levied largely on real estate, there is a greater likelihood that property taxes will be borne by landowners, including those in the rural sector.

Taxes on the export of coffee cannot be shifted to foreign purchasers of coffee; thus they reduce the incomes of coffee growers. Many but not all of these will be rural households; many of the coffee growers with the highest incomes probably reside in urban areas.[60] It is not clear whether the coffee export duties should even be included in a study of tax incidence. To the extent that revenues are used for financing benefits to the coffee sector, they should not be included. By comparison, if they are simply a form of general revenue, they should be included.[61] Inclusion of these taxes in the analysis may also give a picture of incidence that is unstable from year to year, since collections of export taxes depend crucially on changeable conditions in world coffee markets.

Selective excises, the general sales tax, and other indirect taxes are virtually always attributed to consumers of the taxed goods. The data available from budget surveys usually allow the most important excises to be allocated among income classes in proportion to consumption of the taxed goods. The sales tax may simply be allocated among income groups on the basis of aggregate figures such as nonfood consumption. Sales tax burdens are likely to be somewhat greater in relative terms in urban areas than in rural areas because of the greater importance of food in the market basket of rural households; whether food is explicitly exempt or not, much of it is likely to be untaxed in rural areas. Even if food is exempt, food prices may reflect some taxes on agricultural inputs.

Incidence of Colombian Taxes in 1970

Table 3.6 presents three estimates of tax incidence for 1970, based on alternative assumptions about the incidence of the company income tax. Under the assumption that the tax is borne by shareholders, it is attributed to the two highest income classes with barely 4 percent of the nation's households but more than 25 percent of its income. We exclude coffee export duties from the analysis, since Berry and Soligo (1980a) have questioned their treatment in the original study from which these results are drawn. McLure (1975) not only attributed all these duties to rural households, he assumed these households had no other source of income.

The income taxes and other direct taxes are highly progressive, especially under the traditional assumption that the company tax is borne by shareholders. If the company tax is assumed to be borne by consumers, it is essentially proportionate, as are indirect taxes, except at the top of the income distribution, where luxury taxes are important, and at the bottom of the income distribution, where on average households consume more than their income. In total, McLure finds that "The Colombian system ap-

Table 3.6 Taxes as Percent of Income, 1970

Income class	*Direct taxes*				
(pesos per year)	*a,c*	*b*	*Indirect taxes*	*Total a,c*	*All taxes b*
0–5,999	0.1	4.6	7.6	7.6	12.1
6,000–11,999	0.3	3.0	5.3	5.5	8.2
12,000–23,999	0.5	3.5	6.3	6.8	9.9
24,000–59,999	1.5	4.7	6.2	7.7	10.9
60,000–119,999	2.2	5.1	6.6	8.8	11.7
120,000–240,000					
a	11.5	n.a.	5.5	17.5	n.a.
b	n.a.	7.0	6.0	n.a.	12.9
c	4.3	n.a.	6.6	10.9	n.a.
Over 240,000					
a	26.7	n.a.	7.1	34.0	n.a.
b	n.a.	18.8	8.3	n.a.	27.1
c	18.8	n.a.	9.5	28.4	n.a.
Total					
a	6.2	n.a.	6.3	12.5	n.a.
b	n.a.	6.4	6.5	n.a.	12.8
c	3.5	n.a.	6.7	10.1	n.a.
Over 120,000					
a	17.2	n.a.	6.2	23.4	n.a.
b	n.a.	11.2	6.8	n.a.	18.0
c	9.3	n.a.	7.7	17.0	n.a.

n.a. Not applicable.
Note: Coffee export duties excluded.
Incidence assumptions:
a. Corporate income tax is borne entirely by Colombian shareholders.
b. Corporate income tax is shifted entirely to Colombian consumers.
c. Corporate income tax is borne entirely by non-Colombian shareholders.
Source: McLure (1975, p. 181).

pears to exhibit a degree of progressivity ranging from rather mild to fairly strong, depending on whether the corporation income tax is borne by foreigners or Colombian consumers or by Colombian shareholders and on whether coffee export duties are excluded from the analysis or included" (1975, pp. 180–81). Berry and Soligo conclude that "progressivity is substantial and continuous across income levels" (1980a, p. 21).

Distributional Effects of the 1974 Reforms

The 1974 reforms may have shifted an amount of income equal to as much as 1.5 percent of GDP away from the top quintile of the income distribution (Gillis and McLure 1978). Even if this is a substantial overestimate, the reforms probably increased the progressivity of the tax system significantly.[62] By comparison, it seems likely that the counterreforms that occurred between 1974 and 1986, by restoring old loopholes, creating new ones, and by unifying VAT rates, reduced somewhat the progressivity of the tax system. It is unlikely, however, that these eliminated the entire increase in progressivity resulting from the 1974 (and subsequent) reforms.

Distributional Effects of the 1986, 1988, and 1989 Reforms

It is difficult to know whether the 1986 tax reform was revenue-neutral. The substantial reduction in tax rates and the exemption of dividends suggest that it may not have been. If it was not, it reduced the progressivity of the entire tax system, since the income tax is paid by only a small minority of households. Even if the 1986 reforms were revenue-neutral, as claimed, the reduction in the top rate to 30 percent and the elimination of tax on dividends reduced progressivity at the top of the income distribution.

Taxes were completely eliminated for households with incomes below $1 million. This increased progressivity relative to those higher in the income distribution, but reduced it relative to those below. On balance, it appears safe to say that progressivity was reduced significantly by the 1986 reforms. The 1988 reforms also probably reduced progressivity, but only slightly, as did the 1989 repeal of the net wealth tax.

Tax Reform and Marginal Effective Tax Rates

It is difficult to determine from the descriptive analysis presented thus far the net effect of any particular tax structure on investment incentives. Accordingly, the following analysis provides quantitative estimates of the net effects of the major provisions (statutory individual and business income tax rates, depreciation allowances, provisions for inflation adjustment, and net wealth taxes) affecting the taxation of capital income for several recent tax structures.

Specifically, this section presents calculations of the METRS on new investment under several tax regimes. This approach calculates the "tax wedge" between the gross and net returns on a marginal investment of a particular type. This wedge, when expressed as a percentage of the gross return, is termed the METR on an investment of that type. Investments are categorized according to the type of asset purchased, the method of finance, and the characteristics of the saver that provides the investment funds. Taxes at both the business and saver levels are considered. Thus, METRS are summary measures of the net effects of all major tax provisions affecting capital income. They provide an indication of the relative tax burdens facing various types of investments under alternative tax structures.

A number of assumptions—concerning both theoretical issues and empirical magnitudes—must be made in order to perform the METR calculations for Colombia. The methodological approach used in these calculations generally follows King and Fullerton (1984), which should be consulted for further details.

Three basic assumptions are made in the METR calculations. The first is necessary because METR analysis is a partial equilibrium by nature and thus requires that some rate of return in the economy be specified exogenously; the calculations assume that all investment earns the same before-tax rate of return. Thus METRS can be interpreted as reflecting the "impact" effects of the imposition of the tax system on an economy that is in long-run, no-tax equilibrium, before the reallocation of capital that would be induced by introduction of the tax system. The second assumption is the adoption of the rather controversial "new view" of dividend taxation. This view implies that dividend taxes are irrelevant for the calculation of METRS on investments financed from retained earnings, and affect only investment financed with the issuance of new shares. Note, however, that METRS under the alternative traditional view of dividend taxes can be calculated as a weighted average of the METRS on investments financed with retained earnings and with issues of new shares (Zodrow 1991). Third, in the absence of data regarding depreciation rates across assets and sectors, the calculations make the obviously tenuous assumption that depreciation rates are constant across sectors and equal to average rates reported by Hulten and Wykoff (1981) for the United States.[63] Specifically, equipment is assumed to depreciate at a constant exponential rate of 14 percent, while structures are assumed to depreciate at a rate of 3 percent.

Three types of "savers" or sources of funds are considered in each of the calculations: taxable individuals, tax-exempt institutions, and foreign investors. Only Colombian taxes on foreign investors are considered; any taxes imposed by the home countries of such investors are ignored. Such treatment is entirely appropriate if the home country (1) utilizes a territorial system of income taxation, (2) utilizes a residence-based system but provides for "tax sparing" for investment in Colombia, or (3) utilizes a residence-based system but the investing firm is in an "excess foreign tax credit position." However, if the home country employs a residence-based system and the investing firm is not in an excess foreign tax credit position, the taxes of the home country are relevant for marginal investment decisions. In this case, any taxes paid in Colombia may simply replace taxes that would otherwise be paid to the treasury of the home country; that is, under certain circumstances (including accrual taxation by the home country), Colombian income or withholding taxes have no effect on the marginal investment incentives facing foreigners. In other words, the METR calculations may be inaccurate in that they should ideally take account of taxes in the home country rather than Colombian taxes.[64] Three forms of investment finance are considered for each of these savers: debt, new share issues, and retained earnings.

A number of inherent limitations of the METR approach must be noted. In particular, such calculations assume that capital markets (and, implicitly, all other markets as well) function perfectly. Thus, the calculations do not capture the effects of market imperfections in the economy that are important in determining actual tax burdens, especially in certain sectors of the economy. Moreover, the analysis is conducted under the assumption of perfect certainty and thus neglects uncertainty regarding investment returns and ultimate net costs that would typically characterize investment in Colombia (as elsewhere).[65]

Several other problems should also be mentioned. First, the imperfections in administration that plague tax collection are largely ignored in the analysis. Second, the calculations assume the expected rate of inflation is constant over the life of the asset purchased; no attempt is made to adjust the results for actual or predicted fluctuations in inflation rates over time. Third, METRS are not calculated for owner-occupied housing.[66] Fourth, no attempt is made to model the taxation of presumptive income. Fifth, the effects of local property taxes are ignored; the calculations refer only to taxation at the national level. Finally, no allowances are made for any other taxes, including those on foreign trade.[67]

Thus, the following analysis of METRS indicates general tendencies under the appropriate circumstances rather than establishing definitively actual effective tax rates on capital income under alternative tax regimes. This qualification should always be kept in mind when using the METRS presented below to evaluate tax policy toward capital income as it has recently evolved.

The analysis focuses on three tax structures, each of which has been described in detail above. The first, denoted as "1985," corresponds to the tax structure prior to the tax reform of 1986. The second, denoted as "1986," corresponds to the tax structure that would have been in effect had all provisions of the 1986 reform been fully phased in by 1995 (neglecting the changes enacted in 1988). The third tax structure, denoted as "1988," includes all of the provisions for inflation indexing enacted in 1988. The calculations are based on a modeling of these three tax structures, including estimates of the relevant average individual-level tax rates.[68] The analysis also includes a brief discussion of the effects on marginal investment incentives of (1) various aspects of the taxation of capital income during the period 1974–85, and (2) the income tax reforms recommended in the 1988 report discussed above. Results are presented in each case considered below for two rates of inflation—0 and 20 percent. The no-inflation case serves as a convenient benchmark, and the 20 percent inflation rate is typical of recent experience.

Results of the Marginal Effective Tax Rate Calculations

The basic results of the METR calculations are presented in tables 3.7–3.9, which ignore the VAT. The effects including the VAT in the calculations are discussed briefly below. In each case, the METR under the provisions of the 1988 report (T*) is provided as a reference point; these values reflect the METRS that would obtain under full taxation of inflation-adjusted income. These results are interpreted as follows.

New share issues to tax-exempt institutions. Table 3.7A presents METRS for investments financed with new share issues when the saver or supplier of funds is a tax-exempt institution. (Since tax-exempt institutions pay no income or wealth taxes, these tax rates are identical to those on investment financed with retained earnings shown in table 3.8A. Unless otherwise noted, subsequent references to "tax rates" will refer to METRS rather than statutory tax rates.) Consider first the results for inventories and land. For all three tax structures, regardless of the level of inflation, the METR is simply the statutory business tax rate. Thus, the differences in the results reflect only the reduction of the business tax rate in the 1986 reform. This result obtains because the business tax base is real economic income for these two types of assets and there is no individual taxation. In particular, depreciation deductions are not an issue for either asset, and the use of LIFO inventory accounting is assumed to eliminate inflation indexing problems for the cost of goods sold from inventories. (Since no deductions are allowed for the purchase of land, the issue of inflation indexing does not arise.)

The more striking result in table 3.7A is that, when the inflation rate is 20 percent, the METRS for the two depreciable assets under the 1985 and 1986 laws are roughly equal to the statutory corporate tax rate. These results reflect the fact that the accelerated (but unindexed) depreciation allowances prior to 1988 were roughly equal in present value terms to the deductions that would have been allowed under a system that provided for real economic depreciation. This implies that equity-financed investment decisions across assets—and thus across business sectors—were not distorted significantly by the business tax when the inflation rate was in the neighborhood of 20 percent.[69] Of course, such a positive result can be expected to occur only for rates of inflation within a fairly narrow range.

Finally, note that under the 1988 tax system, the METRS are independent of the rate of inflation; this reflects indexation of depreciation allow-

Table 3.7 Marginal Effective Tax Rates on Investment Financed with New Share Issues
(percent)

A. Source of Funds: Tax-Exempt Institutions
(T=30.0)*

	Inflation rate=0			*Inflation rate=0.2*		
Asset	*1985*	*1986*	*1988*	*1985*	*1986*	*1988*
Inventories/land	40.00	30.00	30.00	40.00	30.00	30.00
Equipment	15.59	10.75	10.75	35.42	24.64	9.34
Structures	28.31	20.71	20.71	42.18	30.95	20.71

B. Source of Funds: Foreigners
(T=51.0)*

	Inflation rate=0			*Inflation rate=0.2*		
Asset	*1985*	*1986*	*1988*	*1985*	*1986*	*1988*
Inventories/land	52.00	51.00	51.00	52.00	51.00	51.00
Equipment	32.47	37.53	37.53	48.34	47.25	36.54
Structures	42.65	44.50	44.50	53.75	51.66	44.50

C. Source of Funds: Individuals[a]
(T=40.7)*

	Inflation rate=0			*Inflation rate=0.2*		
Asset	*1985*	*1986*	*1988*	*1985*	*1986*	*1988*
Inventories	65.40	40.70	40.70	56.60	32.69	40.70
	56.99			48.48		
Land	65.40	36.42	40.70	56.60	31.64	40.70
	56.99			48.48		
Equipment	44.66	19.16	19.16	59.04	32.45	17.73
	46.75			51.23		
Structures	51.53	27.08	27.08	58.59	33.76	27.08
	48.20			48.85		

a. In each case, the second figures listed in the 1985 column apply to limited partnerships prior to the unification of the taxation of corporations and limited partnerships in 1986.
Source: Authors' calculations.

ances.[70] However, indexing of the depreciation allowances under prior law implies that the present value of depreciation allowed for tax purposes exceeds the present value of economic depreciation. As a result, the METRS on depreciable assets are below the statutory business tax rate, especially for investment in equipment. Thus the 1988 tax code favors investment in equipment and equipment-intensive production sectors (and to a lesser extent investment in structure-intensive sectors).

By comparison, the 1988 report recommended inflation indexation of the deductions for the best available estimates of economic depreciation, as well as indexed FIFO inventory accounting. This would result in METRS on all four assets (T*) equal to the statutory business tax rate of 30 percent.

Note also that T* corresponds to the tax rates on investment in inventories and land under the 1988 law. This occurs in all the cases analyzed in tables 3.7–3.9; the reason is that depreciation allowances are not an issue for these assets and, for purposes of METR calculations, the only differences between the 1988 law and the recommendations of the 1988 report involve depreciation allowances. Thus, the descriptions below of the taxation of inventories and land under the 1988 law also describe all of the factors that would have determined METRS on all four types of assets under the proposals of the 1988 report.

New share issues to foreigners. Table 3.7B presents results on the taxation of investments funded by new shares issued to foreigners. Under the 1985 law, the METR of 52 percent on investment in inventories and land reflects a corporate tax rate of 40 percent coupled with a withholding tax on dividends paid to foreigners of 20 percent ($0.52 = 1 - 0.6 \times 0.8$). This rate drops slightly to 51 percent under the 1986 and 1988 tax structures, reflecting a reduction in the business tax rate to 30 percent, coupled with an increase in the withholding rate on dividends to the same rate ($0.51 = 1 - 0.7 \times 0.7$). The METRS on investment in equipment and structures follow the same pattern as in table 3.7A, although their absolute levels are higher due to the withholding tax.

New share issues to individuals. Table 3.7C provides METRS on investments financed with new share issues to individuals. Consider first the case of 1985 law and 0 percent inflation. The tax rates for inventories and land now reflect the combined effects of taxation at the business income tax rate of 40 percent, the average marginal income tax rate applied to dividends (34.5 percent with a 10 percent credit for dividends received), and the average tax rate on reported wealth (1.07 percent). The combined income tax rate is 54.7 percent ($0.55 = 1 - 0.6 \times 0.66 - 0.1 \times 0.6$)[71] which, when combined with a wealth tax of 1.07 percent (relative to a fixed before-tax rate of return of 10 percent) yields a METR of 65.4 percent ($0.66 = 0.55 + 1.07/10$). The dramatic effects of inflation on the wealth tax base are illustrated by the fact that an inflation rate of 20 percent virtually eliminates the burden imposed by the wealth tax, as the METR falls to 56.6 percent (in comparison to an income tax burden of 54.7 percent). As above, the METRS on equipment and structures under the 1985 law are roughly equal to the tax rate on inventories and land at a 20 percent rate of inflation but are significantly lower at a 0 percent inflation rate.

The effects of the 1986 reform on the METRS on investment financed with new share issues to individuals arise from the reduction in the business tax rate and the complete elimination of taxation on dividends at the individual level. The results demonstrate that these provisions dramatically reduce the METRS on such investments.

Turning to the results for the 1988 tax structure, the METR on investments in land and inventories is equal to 40.7 percent regardless of the rate of inflation; as described above, this figure reflects the combined effects of business taxation at a rate of 30 percent and individual wealth taxation at a rate of 1.07 percent. (Note that the tax rate on investment in land is higher than under the 1986 law, reflecting the elimination of the provision allowing only 60 percent of the assessed value of land to be included in the wealth tax base.) As noted above, the tax rates on equip-

Table 3.8 Marginal Effective Tax Rates on Investment Financed with Retained Earnings
(percent)

A. Source of Funds: Tax-Exempt Institutions
(T=30.0)*

	Inflation rate=0			Inflation rate=0.2		
Asset	1985	1986	1988	1985	1986	1988
Inventories/land	40.00	30.00	30.00	40.00	30.00	30.00
Equipment	15.59	10.75	10.75	35.42	24.64	9.34
Structures	28.31	20.71	20.71	42.18	30.95	20.71

B. Source of Funds: Foreigners
(T=32.6)*

	Inflation rate=0			Inflation rate=0.2		
Asset	1985	1986	1988	1985	1986	1988
Inventories/land	42.25	32.63	32.63	42.25	32.63	32.63
Equipment	18.76	14.10	14.10	37.84	27.47	12.74
Structures	31.00	23.69	23.69	44.35	33.54	23.69

C. Source of Funds: Individuals
(T=43.3)*

	Inflation rate=0			Inflation rate=0.2		
Asset	1985	1986	1988	1985	1986	1988
Inventories	54.38	43.33	43.33	45.95	35.24	43.33
Land	54.38	39.05	43.33	45.95	34.21	43.33
Equipment	23.87	17.29	17.29	42.01	30.25	15.97
Structures	38.75	29.98	29.98	48.21	36.29	29.98

Source: Authors' calculations.

ment, and to a lesser extent those on structures, are lower than those on inventories and land, reflecting the inflation indexing of depreciation allowances that are already accelerated. If the recommendations of the 1988 report had been adopted, all assets would have been subject to a METR of 40.7 percent.

Finally, note that the results presented in table 3.7C also provide METRS on investments in limited partnerships under the 1985 tax structure. These results reflect the relatively low tax rate of 18 percent applied to the income of limited partnerships prior to 1986. Note that despite the increase in the business-level tax on limited partnerships from 18 to 30 percent under the 1986 reform, the tax burden on equity investment by limited partnerships declined dramatically due to the elimination of individual taxation.

Retained earnings. Table 3.8 considers equity finance in the form of retained earnings. Recall that the new view of the effects of dividend taxation implies that taxation of dividends at the saver level (either foreigners or individuals) has no effect on the METR calculations when the method of finance is retained earnings. Thus, the results for tax-exempt institutions (table 3.8A) are the same as in table 3.7A, and the results for foreign investors (table 3.8B) are virtually identical to those reported for tax-exempt institutions in table 3.7A, the only difference being retained earnings are subject to capital gains taxes at an effective rate of 3.75 percent when the saver is a foreigner.

Table 3.8C provides results for the case of retained earnings finance when the source of funds is individuals. These generally differ from the tax rates reported for investments financed with retained earnings attributable to foreigners only in that the METRS are higher due to the burden of the individual net wealth tax.

Debt finance by tax-exempt institutions or foreigners. Table 3.9A reports METR results on debt-financed investments when the source of funds is either tax-exempt institutions or foreigners; these are identical because interest income to foreigners is not taxable under any of the three versions of Colombian law.

Consider first investment in inventories or land under the 1985 tax structure in the absence of inflation. The tax rates on such marginal investments are zero, reflecting the fact that all returns are paid out as interest, which is not taxable at the level of the saver. At a 20 percent inflation, full deductibility of nominal interest expense implies a very large (in absolute value) negative METR of –133.3 percent. The tax rates on depreciable assets are also very negative at a 20 percent inflation rate and roughly similar to those on inventories and land for the reasons discussed above. Note that at 0 percent inflation, the fact that the present values of depreciation allowances on equipment and structures exceed the present value of deductions for economic depreciation yields negative (rather than simply zero) METRS. Note also that, in contrast to all the previous cases, the METRS on debt-financed depreciable assets increase as the rate of inflation declines, as the advantage of depreciation allowances that are too generous relative to economic depreciation is swamped by the disadvantage of no longer receiving large deductions for the inflationary component of interest expense.

Under the fully phased-in version of the 1986 tax structure, METRS on debt-financed investments in inventories and land are zero independent of the level of inflation since interest expense is fully indexed for inflation. A METR of roughly zero also obtains at a 20 percent rate of inflation for investment in depreciable assets, with overgenerous depreciation allowance resulting in negative METRS for such assets at 0 percent inflation. Under the 1988 tax structure, inflation dependence of the tax rates on depreciable equipment is eliminated; debt-financed investments in inventories and land are subject to a zero METR but, for the reasons described above, the METRS on investment in structures and especially equipment are lower than those on inventories and land, that is, METRS are always negative.

Debt finance funded by individuals. Results for the case of debt-financed investments when the source of funds is individuals are reported in table 3.9B. Consider first the case of investment in inventories or land under the 1985 tax structure in the absence of inflation. The METR of 33.5 percent re-

Table 3.9 Marginal Effective Tax Rates on Investment Financed with Debt
(percent)

A. Source of Funds: Tax-Exempt Institutions and Foreigners
(T=0)*

	Inflation rate=0			*Inflation rate=0.2*		
Asset	*1985*	*1986*	*1988*	*1985*	*1986*	*1988*
Inventories/land	0.00	0.00	0.00	−133.33	0.00	0.00
Equipment	−40.68	−27.50	−27.50	−140.96	−7.65	−29.52
Structures	−19.49	−13.27	−13.27	−129.69	1.35	−13.27

B. Source of Funds: Individuals[a]
(T=24.7)*

	Inflation rate=0			*Inflation rate=0.2*		
Asset	*1985*	*1986*	*1988*	*1985*	*1986*	*1988*
Inventories/land	33.50	24.66	24.66	−51.19	24.66	24.66
	33.50			17.85		
Equipment	2.09	1.00	1.00	−57.08	18.07	1.00
	22.10			13.73		
Structures	18.46	13.24	13.24	−48.38	25.82	13.24
	27.95			17.97		

a. In each case, the second figures listed in the 1985 column apply to limited partnerships prior to the unification of the taxation of corporations and limited partnerships in 1986.
Source: Authors' calculations.

flects taxation of interest receipts at the individual level (at a rate of 22.8 percent) and individual wealth taxation (at a rate of 1.07 percent) (0.34 = 1 – 0.77 + 1.07/10). The METR declines dramatically with inflation since nominal interest expense is fully deductible at the business rate of 40 percent but the inflationary component of interest receipts is partially excluded at the individual level (60 percent) and the relevant tax rate on interest receipts is only 22.8 percent. Tax rates on depreciable assets are roughly similar to those on inventories and land at a 20 percent inflation rate and, for the same reason outlined above, increase as the rate of inflation declines.

Under the 1986 tax structure, interest income and expense are fully indexed, so that the METRS reflect only individual taxation of real interest receipts (at the reduced rate of 14.0 percent), and taxation of individual wealth (at a rate of 1.07 percent). Thus, the tax rate on inventories and land, irrespective of the rate of inflation, is 24.7 percent (0.25 = 1 – 0.86 + 1.07/10). Similar (reduced) METRS obtain for investments in depreciable assets at an inflation rate of 20 (0) percent.

Under the 1988 tax structure, the METRS on debt-financed investment are independent of inflation. The METR on investment in inventories is equal to the 24.7 percent figure cited above; lower rates obtain for structures and especially equipment.

Finally, results for debt-financed investments in limited partnerships under the 1985 tax structure (when the source of funds is individuals) are also presented in table 3.9B. Note that at an inflation rate of 20 percent, the METRS are relatively high for limited partnerships because the benefit of full deduction of nominal interest expense is worth considerably less when the business tax rate is 18 rather than 40 percent.

A Note on Marginal Effective Tax Rates Inclusive of the Value Added Tax

The results presented in tables 3.7–3.9 neglect the effects of the VAT on the METRS on investment in equipment and structures. They are thus applicable to imports of equipment (and perhaps some structures) used in basic industries, since such purchases are exempt from tax. Moreover, these results are also generally applicable to purchases from domestic owners if, as is sometimes argued, the application of the VAT to capital goods merely compensates for an undervalued exchange rate.[72] However, if the VAT does raise the cost of domestic capital goods to Colombian businesses (and, as theory suggests, is not shifted forward in the form of higher output prices), then the effects of the VAT should be included in the METR calculations.

The most important effect of this modification is that inclusion in the calculations of a 10 percent VAT (which is not creditable against the Colombian income tax) increases the reported METRS in a way that does not vary significantly with inflation; the increases are greater for equipment than for structures.[73] Under the 1985 and 1986 tax structures and 20 percent inflation, the differential impact of the VAT tends to convert a tax system that is roughly neutral across assets to one that discourages investment in equipment and to a lesser extent in structures, relative to investment in inventories and land. By comparison, under the 1988 tax structure the introduction of the VAT generally offsets the distortions caused by the income and wealth taxes, resulting in a tax system that is roughly neutral with respect to the allocation of investment.

Comments on Marginal Effective Tax Rates under the 1974 Law and between 1974 and 1985

This section provides a rough indicator of the magnitudes of the effects of various changes that were enacted during the period 1974–85 by comparing the METRS for each type of asset under the "1985" tax structure and under a "revised" version of that same structure with a single provision altered. The results are presented in table 3.10.

There were three major changes in the taxation of capital income between 1974 and 1985.[74] In reverse chronological order, in 1983 partial inflation indexing of interest receipts at the individual level was introduced. Table 3.10A indicates that elimination of this provision significantly increases the METR on debt-financed corporate investment when the source of funds is individuals and the inflation rate is 20 percent. Earlier in 1983 the 10 percent credit for corporate dividends received by individuals was enacted.[75] Since dividend taxes affect only investment financed with the issuance of new shares (under the new view of the effects of such taxes),

Table 3.10 Effects of Changed Provisions, 1974–85

A. Effects of Partial Exclusion of Inflationary Component of Interest Receipts
(METR debt-financed investments made by individuals)

	Without partial exclusion		With partial exclusion	
Asset	P=0	P=0.2	P=0	P=0.2
Land/inventories	33.5	–23.8	33.5	–51.2
Equipment	2.1	–29.7	2.1	–57.1
Structures	18.5	–21.0	18.5	–48.4

B. Effects of Tax Credit for Dividends
(METRs on investments financed by new share issues to individuals)

	Without dividend credit		With dividend credit	
Asset	P=0	P=0.2	P=0	P=0.2
Land/inventories	71.4	62.4	65.4	56.6
Equipment	53.1	65.5	44.7	59.0
Structures	58.4	64.2	51.5	58.6

C. Effects of 1974 Depreciation Schedule
(METRs on investments financed by new share issues or retained earnings attributable to tax-exempt institutions)

	With 1974 schedule		With 1985 schedule	
Asset	P=0	P=0.2	P=0	P=0.2
Equipment	28.6	56.8	15.6	35.4
Structures	33.7	45.6	28.3	42.2

Source: Authors' calculations.

the relevant comparison is with table 3.7C. Table 3.10B indicates that eliminating the dividend credit results in an increase in METRS of roughly 6 percentage points for each asset type and is independent of the rate of inflation. The final change considered is the introduction in 1976 of the accelerated depreciation schedules that replaced the double declining balance schedules enacted in 1974. Table 3.10C shows METRS for the case of new share issues or retained earnings when the saver is a tax-exempt institution (table 3.7A). At 20 percent inflation, the METRS exceed the statutory rate of 40 percent (especially for equipment), indicating the present value of depreciation allowances under a double declining balance schedule is less than the present value of real economic depreciation. This contrasts markedly with the results under the 1985 depreciation schedules which roughly approximate economic depreciation in present value terms at 20 percent inflation. In contrast, at 0 inflation, the METRS fall below the statutory rate as depreciation (according to a double declining balance schedule) is too generous relative to economic depreciation.

The effects of simultaneously "backing out" the three changes noted above from the 1985 tax structure to construct an approximate "1974" tax system are shown in table 3.11 (investments financed by individuals). Consider first the METRS for retained earnings finance (relevant only for corporations). These differences reflect only the effects of the differences in depreciation schedules under the two tax structures; the interpretation is analogous to that presented above.

Table 3.11 Comparison of Marginal Effective Tax Rates
(percent)

	Corporations				*Limitadas*			
	1974		*1985*		*1974*		*1985*	
Asset	*P=0*	*P=0.2*	*P=0*	*P=0.2*	*P=0*	*P=0.2*	*P=0*	*P=0.2*
Debt finance								
Inventories/land	33.5	−23.8	33.5	−51.2	33.5	45.2	33.5	17.8
Equipment	18.9	−2.2	2.1	−57.1	27.8	50.5	22.1	13.7
Structures	25.4	−16.7	18.5	−48.3	30.5	47.0	28.0	18.0
New share issues								
Inventories/land	71.4	62.4	65.4	56.6	57.0	48.5	57.0	48.5
Equipment	61.6	79.5	44.7	59.0	50.7	57.8	46.7	51.2
Structures	63.9	66.6	51.5	58.6	51.9	50.3	48.2	48.8
Retained earnings								
Inventories/land	54.4	54.4	46.0	46.0				
Equipment	35.6	61.8	23.9	42.0				
Structures	45.7	51.6	38.8	48.2				

Note: Under certain provisions of the 1974 and 1985 law. Source of funds is individuals.

Consider next debt-financed corporate investments when the source of funds is individuals. In comparison to the 1985 law, the METRS under the 1974 structure reflect both slower depreciation for tax purposes and (at 20 percent inflation) full taxation of nominal interest income. The results for the case of 0 inflation indicate that the former factor acts to increase METRS considerably (somewhat) for equipment (structures). In the case of 20 percent inflation the latter factor significantly increases METRS (by making them considerably less negative). Note that the same pattern occurs for debt-financed investments made by *limitadas*.

Finally, consider the METRS for investments financed by issuing new shares to individuals. In comparison to the 1985 law, tax rates under the 1974 structure for investment by corporations reflect both slower depreciation for tax purposes and the absence of a tax credit for dividends received. The effect of the dividend credit is isolated in the results for investment in inventories; elimination of the credit increases the METRS on such investment by roughly 6 percentage points. The increases in METRS for structures and especially equipment are larger, reflecting less rapid depreciation allowances. The increases in METRS under the 1974 tax structures for investment through *limitadas* reflect only differences in depreciation allowances.

Evaluation

This section highlights the major policy implications of the METR analysis and provides an evaluation of the changes in the taxation of capital income that occurred in 1986 and 1988.

Inflation sensitivity: Depreciation allowances. METRS clearly varied with inflation under the 1985 and 1986 tax structures, regardless of the method of finance or the source of funds, primarily because depreciation allowances were not indexed for inflation. However, at 20 percent inflation (typical of recent history in Colombia), the accelerated depreciation allowances provided prior to 1988 were roughly equivalent in present value terms to those allowable under a system that provided for real economic depreciation. Thus, the 1985 and 1986 income tax systems were roughly economically neutral across assets and business sectors at 20 percent inflation. These tax structures would not cause inefficient misallocations of resources across assets if such a rate of inflation were stable and fully anticipated.[76]

This positive result does not obtain at inflation rates significantly different from 20 percent. For example, at 0 inflation, the accelerated depreciation allowances under the 1985 and 1986 laws were too generous relative to real economic depreciation and caused a tax distortion favoring investment in depreciable assets, especially equipment. Such tax-induced distortions are generally undesirable, as they result in an inefficient allocation of scarce capital resources. This misallocation in turn implies a reduction in the overall productivity of investment and thus a lower level of output.

Moreover, the dependence of METRS on inflation introduces unnecessary uncertainty into the investment decision with deleterious effects on the level of investment. Uncertainty about the course of future inflation may also distort decisions regarding debt maturity, as both borrowers and lenders are likely to reduce uncertainty by choosing short-term over long-term debt.[77] Finally, distortions of financing decisions will occur if inflationary expectations differ across borrowers and lenders.

The introduction of inflation indexing of depreciation allowances and of the base of the net wealth tax eliminated inflation sensitivity. However, depreciation allowances were still based on the pre-1988 schedules rather than on estimates of real economic depreciation (as recommended by the 1988 report). As a result, the 1988 law did not satisfy the criterion of economic neutrality. Instead, investment in equipment was favored over investment in structures, which was in turn favored over investment in inventories and land.

Preferential treatment of land under the net wealth tax. The exclusion of 60 percent of the assessed value of land from the base of the individual tax on net wealth under the 1986 tax structure lowered the METR on investment in land when the source of funds was individuals. This tax distortion resulted in a bias toward investment in land over investment in other capital assets. A positive aspect of the 1988 reform was the elimination of this distortion.[78] For ease of exposition, the discussion of remaining issues will ignore inflation sensitivity of depreciation allowances and preferential treatment of land under the net wealth tax.

Inflation sensitivity: Interest income and expense. One critical feature demonstrated by the results of the present analysis is that METRS on debt-financed investment (in inventories) were highly sensitive to the rate of inflation under the 1985 law. Full deductibility of interest expense at the business level, coupled with either partial taxation (at relatively low rates) of interest income at the saver level for individuals or tax exemption for tax-exempt institutions and foreigners, resulted in extremely negative METRS at 20 percent inflation. Such negative tax rates encourage economically unprofitable investments by making it profitable to borrow at an interest rate higher than the return on the investment. Negative METRS provide a tax bias toward debt finance, and almost certainly played a role in the increasing reliance on debt finance over the last thirty-five years. The ratio of corporate indebtedness to the sum of corporate debt and equity increased from roughly 25 percent in 1950 to 70 percent in 1980 (McLure, Mutti, Thuronyi, and Zodrow 1990, chapter 7). Such a tax bias distorts the allocation of risk bearing in the economy, and increases the likelihood of bankruptcies during a cyclical downturn. In addition, the increase in tax differentials across assets and business sectors that occurs as the rate of inflation declines is more pronounced for debt than for equity finance when full nominal interest is deductible.

One of the most beneficial aspects of the 1986 reform (when fully phased in)—and one maintained in the 1988 reforms—was the introduction of inflation indexing of interest deductions at the business level. This was coupled with full inflation indexing of interest receipts at the individual level. This reduced a significant tax bias favoring debt-financed investment and eliminated the negative effect on tax rates of increases in inflation attributable to the increased attractiveness of full interest deductibility at the business level.

Taxation and financial policy. METRS on equity investment were reduced under the 1986 reform by the elimination of taxation of dividends at the individual level and by rate reduction at the business level. When coupled with the increases in tax rates on the income from debt-financed investment described above, the 1986 law dramatically reduced the tax bias favoring debt over equity finance under the 1985 law.

Elimination of dividend taxation at the individual level also affected the relative investment incentives facing new firms and established enterprises. New firms commonly raise a relatively large portion of their equity funds through the issuance of new shares; by comparison, more established firms rely on retained earnings. The 1986 reform eliminated the tax advantage favoring retained earnings finance over new share issues for investments financed by individuals. It thus virtually eliminated a significant tax advantage under the 1985 tax structure favoring investment by established firms over investment by new enterprises.

METRS on debt-financed investments funded by tax-exempt institutions increased under the 1986 reform (at 20 percent inflation) and decreased on equity-financed investments. The only differences from the cases in which the savers are individuals are that the tax rates on retained earnings and new share issues are identical for tax-exempt institutions, and that none of the reduction in the tax bias favoring debt over equity finance comes from changes in the taxation of dividends at the saver level. The relatively low METRS on debt-financed investments funded by tax-exempt institutions indicate the desirability of limiting the investment opportunities available to such institutions, as was done in the 1986 reform. In the absence of such limitations, tax-exempt institutions created to avoid taxes may become a source of unfair competition to taxable enterprises.[79]

The tax differential favoring debt finance over new share issues to foreigners was virtually unchanged between 1985 and 1986 (and 1988); this occurred because the reduction in the business tax rate in 1986 was coupled with an increase in the withholding rate, so that in both cases investment financed with new share issues faced a METR of roughly 50 percent, while the tax rate on debt-financed investment was zero. However, the tax rate on investments financed with retained earnings was reduced when the business tax rate was reduced in 1986. As a result, the relatively more important tax bias favoring debt finance over retained earnings was reduced somewhat, while the tax bias favoring retained earnings over new share issues was increased, meaning investment by firms already established in Colombia over new foreign enterprises was favored even more. In addition, the relatively high tax rate on new shares issued to foreigners may act as a deterrent to such investment.[80]

The treatment of limited partnerships. Limited partnerships enjoyed a preferential tax rate of 18 percent (relative to a corporate tax rate of 40 percent) under the 1985 tax structure. However, they also were subject to "pass-through" treatment, as all annual earnings were attributed to partners; thus, such earnings did not receive the benefit of tax deferral from individual taxation accorded to corporate retained earnings. As a result, the interpretation of the 1986 reform on investment through limited partnerships is somewhat complicated.

Under the 1985 tax structure, METRs on debt-financed investment in limited partnerships were lower than those on equity-financed investments. The tax code thus provided an incentive for debt financing by limited partnerships although this incentive was smaller than the differential facing corporations. This incentive was narrowed by the unification of tax treatment of limited partnerships, corporations, and mixed public-private enterprises in the 1986 reform. It is clear that under the 1985 tax law investment in limited partnerships financed with new share issues was tax advantaged relative to new issues made by corporations (although the METRs on investment in limited partnerships financed with new issues were roughly equal to the tax rates on investment in corporations financed with retained earnings). In any case, the misallocations of investment caused by differential treatment of limited partnerships and corporations were eliminated under the 1986 law.

Inflation sensitivity: The net wealth tax. The only remaining source of inflation sensitivity under the 1986 tax structure was due to the absence of inflation indexing of the individual net wealth tax base. The associated erosion of the net wealth tax base at 20 percent inflation virtually eliminated its burden on equity-financed investments. This feature was eliminated with full inflation indexing of the wealth tax base in 1988. As a result, there was no inflation sensitivity under the 1988 tax structure, as the burden of the individual wealth tax was independent of the rate of inflation.

Inflation sensitivity: Net effects. The above discussion indicates that METRS under the tax structures in 1985 and 1986 were sensitive to inflation. Under the 1986 law, this inflation sensitivity occurred because depreciation allowances and the base of the net wealth tax were not indexed. In addition, under the 1985 tax structure, deductions for interest expense were not indexed, and interest receipts at the individual level were only partially indexed.

The net effects of these provisions are summarized in qualitative terms in table 3.12, which shows that, for equity-financed investments in depreciable assets, METRS increased with inflation under both the 1985 and 1986 laws. This effect occurred because depreciation allowances were not indexed; it was not reversed by inflation-induced erosion of the net wealth

Table 3.12 Inflation Sensitivity of the Tax Structures

Case analyzed	*1985*	*1986*	*1988*
Equity finance			
METR, depreciable assets Saver is tax exempt Institution of foreigner	+	+	0
METR, depreciable assets Saver is individual	+	+	0
METR, inventories/land Saver is tax exempt Institution or foreigner	0	0	0
METR, inventories/land Saver is individual	–	–	0
Debt finance			
METR, depreciable assets Saver is tax exempt Institution or foreigner	–	+	0
METR, depreciable assets Saver is individual	–	+	0
METR, inventories/land Saver is tax exempt Institution or foreigner	–	0	0
METR, depreciable assets Saver is individual	–	+	0
METR, inventories/land Saver is tax exempt Institution or foreigner	–	0	0
METR, inventories/land Saver is individual	–	0	0

Note: +(–,0) means METR increases (decreases, is constant) as inflation increases.
Source: Authors.

tax base when the saver was an individual. In contrast, for investments in nondepreciable assets (inventories and land) under the 1985 and 1986 laws, METRS were sensitive to inflation only if the saver was an individual who benefited from the erosion of the wealth tax base.

For debt-financed assets, METRS declined with inflation under the 1985 law. As inflation increased, the decline in the METR due to full deductibility of nominal interest expense swamped the additional tax burden attributable to unindexed depreciation allowances; in the case of inventories, only the former effect was relevant. Under the 1986 reform, interest income and expense were fully indexed, so that unindexed depreciation allowances were the only remaining source of inflation sensitivity for debt-financed investments. Thus, METRS for investment in depreciable assets increased with the rate of inflation and were invariant with inflation for investment in inventories and land. Finally, note that all inflation sensitivity was eliminated when depreciation allowances and the base of the net wealth tax were indexed under the 1988 reform.

Chronological Summary

The above analysis reveals five major problems with the taxation of capital income in 1974. First, METRS were characterized by a high degree of inflation sensitivity due to a lack of inflation indexing of depreciation allowances, interest income and expense, and the wealth tax base. Second, full deductibility of nominal interest expense resulted in negative METRS on debt-financed investments. Third, due to the taxation of dividends at the individual level (and full deductibility of nominal interest), debt finance was tax advantaged relative to equity finance, especially the issuance of new shares. Fourth, at rates of inflation extant at that time (17 to 25 percent during 1974–76), the unindexed deductions allowed for depreciation were smaller in present value terms than those associated with real economic depreciation. Finally, differential taxation of corporations and limited partnerships resulted in a complex set of distortions across organizational forms.

Each of these problems has been addressed in recent tax reforms. Three major changes occurred between 1974 and 1985. First, the introduction of accelerated depreciation for equipment in 1976 resulted in depreciation deductions for tax purposes that were roughly equal in present value terms to economic depreciation at 20 percent inflation. Second, the enactment of a 10 percent dividend credit in 1983 reduced slightly the tax advantage of debt finance over the issuance of new shares. Third, partial inflation indexing of interest receipts at the individual level was introduced in 1983. This change can be viewed as positive if it is perceived as an intermediate step toward full indexing of interest income and expense. However, in isolation, such a change was not desirable, since it reduced METRS on debt-financed investment to very negative levels.

The effects of the 1986 reforms were generally positive. Full inflation indexing of interest income and expense was introduced, which greatly reduced the tax advantage of debt finance, and reduced (but did not eliminate) inflation sensitivity of the tax structure. The elimination of dividend taxation at the individual level further reduced the tax advantage of debt over equity finance and, within categories of equity finance, largely eliminated the tax advantage favoring retained earnings over the issuance of new shares. In addition, this reform eliminated a tax advantage favoring established firms over new enterprises, since new firms rely disproportionately on new share issues rather than the retained earnings available to more established firms. The unification of the taxation of corporations and limited partnerships eliminated a complex set of distortions regarding the choice of business organizational form. Finally, the lower rates provided under the 1986 reform generally reduced the degree of distortions remaining in the tax structure.

The 1988 reforms also had several desirable effects. Inflation indexing of depreciation allowances and of the net wealth tax base eliminated inflation sensitivity of METRS. Elimination of the provision allowing only 60 percent of the assessed value of land to be included in the wealth tax base eliminated a distortion of investment decisions and was also desirable on distributional grounds. However, the newly indexed depreciation allowances were too generous, as they were based on allowances under prior law, rather than conceptually correct allowances based on the best available estimates of economic depreciation. As a result, investment in depreciable assets, especially equipment, was tax advantaged over investment in inventories and land. This provision of the 1988 law should be reconsidered in future discussions.

Finally, to the extent that the VAT should be included in the METR calculations, the combined effect of the income/wealth taxes under the 1988 reforms and the VAT was economically neutral across assets and business sectors. Since this result was both accidental and rather tenuous, it does not provide a strong argument against future reform. Rather, neutrality with respect to investment decisions should be achieved directly and with greater certainty by basing the income tax on real economic income and by exempting all capital goods from the VAT.

Simplification

The individual income tax system that prevailed in the 1960s resembled the United States, with a few modifications. The most important modification was the taxation of income from owner-occupied housing, an approach long advocated unsuccessfully by tax scholars in the United States. This approach was used in the United Kingdom until it was abandoned in the 1960s. Because of efforts to "fine-tune" tax liabilities to individual circumstances, some provisions of Colombian law were even more complicated than their counterparts in industrial countries. Many individuals who had no tax liability were required to file tax returns. Since tax was officially assessed, rather than self-assessed, the administrative burden was overwhelming.

It seems likely in retrospect that this system when introduced was too complex for a country at Colombia's level of development. This judgment is borne out by subsequent events, which have seen progressive (if not monotonic) simplification of the individual income tax. This section describes the simplifications of the 1986 reform concentrating on a few clear-cut examples.

The 1986 reforms eliminated personal exemptions and the privilege of ceding income to the spouse. As a result, tax liability is now based on individual income, with no splitting, and more accurate withholding is possible. There is no longer any need to be concerned with the marital status or family size of employees.

The 1986 reform repealed the tax credits for expenditures on medical care and education and the special standard deduction. In combination with the elimination of credits for personal exemptions and the ceding of income, this made it possible to make withholding an accurate final tax for many taxpayers and to eliminate filing of returns (discussed below).

The 1986 reforms recognized that it is impossible to administer the taxation of imputed income from owner-occupied housing, so they eliminated both the taxation of such income and the credit for renters. In addition, deductions for mortgage interest were limited to the interest on loans of no more than $5.9 million; the deduction could not exceed $1 million. Property taxes were no longer deductible, except when related to the earning of income.

The elimination of many deductions and many credits under prior law offset tax liability made it possible in the 1986 law to raise the zero bracket amount from $180,000 to $1 million. Thus it was possible to eliminate the filing requirement and make withholding a final tax for those with total incomes below $4 million, for 80 percent of income from labor, for coverage of all income by withholding, for net wealth below $6 million, and with no obligation for sales tax.

The 1986 reforms completely eliminated double taxation of dividends by exempting such dividends, rather than by adopting either of the conceptually correct approaches, the withholding or imputation method involving a shareholder credit or the deduction for dividends paid (or different rates for retained earnings and distributed income).

Simplification of inflation adjustment in the measurement of income from business and capital cannot be directly observed. (Inflation adjustment of nominal amounts, being done by the fiscal authorities and reported to taxpayers annually, involves little loss of simplicity and is not discussed.) In this case it has been deemed worthwhile to add complexity to the system to avoid the very real inequities and economic costs that an unindexed system entails. It is important to note that this complexity is required to achieve an accurate measure of real income; it does not reflect an attempt to fine-tune the system to the individual circumstances of taxpayers.

Some of the reforms of the individual income tax system during the past three decades have not simplified compliance and administration. Episodes in which the system has been made more complex include the brief use of vanishing deductions. Moreover, many reforms have permanently increased complexity. Among these are the elimination of exemptions for some kinds of income and the introduction of inflation adjustments for interest income and expenses. However, the combination of inflation adjustment and the elimination of exemptions often reduces complexity by reducing the "games" taxpayers can play.

There is inevitably a conflict between simplicity and efforts to fine-tune tax burdens. For the most part, the reforms that have increased complexity brought the measurement of taxable income into greater conformity with real economic income, a worthwhile price to pay to avoid the inequities and distortions that result from a great divergence of those two concepts.

Most efforts to reduce complexity involved prior fine-tuning of the tax system in order to take account of family circumstances, usually in the name of equity. In the case of the elimination of dividends a "rough-and-ready" form of integration was chosen over approaches that are conceptually more attractive but much more complicated, reflecting the judgment that any gains in equity were not worth the increased complexity such fine-tuning—or the conceptually preferable treatment of dividends—entails.

Lessons

This review of tax reform episodes in Colombia reveals many lessons.[81] Some may be unique to the Colombian setting, but most are probably applicable to other countries.

The improvement of the tax system has been marked by a process of groping which has been slow and sometimes reversed by subsequent counterreforms. Yet, on balance, the system has improved. This is seen in a number of areas: the elimination of tax incentives, the unification of the taxation of corporations and partnerships, the simpler tax treatment of the family, the elimination of deductions for personal expenditures, and the systematic treatment of income measurement problems caused by inflation. It is important to be patient: progress is not always steady or incremental.

1. *Tax reform requires a long gestation period.* For example, it was five years between the time of the Musgrave Report and the introduction of the 1974 reforms. This is a good lesson for those who would become impatient when reform proposals are not adopted quickly.[82] Yet there are instances of badly needed reforms that have been adopted relatively quickly. Among them are the introduction of withholding, self-assessment, advance payments, vanishing deductions, the presumptive income tax, and the 1988 reforms. These examples should also act as an antidote to pessimism.

2. *Tax reform studies serve an important educational purpose, in addition to informing policymakers; thus they should be widely disseminated.*[83] This lesson is directly related to several others, notably the creation of a cadre of local advocates for tax reform, the atonement nature of tax reform, and the fact that tax reform often occurs with lags. The easy availability of tax reform studies contributes in all these areas.[84] The 1974 and 1986 reforms were put together by local experts, who in most cases had previously worked with the foreign missions and/or had studied their reports (Perry and Cárdenas 1986, pp. 11–13, 256–57, 292–94). This suggests another lesson: in order to institutionalize tax policy it is necessary to develop a cadre of local experts who understand local conditions and special circumstances and are knowledgeable about both the appropriate goals of tax policy and the existing system, and who have an interest in reform. Otherwise, a country will continue to rely on outside advisers to formulate tax policy and appraise proposals to modify the system.[85]

3. *Tax reform is often an uphill battle; similarly "counterreforms" motivated by strong political interests may derail a reformed system.* The important procedural features of the 1974 reforms were declared unconstitutional, severely limiting the effectiveness of the substantive reforms made. Also, reform can deteriorate as it did following the dramatic 1974 reforms, especially the 1979 provisions that limited taxation of capital gains to one-half the tax rate on ordinary income and allowed the optional exemption of all capital gains from tax, and the 1976 liberalization of depreciation allowances (perhaps justifiable as an ad hoc means of dealing with inflation), and in the continued success of cattle raisers in remaining tax free.

4. *Reform occurs when the country can no longer afford to put it off.* Reform has occurred most often during times of fiscal crisis, and counterreforms have generally occurred when revenues have been relatively plentiful. The implication of this is not that tax reformers should pray for such crises, but that they should be prepared for them and the possibility of achieving reforms that would not otherwise occur. This highlights the importance of having a ready supply of well-considered proposals for enactment, with at most minor modifications.[86]

5. *In the tax field, as in all walks of life, advice is more likely to be accepted and heeded when it is sought, rather than when it is offered gratuitously.* It is interesting—and a bit depressing—to note how little attention the impressive report of the Taylor Mission received, compared to the report of the Musgrave Mission.[87] Whereas the Taylor Mission was conducted under the auspices of international agencies, despite the apparent cooperation of the government, it was not a Colombian initiative. By comparison, the Musgrave Mission was conducted at the request of the president of the government and its commission included Colombian members. This may be why many of the recommendations of the 1988 report on the taxation of income from business and capital were quickly accepted and implemented. It was prepared at the request of the Ministry of Finance and financed by

the government to assist it in choosing how best to utilize the emergency powers provided by the 1986 tax reform act.[88]

Of course, when advice is offered by outsiders with "clout," particularly the International Monetary Fund or the World Bank as part of a country consultation or Structural Adjustment Loan, the advice may be more acceptable. However, Colombia has experienced little outside coercion to improve its tax system, both because its system is one of the best among developing countries and because it has had the good fortune (based, not coincidentally, on a history of relatively sensible economic policy) not to have had to come hat in hand to international agencies. Even so, it should be noted that Colombia has often gladly accepted the advice of outsiders when it was offered by international agencies, as well as when solicited by the government.[89]

The tax system has been heavily influenced by external intellectual forces, including extensive borrowing from tax systems in industrial countries. When the modern system was first introduced in 1935, more attention was given to recent reforms in industrial countries, and the continuing influence of the "conventional wisdom" prevailing in academic circles. Of course, neither the tax systems of industrial countries nor the conventional wisdom has remained static; in a sense, the Colombian government has tried very hard to hit a moving target in tax policy. For example, investment incentives have come and gone; dividends have been taxed and then exempted; concerns about equity have led to fine-tuning of the system and then have been replaced with greater concern for economic neutrality and simplicity; inflation adjustment has first been rejected and then endorsed. Of course, the target has not quit moving. It may be that Colombia will eventually aim at the SAT discussed in the 1988 report.

During the period covered by this paper, advocates of tax reform have become much less ambitious in their expectations for tax policy. This can easily be seen in three of the areas just mentioned; presumably there are more.

- Much less is expected of economic planning and governmental intervention in the economy. This is most clearly seen in the elimination of virtually all tax incentives. Economic neutrality has replaced planning and intervention as a prime objective of tax policy.[90]
- Tax reformers are much less sanguine about the ability of the government to keep inflation low. As a result, emphasis has shifted from a strategy of "feet to the fire" to the acceptance of inflation adjustment. In part this reflects greater appreciation of the inequities and economic distortions caused by the combination of inflation and an unindexed tax system.
- A more realistic view of administrative capabilities has forced tax policy to be less ambitious in some areas.[91] The most important of these involve the personal exemptions, deductions, credits, and income splitting provided under prior law. To the extent that fine-tuning of tax burdens has been abandoned, equity—or at least the appearance of equity—may have been given lower weight.

6. *Tax policy for developing countries must be based on careful consideration of conditions prevailing in the country.* Tax policy that is in a constant state of flux, based on adoption of still-evolving fiscal ideas and fads, is undesirable, and must be maddening to those who have had to live through it. The latest fiscal fad must not simply be imported from industrial countries without appropriate modification.[92] Since hardly anyone can disagree with platitudes like this, it may be worthwhile to cite several examples.

First, it should have come as no surprise that Colombia would have difficulty administering an income tax with many "bells and whistles" intended to achieve equity among households that have similar incomes but are different in other respects. Whereas at a conceptual level such differences are potentially important, as a practical matter they may be hard to detect or verify. Under such circumstances "rough justice" may be preferable to a system that attempts to achieve "exact justice" but that cannot be administered.[93]

Second, foreign advisers commonly advised against the introduction of inflation adjustment in the measurement of income from business and capital. This may reflect a failure to appreciate the extent of inequities and economic distortions that can result from the combination of even mild rates of inflation and an unindexed tax system. This may result, in turn, from not being accustomed to thinking much about the problem. After all, foreign advisers have commonly come from industrial countries with relatively low rates of inflation. Recent tax policy reflects a more complete appreciation of the costs of an unindexed system in a country with a higher rate of inflation—though one well below rates typical of many developing countries.

The goals of equity and simplicity are often in conflict. This was shown clearly in the section on simplification. In earlier times that conflict has tended to be resolve itself in favor of equity, while more recently simplification has been in ascendancy.

7. *If it cannot be administered, the beauty of the legislation is lost in the morass of administrative failure.*[94] On paper at least, Colombia has one of the best tax systems for a developing economy. Whether the system is that good in practice, however, is open to some doubt. This underlines how unfortunate it was that the procedural portion of the 1974 law was declared unconstitutional and how important the procedural steps taken in the wake of the 1986 act can be (McLure and Pardo 1992).

It is difficult, in the absence of an in-depth study, to determine the skill and honesty with which a tax system is being administered; after all, if ineptitude and malfeasance leaped out at the investigator, they could presumably be detected and perhaps eliminated by the fiscal authorities. Thus, one must rely to some extent on impressions. In 1982 the World Bank reviewed the Colombian tax system and recommended the formation of a commission to study fiscal reform (McLure 1982). Such a commission would provide a detailed study of tax administration and help to focus public attention on needed reforms. Some means of producing administrative reform would probably be useful in all countries with the ambition of having a well-functioning income tax system.

Notes

1. See Gittes (1967) for the view that the recommendations of the Shoup mission to Venezuela affected the content of tax reform in that country; Gittes argues that some reform would have occurred in any event.

2. See OAS and IDB (1965, chapter 11), McLure (1971, 1975), and Gillis and McLure (1978). Much of the literature on income distribution, including especially studies by the national statistical office (Departamento Administrativo Nacional de Estadística), is brought together in United Nations (1988). For work on the distributional impact of government spending and other nontax policies, see Urrutia and de Sandoval (1971), Jallade (1974), Selowsky (1979), and the papers in Berry and Soligo (1980b). The usefulness and methodology of such studies have been questioned in Bird and De Wulf (1973), De Wulf (1975), McLure (1977, 1990b), and Bird (1980).

3. See Bilsborrow and Porter (1972). Their work influenced the mission's staff paper on incentives (see McLure 1971).

4. Canada is an important exception. Following the report of the Lortie Commission, Canada tried inflation adjustment briefly for a limited set of investments before repealing it because of complexity and the feeling that the threat of inflation had subsided. On the Lortie Report, see Canadian Tax Foundation (1982). John Bossons has been exceptionally active in the study of inflation adjustments for tax purposes.

5. Bird (1970, pp. 73, 79) makes an explicit appeal for using base broadening to increase taxes on upper-income Colombians; he rejects using higher tax rates for this purpose.

6. The view that the tax system could and should be used actively as an instrument of economic planning is perhaps nowhere more clearly stated than in Bird (1970, p. 54) "Tax policy should be deliberately designed so far as possible to correct major distortions in the Colombian price system." In a footnote to that sentence, Bird goes on to say (p. 247), "The traditional goal of 'fiscal neutrality' thus has no place in a development-oriented fiscal system. . . . Such strict neutrality seems neither possible nor desirable in this imperfect world, though of course *deliberate interference is to be avoided unless we are fairly sure of what we are doing*." (Emphasis added.)

7. In actuality the 1988 reforms employed replacement cost accounting for inventories, rather than indexation per se. Nonetheless, this was done as an explicit substitute for inflation adjustment.

8. On the other hand, the 1988 report did consider international issues in the tax treatment of income from business and capital, including indexation for inflation and the possibility of replacing the income tax with a consumption based direct tax. See McLure, Mutti, Thuronyi, and Zodrow (1990, pp. 41–44, 118–35, 219–22, and 303–12). By comparison, both the Taylor and Musgrave Reports devoted relatively little attention to this topic.

9. This discussion draws heavily on McLure (1989a), as well as on Perry and Cárdenas (1986).

10. Since the excess profits tax was abolished in 1974, and is thus primarily of historical interest, it is not described here. See, however, OAS and IDB (1965, pp. 35–37) and Musgrave and Gillis (1971, p. 77).

11. The basic structure of the tax is described in OAS and IDB (1965, pp. 37–40). For a discussion of its justification see McLure, Mutti, Thuronyi, and Zodrow (1990, pp. 167–68).

12. See also Carrizosa (1986, pp. 55–66).

13. Musgrave and Gillis (1971, p. 77), argue that "the excess profits tax is a major contributor to the high and erratic marginal rate structure. . . . Any function the excess profits tax may have served in the past is now outweighed by its evident inequities, distortions, incentives to waste "cheap tax pesos," and disincentives to growth and efficiency." For another negative appraisal of the excess profits tax, see Bird (1970, pp. 83–84).

14. The Taylor Mission's recommendation may have reflected both political and economic reality; after all, the taxation of dividends had only been introduced in 1953. The reasons given by the Musgrave Mission were economic: revenue loss, reduced progressivity, the possibility that double taxation did not occur because the corporate tax was shifted to consumers and wage earners, and the Colombian tradition of taxing income from capital more heavily than income from labor. Bird, (1970, p. 82), also did "not see any need for relief from 'double taxation'," in part because the corporate income tax may be shifted. For the contrary view that shifting of the corporate tax does not detract from the need for integration of the corporate and individual taxes, see Mieszkowski (1972) and McLure (1975).

15. For similar views see Bird (1970, pp. 68–69, 72, 84–85, 250).

16. It is interesting to note, however, that OAS and IDB (1965, p. 88), described incentives for basic industries for iron and steel fabricators as "general in scope."

17. See Musgrave and Gillis (1971, pp. 93–94). Bilsborrow and Porter (1972, p. 417), found that the amount of additional investment stimulated by the incentives was less than the revenue loss involved. See also Bird (1970, pp. 132–33). On page 144 Bird gives the following scathing summary assessment: "The objects of most existing incentive legislation are poorly defined. Their administration is complex and appears to be inefficient. The cost of these incentives is never considered and evaluated as an expenditure, as should be done. In any subsequent incentive legislation, it would be highly desirable to include as essential features a clear, well defined law, a mandatory annual revenue-cost estimate, and a centralized and simple administrative mechanism. All concessions should be granted for a limited time period and should be reexamined periodically. Few rich countries today satisfy these counsels of perfection; but poor countries like Colombia can less afford the waste of resources implied by not satisfying them and must lead the way in devising workable incentives."

18. Bird (1970, pp. 110–11) favored a similar approach.

19. The tax had actually been enacted in 1963, but its introduction was postponed.

20. See Perry and Orozco de Triana (1990). Tait (1988, pp. 6–8) briefly describes international practice in these areas. For further details of such practice, see Due (1988, chapter 6).

21. Space does not permit a full discussion of the political process leading to the 1974 reforms. For further discussion see Perry and Cárdenas (1986) and especially Urrutia (1989). The basic decrees were No. 2053 of September 30 and No. 2247 of October 21.

22. For a full discussion of the political process leading to the 1974 reforms see Perry and Cárdenas (1986) and especially Urrutia (1989).

23. For further description and analysis of the working of this rule, see Gillis and McLure (1977, pp. 36–41).

24. See also Perry and Cárdenas 1986, chapter 3.

25. See McLure (1982) for a recitation of the general outlines of several illegal tax fraud schemes and the call for a high-level commission to examine tax administration.

26. The monetary correction is an adjustment made to the principal of certain deposits to maintain their real value.

27. Thus, the diagnosis of the present system in the "Exposición de Motivos" contains the following section headings: equity, neutrality, progressivity and high nominal level of rates, effects of inflation, simplification, and administrative capacity.

28. Dollar signs, when used alone, denote Colombian pesos. U.S. dollars are indicated by "US$." It should be remembered that many of the personal tax benefits allowed taxpayers under prior law took the form of credits, rather than deductions. The reduction in tax liabilities due to the increase in the zero bracket amount is not as dramatic as it might appear, since many of these personal tax benefits were eliminated.

29. For discussions of this issue, see Brazer (1980), Munnell (1980), McIntyre and Oldman (1977), and the papers in Penner (1983).

30. To avoid exempting dividends paid from tax-exempt income, tax free distributions were limited to 7/3 of the cumulative amount of tax paid by the entity. To be consistent with the exemption of dividends, the value of shares was removed from the figure for net wealth used in calculating presumptive income. For further discussion of these provisions, see McLure, Mutti, Thuronyi, and Zodrow (1990, chapters 5 and 6).

31. On the wisdom of such a policy, see Sinn (1987). For a recent discussion of alternative views of the effects of dividend taxes, see Zodrow (1991).

32. See Chica (1984/85), Carrizosa (1986), and reviews of these arguments by McLure and Zodrow (1989).

33. World Bank staff estimate based on data in Comisión Nacional de Valores, Informe de Labores, 1987–88, reported in Colombian Economic Memorandum, April 21, 1989.

34. This was already available to some extent prior to the 1986 changes, due to subsequently enacted exceptions (for real estate, corporate shares, and partnership interests) to the 1974 rule that revaluations be made earlier, not at the time of disposition of assets.

35. See McLure, Mutti, Thuronyi, and Zodrow (1990, chapters 7–9) for details of the consumption-based direct tax and the Chilean system. Perry and Cárdenas (1986, pp. 178–81) briefly discuss both inflation adjustment and a tax system with the basic features of the SAT.

36. The METR measures the extent to which the after-tax rate of return falls below the before-tax return. The concept and its calculation are described fully in the section on Tax Reform and Marginal Effective Tax Rates.

37. For a complete discussion, see McLure, Mutti, Thuronyi, and Zodrow (1990, chapter 7).

38. These issues are also covered in McLure, Mutti, Thuronyi, and Zodrow (1990, pp. 303–23).

39. This description and appraisal of the 1988 law is taken from McLure, Mutti, Thuronyi, and Zodrow (1990, chapter 11).

40. These are described in Legislative Decrees No. 2686 and 2687 of December 26, 1988 and No. 2633 of December 21, 1988 and in Law 84 of December 21, 1988.

41. There is a relief provision for taxpayers whose assets have a value of less than 50 percent of that resulting from the application of the regular rules for inflation adjustment.

42. The decree specified separate procedures for determining the replacement costs of inventories of unprocessed goods, finished goods, and goods in process. Note that whereas net wealth is based on replacement costs, cost of goods sold in a given year is calculated before revaluing inventories to their replacement values.

43. For further explanation, see McLure, Mutti, Thuronyi, and Zodrow (1990, chapter 7) and references cited there.

44. The report had been available in English in September.

45. As an interim measure, the value of owner-occupied houses, up to $10 million, was removed from the base of the net wealth tax, beginning in 1989.

46. Decree 1321 of June 20, 1989.

47. See also Bird (1970, p. 41), OAS and IDB (1965 pp. 2–4), and Musgrave and Gillis (1971, chapter 3).

48. Perry and Cárdenas (1986, p. 77) do not indicate whether these figures include only revenues of the central government or those of lower-level governments as well. The figures fall between those reported for 1950 to 1962 by Taylor (OAS and IDB, 1965, p. 4) for central government and for all governments. They are best seen as indicative of trends.

49. Bird (1970, pp. 33, 38) notes that the 1960 reforms, which were very expensive in terms of lost revenue, were introduced at the height of the export/import cycle. Although the reforms were affordable when first introduced, the subsequent drop in economic activity soon reduced revenues to unacceptable levels.

50. Perry and Cárdenas (1986, p. 78) note that imports have not always fallen when exports have fallen; in such cases, the continuation of a high level of imports has sustained revenues.

51. Note that these figures from Musgrave and Gillis (1971, chapter 3) are different in coverage and source from figures given earlier in this section.

52. The following figures are from Tanzi (1987). The concepts used in this later period are not necessarily the same as those underlying the figures reported by Musgrave and Gillis (1971).

53. Bird (1970, p. 239), commenting on the Taylor Mission's figures, states, "The degree of accuracy to be attributed to the final figures is obviously not too high. Nevertheless, there are no conceivable adjustments which would alter the picture of great inequality." Though most studies focus on the urban sector, some also cover the rural sector; the rural data are generally thought to be less reliable than those for the urban sector, which are the primary focus of the United Nations study (1988). Bird (1970, pp. 15–17) notes that labor's relatively small share of total income also suggests that the distribution of income is highly skewed.

54. Inclusion of income in kind does not significantly change this pattern. The finding that the distribution of income has become more unequal during the 1970s would suggest a continuation of the pattern for somewhat earlier periods reported in Bird (1970, p. 15) and in Berry and Soligo (1980a, pp. 14–17). The corresponding figures for 1982 reported in the U.N. study (1988) are 2.5 and 40.3 percent, suggesting somewhat less worsening of the relative position of the lowest two deciles and a slightly lower share of the top decile; however, the U.N. study warns (p. 61) that not too much should be made of this minor change, due to lack of reliability of data.

55. See also Berry and Soligo (1980b, p. 5).

56. For more on the difficulties of incidence analysis see Bird and De Wulf (1973) and McLure (1977, 1990b).

57. McLure (1975) does so for Colombia.

58. For evidence on this phenomenon from Jamaican household data and further references to the literature, see Miller and Stone (1985).

59. For development of this theme, see Harberger (1973, 1983) and McLure (1979). Since a creditable tax imposes no net additional burden on foreign multinationals, Colombian firms competing with foreign firms may be forced to absorb the tax.

60. See Berry and Soligo (1980a) for a criticism of McLure (1975), which attributes coffee export duties entirely to rural households.

61. See Bird (1970, pp. 211–18) for a description of tax policy applied to the coffee sector early in the period under examination and Bird (1984) for a more recent assessment that very little of the revenue from the coffee tax actually finances direct benefits to the coffee sector.

62. See also Perry and Cárdenas (1986, chapter 7).

63. It is unclear whether one would expect average depreciation rates in Colombia to be faster or slower than in the United States. On the one hand, depreciation rates may be relatively slow because assets tend to be used for longer periods of time and repaired more frequently. On the other hand, depreciation rates may be relatively fast because labor is less skilled and/or maintenance is less adequate than in the United States. In the absence of empirical data on depreciation rates, there seems to be little alternative to using U.S. estimates.

64. For a detailed discussion of these issues see McLure, Mutti, Thuronyi, and Zodrow (1990, chapter 5).

65. For a METR analysis that incorporates considerations of risk, see Hamilton, Mintz, and Whalley (1991).

66. The approach used to calculate METRS in this chapter is not well suited to calculating tax rates on investment in housing; in particular, the assumption of individual arbitrage (to be discussed below) implies that, in the absence of the taxation of imputed rents, the METR on debt-financed investment in owner-occupied housing is zero by definition.

67. For details on additional assumptions made in the analysis, see McLure, Mutti, Thuronyi, and Zodrow (1990, chapter 4). For a discussion of how taxes on foreign trade might be included in the calculation of METRS, see Guisinger (1988).

68. For details regarding this procedure, see McLure, Mutti, Thuronyi, and Zodrow (1990).

69. Sectoral METRS are calculated as weighted averages of the asset METRS; for details and results see McLure, Mutti, Thuronyi, and Zodrow (1990).

70. In the case of equipment, the slight variation in tax rates with inflation reflects only the fact that annual inflation indexing of the deductions allowed under the three-year write-off for equipment is only an approximation to the conceptually correct continuous time inflation indexing. The following discussion will neglect these slight differences in the METRS on investment in equipment.

71. The dividend tax credit is reflected in the 0.1×0.6 term, as a 10 percent tax credit is allowed for the corporate distribution of 0.6.

72. This line of reasoning is consistent with the METR methodology, which ignores issues of incidence and general equilibrium adjustments. One possible interpretation that takes such issues into account follows. It seems reasonable that the prices of capital goods (other than land) are fixed internationally, in which case both the VAT on domestic purchases and any exchange rate undervaluation would appear as higher prices to purchasers of the capital goods. To the extent that businesses are successful in shifting such costs forward in the form of higher prices, the results in tables 3.7–3.9 (which neglect the VAT) are more relevant. To the extent such

costs cannot be shifted forward, results that would include a VAT at a rate of 10 percent would be more relevant.

73. For numerical results in this case, see McLure, Mutti, Thuronyi, and Zodrow (1990).

74. These three changes are exclusive of changes in capital gains taxation that occurred over the period. Since the effective rate of taxation of capital gains is generally small, changes in the rules for taxing such gains do not have a large impact on the METR calculations.

75. As described in the section on episodes of tax reform, more generous credits were available under certain circumstances; however, the analysis focuses on the effects of the widely available 10 percent credit.

76. Note that the discussion in the text equates efficiency in the taxation of capital income with uniform METRS. Of course, the optimal taxation literature has derived theoretical conditions under which efficiency requires differential taxation (Stern 1987a, 1987b). However, as argued by Deaton (1987), the data required to implement "optimal" differential taxation are generally not available in most developing countries, including Colombia. In addition, political realities suggest that the implementation of differential taxation is more likely to reflect the strengths of various special interest groups than theoretical considerations based on optimal taxation principles. Accordingly, we assume that a system with neutral taxation of capital income is close to an efficient tax system for Colombia. For further elaboration of this argument, see the chapter by Thirsk in this volume.

77. The earlier discussion of decapitalization suggests this has occurred in Colombia. For further discussion of this issue and data on the term structure of debt in recent years, see McLure, Mutti, Thuronyi, and Zodrow (1990, chapter 7).

78. The benefits of preferential treatment of land under the net wealth tax are likely to be capitalized into land values. In this case, such treatment would have no effects on marginal investment decisions. As noted previously, such price adjustments are not considered in the METR methodology. In any case, elimination of this "distortion" under the 1988 reform was probably desirable on distributional grounds.

79. See also McLure (1982).

80. Note again that the METR calculations do not consider the effects of home country taxes on marginal investment decisions made by foreigners.

81. This draws on McLure (1989a), Thirsk (1988), and Gillis (1989), as well as the material above. Thirsk mentions several others factors that presumably are relevant outside Colombia, but are not discussed, such as the obvious political difficulty of reforming provisions benefiting the rich and powerful; he uses "the case of cows" to illustrate this point, noting that many members of congress raise cattle. He also notes that inadequate design of one tax may jeopardize the entire system. Finally, how tax reform is perceived may depend on the macroeconomic climate in which it is introduced.

82. One can hardly resist the temptation to quote the following famous—if self-serving—passage from Keynes (1936, pp. 383–84): "The ideas of economists and political philosophers, both when they are right and when they are wrong, are more powerful than is commonly understood. Indeed the world is ruled by little else. Practical men, who believe themselves to be quite exempt from any intellectual influences, are usually the slave of some defunct economist. . . . I am sure that the power of vested interests is vastly exaggerated compared with the gradual encroachment of ideas. Not, indeed, immediately, but after a certain interval. . . . The ideas which civil servants and agitators apply to current events are not likely to be the

newest. But, sooner or later, it is ideas, not vested interests, which are dangerous for good or evil."

83. In this regard, the Colombian experience has been quite remarkable. The Spanish version of the Musgrave Report was published in May 1969 by the Banco Popular, which also published a Spanish translation of Gillis and McLure (1977), a report originally prepared for the World Bank. The English version of McLure (1982) circulated in Colombia, although it was never released by the World Bank or the author. Finally, the Ministry of Finance published the English version of McLure, Mutti, Thuronyi, and Zodrow (1990) before the Spanish version. The government also approved English publication of both the Musgrave Report and the 1988 report by foreign academic presses.

84. Several of the reports mentioned in the previous footnote have been used in university courses in Colombia. It is worth noting that Enrique Low, a staff member of the Musgrave Mission, was the senior coauthor of a textbook on fiscal policy in Colombia, Low and Gómez (1986).

85. On the other hand, they can be expected to offer advice that has no tinge of personal vested interest—a high standard indeed for a local expert who must pay the taxes he or she recommends. Note that an early criticism of the 1988 report was that there was no Colombian participation in its formulation, which seems somewhat inappropriate given the intimate involvement of two directors general of internal taxation. Yet it is true that there was less involvement at the staff level than in either the Taylor or Musgrave Missions.

86. See also Gillis (1989).

87. Of course, the work of the Taylor Mission and Richard Bird fed directly into the deliberations of the Musgrave Mission.

88. Of course, in this case the government confronted the imminent expiration of the extraordinary powers responsible for introducing the 1988 reforms. On the other hand, those emergency powers were enacted with the expectation that the 1988 report would be prepared to guide the government during implementation.

89. Note that both Gillis and McLure (1977) and McLure (1982) were prepared under the auspices of the World Bank.

90. In commenting on this paper Richard Bird has suggested that this may be a cyclical phenomenon and that incentives may be revived in the near future. This could be true, for example, to the extent that the highly lucrative trade in cocaine is a form of the "Dutch disease," the inability to export anything else, because of the strength of the peso. Sachs and Morales (1988, pp. 41–42), have made such an argument for export incentives in Bolivia, the first link in the chain of cocaine traffic. Be this as it may, the judgment in the text seems to be appropriate.

91. Thus, Colombia seems to have accepted the wisdom of the following maxim from Bird (1989, p. 326): "The best way to cope with the administrative problem is to design tax reforms for developing countries in full recognition of the severe limitations imposed by administrative realities. The administrative dimension is central, not peripheral, to tax reform."

92. Bird (1970, p. 203) noted almost twenty years ago that "Colombia thus has a long tradition of foreign tax missions, as well as a surprising record of listening to them, not always with desirable results. One reason for the relatively poor success record of foreign advisers in the past has been the inadequate consideration of local conditions affecting the transferability of tax techniques." Bird (1970, p. xiii), also wrote: "The tax policy appropriate for a given country at a given time is determined by the economic, political, and social circumstances of the country. It follows that,

as these circumstances change, the appropriate tax policy will also change. Tax reform is therefore a never-ending process, not something that can be brought about once-and-for-all and then forgotten."

93. Bird (1970, p. 62) notes, "The aim must be practicality, not perfection."

94. Bird (1970, p. 200) notes, "In the hands of incompetent administration, good tax policy and bad tax policy may end up looking remarkably alike." Or, as Casanegra has written, "tax administration *is* tax policy" (quoted in Bird 1989, p. 315).

References

Berry, R. Albert, and Ronald Soligo. 1980a. "The Distribution of Income in Colombia: An Overview." In R. Albert Berry and Ronald Soligo, eds., *Economic Policy and Income Distribution in Colombia*. Boulder, Colo.: Westview Press.

———, eds. 1980b. *Economic Policy and Income Distribution in Colombia*. Boulder, Colo.: Westview Press.

Bilsborrow, R. E., and R. C. Porter. 1972. "The Effects of Tax Exemptions on Investment by Industrial Firms in Colombia." *Weltwirtschaftliches Archiv* 108(3): 396–425.

Bird, Richard M. 1970. *Taxation and Development: Lessons from the Colombian Experience*. Cambridge, Mass.: Harvard University Press.

———. 1980. "Income Redistribution through the Fiscal System: The Limits of Knowledge." *American Economic Review* 70(2): 77–81.

———. 1984. "Director, Intergovernmental Finance in Colombia." Harvard University International Tax Program, Cambridge, Mass.

———. 1989. "The Administrative Dimension of Tax Reform in Developing Countries." In Malcolm Gillis, ed., *Lessons from Fundamental Tax Reform in Developing Countries*. Durham, N.C.: Duke University Press.

Bird, Richard M., and Luc Henry De Wulf. 1973. "Taxation and Income Distribution in Latin America." *International Monetary Fund Staff Papers* 20: 639–82. International Monetary Fund, Washington, D.C.

Brazer, Harvey E. 1980. "Income Tax Treatment of the Family." In Henry J. Aaron and Michael J. Boskin, eds., *The Economics of Taxation*. Brookings Institution, Washington, D.C.

Cámara de Comercio de Bogotá. 1987. *Nueva Reforma Tributaria*. Segunda Edición. Bogotá.

Canadian Tax Foundation. 1982. "Inflation, Indexation, and the Tax System, Where Do We Go from Here?" Proceedings of the Thirty-fourth Tax Conference. Toronto: The Canadian Tax Foundation.

Carrizosa S., Mauricio. 1986. "Hacia la Recuperación del Mercado de Capitales en Colombia." Bolsa de Bogotá, Bogotá.

Chica, Ricardo A. 1984/85. "La Financiación de la Inversión en la Industria Manufacturera Colombiana: 1970–1980." *Desarrollo y Sociedad (Colombia)*15/16: 192–285.

Deaton, Angus. 1987. "Econometric Issues for Tax Design in Developing Countries." In David Newbery and Nicholas Stern, eds., *The Theory of Taxation for Developing Countries*. New York: Oxford University Press for the World Bank.

De Wulf, Luc Henry. 1975. "Fiscal Incidence Studies in Developing Countries: Survey and Critique." *International Monetary Fund Staff Papers* 23: 61–131. International Monetary Fund, Washington, D.C.

Due, John F. 1988. *Indirect Taxation in Developing Economies*. Revised edition. Baltimore, Md.: Johns Hopkins University Press.

Gillis, Malcolm. 1989. "Tax Reform: Lessons from Postwar Experience in Developing Countries." In Malcolm Gillis, ed., *Lessons from Fundamental Tax Reform in Developing Countries*. Durham, N.C.: Duke University Press.

Gillis, Malcolm, and Charles E. McLure, Jr. 1971. "Coordination of Tariffs and Internal Indirect Taxes." In Richard A. Musgrave and Malcolm Gillis, eds., *Fiscal Reform for Colombia: The Final Report and Staff Papers of the Colombian Commission on Tax Reform*. Cambridge, Mass.: Harvard University International Tax Program.

———. 1977. "La Reforma Tributaria Colombiana de 1974." Banco Popular, Bogotá.

———. 1978. "Taxation and Income Distribution: The Colombian Tax Reform of 1974." *Journal of Development Economics* 5: 233–58.

Gittes, Enrique F. 1967. "Income Tax Reform: The Venezuelan Experience." *Harvard Journal on Legislation* 5(1): 125–73.

Goode, Richard A. 1960. "New System of Direct Taxation in Ceylon." *National Tax Journal* 13: 329–40.

Guisinger, Stephen. 1988. "Total Protection: A New Measure of the Impact of Government Interventions on Investment Profitability." University of Texas at Dallas, Department of Economics.

Hamilton, Bob, Jack Mintz, and John Whalley. 1991. "Decomposing the Welfare Costs of Capital Tax Distortions: The Importance of Risk Considerations." Working Paper Number 3628. National Bureau of Economic Research, Washington, D.C.

Harberger, Arnold C. 1973. "The Panamanian Income Tax System: A Heterodox View." University of California at Los Angeles, Department of Economics, Los Angeles.

———. 1983. "The State of the Corporate Income Tax: Who Pays It? Should It Be Repealed?" In Charles E. Walker and Mark A. Bloomfield, eds., *New Directions in Federal Tax Policy for the 1980s*. Cambridge, Mass.: Ballinger.

Hulten, Charles E., and Frank C. Wykoff. 1981. "The Measurement of Economic Depreciation." In Charles E. Hulten, ed., *Depreciation, Inflation and the Taxation of Income from Capital*. Washington, D.C.: Urban Institute Press.

IMF (International Monetary Fund). 1988. *International Financial Statistics Yearbook*. Washington, D.C.

Jallade, Jean-Pierre. 1974. *Public Expenditures and Income Distribution in Colombia*. Baltimore, Md.: Johns Hopkins University Press for the World Bank.

Keynes, John Maynard. 1936. *The General Theory of Employment, Interest, and Money*. New York: Harcourt, Brace & World.

King, Mervyn A., and Don Fullerton, eds. 1984. *The Taxation of Income from Capital*. Chicago: The University of Chicago Press.

Low M., Enrique, and Jorge Gómez R. 1986. *Política Fiscal*. Bogotá: Universidad Externado de Colombia.

McIntyre, Michael J., and Oliver Oldman. 1977. "Treatment of the Family." In Joseph A. Pechman, ed., *Comprehensive Income Taxation*. Washington, D.C.: Brookings Institution.

McLure, Charles E., Jr. 1971. "The Incidence of Taxation in Colombia." In Richard A. Musgrave and Malcolm Gillis, eds., *Fiscal Reform for Colombia: The Final Report and Staff Papers of the Colombian Commission on Tax Reform*. Cambridge, Mass.: Harvard University International Tax Program.

———. 1975. "The Incidence of Colombian Taxes: 1970." *Economic Development and Cultural Change* 24(1): 155–83.

———. 1977. "Taxation and the Urban Poor in Developing Countries." *World Development* 5: 169–88.

———. 1979. "The Relevance of the New View of the Incidence of the Property Tax in Developing Countries." In Roy Bahl, ed., *The Taxation of Urban Property in Less Developed Countries*. Madison, Wisc.: University of Wisconsin Press.

———. 1982. "Income and Complementary Taxes." Available from author. Processed.

———. 1989a. "Analysis and Reform of the Colombian Tax System." In Malcolm Gillis, ed., *Tax Reform in Developing Countries*. Durham, N.C.: Duke University Press.

———. 1990a. "Reform in an Inflationary Environment: The Case of Colombia." In Michael Boskin and Charles E. McLure, Jr., eds., *World Tax Reform*. San Francisco: Institute for Contemporary Studies Press.

———. 1990b. "Income Distribution and Tax Incidence under the VAT." In Malcolm Gillis, Gerardo P. Sicat, and Carl S. Shoup, eds., *Value Added Tax in Developing Countries*. Washington, D.C.: World Bank.

McLure, Charles E., Jr., and Santiago Pardo R. 1992. "Improving the Administration of the Colombian Income Tax." In Richard M. Bird and Milka Casanegra de Jantscher, eds., *Improving Tax Administration in Developing Countries*. Washington, D.C.: International Monetary Fund. Also as "Mejora de la Administración del Impuesto sobre la Renta Colombiano (1986–88)." In Richard M. Bird and Milka Casanegra de Jantscher, eds., *La Administración Tributaria en los Países del CIAT*. Madrid: Ministerio de Economía y Hacienda.

McLure, Charles E., Jr., and George R. Zodrow. 1989. "Tax Reform and Financial Policy in Colombia." In Carlos Caballero Argaez, ed., *Macroeconomía*. Mercado de Capitales y Negocio Financiero. Bogotá: Asociación Bancaria de Colombia.

McLure, Charles E., Jr., Jack Mutti, Victor Thuronyi, and George R. Zodrow. 1990. *The Taxation of Income from Business and Capital in Colombia*. Durham, N.C.: Duke University Press.

Mieszkowski, Peter. 1972. "Integration of the Corporate and Personal Income Taxes: The Bogus Issue of Shifting." *Finanzarchiv* 31: 286–97.

Miller, Barbara D., and Carl Stone. 1985. "The Low-Income Household Expenditure Survey: Description and Analysis." Staff Paper No. 25 of the Jamaica Tax Structure Examination Project. Syracuse University, Maxwell School, Syracuse, New York.

Ministerio de Hacienda y Crédito Público. 1984. "Impuesto sobre la Renta en Colombia: Compilación Cronológica de Normas Vigentes del Impuesto sobre la Renta, Complementarios, y Procedimiento." Bogotá.

———. 1987. "Hacienda." Bogotá.

Munnell, Alicia H. 1980. "The Couple versus the Individual under the Federal Personal Income Tax." In Henry J. Aaron and Michael J. Boskin, eds., *The Economics of Taxation*. Washington, D.C.: Brookings Institution.

Musgrave, Richard A., and Malcolm Gillis, eds. 1971. *Fiscal Reform for Colombia: The Final Report and Staff Papers of the Colombian Commission on Tax Reform*. Cambridge, Mass.: Harvard University International Tax Program.

OAS and IDB (Organization of the American States and Inter-American Development Bank). 1965. *Fiscal Survey of Colombia*. Baltimore, Md.: Johns Hopkins University Press.

Penner, Rudolph G., ed. 1983. "Taxing the Family." American Enterprise Institute for Public Policy Research, Washington, D.C.

Perry, Guillermo, and Mauricio Cárdenas S. 1986. *Diez Años de Reformas Tributarias en Colombia*. Bogotá: FEDESARROLLO.

Perry, Guillermo, and Alba Lucia Orozco de Triana. 1990. "The VAT in Colombia." In Malcolm Gillis, Gerardo P. Sicat, and Carl S. Shoup, eds., *Value Added Taxation in Developing Countries*. Washington, D.C.: World Bank.

Restrepo, J. C. 1983. "La Economía Colombiana y el Desarrollo de la Sociedad Anónima en los Ultimos Treinta Años." *En Comisión Nacional de Valores, Boletín Mensual* 14(Mayo), *Superintendencía de Sociedades Anónimas, Boletín Mensual* (1981–1983), Nos. 3, 4, y 6.

Sachs, Jeffrey, and Juan Antonio Morales. 1988. *Bolivia: 1952–1986*. San Francisco: Institute for Contemporary Studies Press.

Selowsky, Marcelo. 1979. *Who Benefits from Government Expenditure? A Case Study of Colombia*. New York: Oxford University Press for the World Bank.

Sinn, Hans-Werner. 1987. *Capital Income Taxation and Resource Allocation*. New York: North Holland.

Slitor, Richard. 1971. "Reform of the Business Tax Structure: Analysis of Problems and Alternative Remedial Proposals." In Richard A. Musgrave and Malcolm Gillis, eds., *Fiscal Reform for Colombia: The Final Report and Staff Papers of the Colombian Commission on Tax Reform*. Cambridge, Mass.: Harvard University International Tax Program.

Stern, Nicholas. 1987a. "Aspects of the General Theory of Tax Reform." In David Newbery and Nicholas Stern, eds., *The Theory of Taxation for Developing Countries*. New York: Oxford University Press for the World Bank.

———. 1987b. "The Theory of Optimal Commodity and Income Taxation: An Introduction." In David Newbery and Nicholas Stern, eds., *The Theory of Taxation for Developing Countries*. NewYork: Oxford University Press for the World Bank.

Tait, Alan A. 1988. *Value Added Tax: International Practice and Problems*. Washington, D.C.: International Monetary Fund.

Tanzi, Vito. 1977. "Inflation Lags in Collection and the Real Value of Tax Revenue." *International Monetary Fund Staff Papers* 24(1): 154–67. Washington, D.C.

———. 1987. "Quantitative Characteristics of the Tax Systems of Developing Countries." In David Newbery and Nicholas Stern, eds., *The Theory of Taxation for Developing Countries*. New York: Oxford University Press.

Thirsk, Wayne. 1988. "Some Lessons from the Colombian Experience." World Bank, Public Economics Division, Washington, D.C.

United Nations, Economic Commission for Latin America. 1957. *The Economic Development of Colombia*. Geneva: United Nations Department of Economic and Social Affairs.

———. 1988. "La Distribución del Ingreso en Colombia: Antecedentes Estadísticos Características Socioeconómicas de los Receptores." Comisión Económica para América Latina y el Caribe, Santiago.

Urdinola Uribe, Antonio. 1987. "Tributación y Deuda: Aspectos de la Reforma Tributaria de 1986." Nueva Reforma Tributaria, 1986. Bogotá: Asociación Nacional de Industriales.

Urrutia M., Miguel. 1988. "Comments: 'Fiscal Policy and Equity in Developing Countries' by Charles E. McLure, Jr." In Elliot Berg, ed., *Policy Reform and Equity*. San Francisco: Institute for Contemporary Studies Press.

———. 1989. "The Politics of Fiscal Policy in Colombia." In Miguel Urrutia, Schinichi Ichimura, and Setsuko Yukawa, eds., *The Political Economy of Fiscal Policy*. Tokyo: The United Nations University.

Urrutia, Miguel M., and Clara Elsa de Sandoval. 1971. "Política Fiscal y Distribución del Ingreso en Colombia." *Revista del Banco de la República* 44: 1072-87.

Zodrow, George R. 1991. "On the 'Traditional' and 'New' Views of Dividend Taxation." *National Tax Journal* 44(4): 497–511.

4

Reforming the Tax System in Indonesia

Mukul Asher

During the 1980s, Indonesia carried out a series of major economic reforms as part of an overall structural adjustment program designed to enhance the macroeconomic stability and international competitiveness of the country's economy. This chapter analyzes the tax reform experience of Indonesia in the areas of both tax policy and administration during the 1980s. It discusses the reasons why reform was undertaken, attempts to evaluate its success in fulfilling its objectives, and offers some general lessons.

Motivations for Tax Reform

Indonesia's tax structure relied heavily on revenue from the oil sector from 1973 onward. From 1978 to 1983, the ratio of oil revenue to gross domestic product (GDP) went up from 9.6 to 12.9 percent, while the ratio of nonoil tax revenue (NOTR) to GDP went down from 7.4 to 6.0 percent. Thus, during this period, NOTR accounted for only one-third of total tax revenues and financed barely more than a quarter of total expenditures.

The system was plagued by uneven enforcement and compliance, partly due to the complexity and ambiguity of the tax laws. In 1980 only 1.2 million individuals out of an estimated 11 million taxable families paid income taxes. In the corporate sector, income tax collected in the same year was estimated to be less than 50 percent of taxes due. Tax arrears were also large. There was substantial subjective interpretation of the income tax, and liability was often negotiated rather than determined objectively.

There was a wide dispersion in effective rates within activities, sectors, and products, leading to inefficient allocation and inequities between individual taxpayers. The progressivity of the individual income tax was circumvented by its narrow base, reduced by nontaxation of most fringe benefits, capital gains, and pensions. For private businesses, audited and nonaudited profits were taxed differently and public companies with a higher percentage of shares held by the public received preferential tax treatment.

Tax Reform Measures

The tax reform measures unveiled in 1983 had four broad objectives: increase NOTR yield, streamline the tax laws and improve administrative efficiency, reduce tax-induced distortions in the allocation of resources and achieve economic neutrality, and ensure that the poor were not made worse off as a result of the reforms.

The following fundamental changes were made to the tax structure in an attempt to fulfill these objectives. First, to maximize compliance the new tax laws were designed to be substantially free of ambiguity. The income tax base was broadened, nominal tax rates were substantially reduced, and a common rate structure for taxation of capital and wage incomes was established. Exemptions were abolished for high-income groups and those remaining were set at levels aimed at keeping the poor from being liable. Finally, a value added tax (VAT) and a luxury sales tax were introduced.

The following paragraphs present a preliminary evaluation of the tax reforms in meeting stated objectives and insights gained from the reform process.

Revenue gains from tax changes reached the ambitious targets set in the Development Plan, primarily through the introduction of the VAT. Improvements in tax administration, on the other hand, have proven more elusive. Audit and enforcement functions still need considerable strengthening.

In general the tax reform measures have enhanced the neutrality of the tax system. The VAT eliminated differential tax treatment of domestically produced and imported goods, circumvented the cascading effects of the prereform sales tax, and spared business inputs and exports from taxation. Income tax reform eliminated differential tax treatment of various assets and sectors. Furthermore, the income tax was made neutral according to type of business organization and form of ownership and control.

The tax reform paid special attention to shielding the poor by increasing the basic exemption level for the income tax. The introduction of a VAT did not have a measurable impact on prices, which also suggests that the VAT might not have had any adverse impact on the relative position of the poor. Thus, while final judgment on the distributional impact of the tax reform must await careful study, a qualitative analysis suggests no cause for alarm.

The Indonesian experience demonstrates that a major overhaul of the tax system can be accomplished in a relatively short period of time, given careful preparation. Improving tax administration, on the other hand, is a slow and painstaking task. It is also clear that value added taxation is superior to income taxes in terms of raising additional revenues in a relatively nondistorting fashion. The Indonesian experience also lends credence to the view that the VAT is administratively feasible in developing countries (particularly at the preretail level). It is capable of generating considerable additional tax revenues and minimizing taxation of inputs and exports. However, the success of a VAT critically depends upon its design and implementation.

This chapter concludes that the tax reform program represented a substantial improvement over the prereform situation by strengthening the tax system and making it responsive to growth requirements.

Reforming the Tax System

According to the *World Development Report 1989* (World Bank 1989), Indonesia is a low-income country (per capita income in 1987 of US$450).[1] Indonesia is also a member of the Organization of the Petroleum Exporting Countries (OPEC).

During the 1980s, major reforms were carried out in several parts of the economy as part of an overall structural adjustment program designed to enhance the economy's macroeconomic stability and international competitiveness. These reforms went considerably beyond tax policy and administration, which is the focus of this study, to include changes in exchange rate management; the financial sector, including capital markets; industrial policy, including deregulation measures aimed at both domestic and foreign investment; and trade policy, including customs administration.

The tax reform process was initiated by the government in early 1981, and the resulting program was enacted by parliament in December 1983, with different dates of implementation for the various taxes. For example, the new income tax was implemented in January 1984, the VAT and the luxury sales tax (the VAT package) in April 1985, and the land and building tax and the new stamp duty in January 1986. Several modifications, particularly in the VAT package, have been introduced since their original implementation. There have also been changes in taxes not covered by the tax reform program, such as import duties and export taxes. Because the tax reforms of the 1980s left petroleum tax arrangements unchanged, they are not discussed here.

Historical Review of Tax Reform and Fiscal Policy

The New Order government of President Suharto has, since its inception in 1966, viewed poor fiscal control as a major reason for the poor economic performance under the previous government. Therefore, it gave high priority to pursuing fiscal policy that would be consistent with macroeconomic stability. It set itself the task of pursuing a "balanced budget" policy. In the Indonesian context, this meant balancing total expenditure with total receipts from all sources, including project and program aid. Of course, this does not correspond to the usual definition of a balanced budget, which excludes project and program aid receipts. Whereas the balanced budget policy did introduce some distortions, it nevertheless represented a substantial improvement over previous fiscal policy. The fiscal policy of the New Order government hoped to curb current expenditure and improve domestic resource mobilization. All these features have essentially remained unchanged since the 1980s, so there has been continuity in fiscal policy objectives although the manner of implementation has varied. The tax reforms of the 1980s should be viewed from this perspective.

It is useful to describe the prereform tax and revenue system and the earlier reform attempts. The receipts of the central government are divided into domestic revenue and development receipts. The latter consist of program and project aid, essentially foreign loans. During the prereform period from 1978 to 1983, development receipts financed about one-third of annual development expenditures (table 4.1).

It is traditional in an oil-exporting country to divide tax revenue into NOTR and income tax revenue from the oil and gas sector because each has a different impact on the economy. Tax revenue from the oil sector consists largely of the conversion of foreign currency revenue from the export of oil, which the government receives according to production sharing contracts with the companies in the oil sector. Thus, the origin of oil sector revenue is external and its conversion from foreign currency into rupiah leads to an increase in reserve money that is potentially inflationary. In

Table 4.1 Central Government, Selected Revenue Indicators
(percent)

Fiscal year beginning	Total expenditure as a share of GDP	Total domestic revenue as a share of GDP[a]	Total tax revenue as a share of GDP	Oil revenue tax as a share of GDP[b]
1978	22.1	17.8	17.0	9.6
1979	23.5	19.5	19.0	12.5
1980	24.0	20.9	20.3	14.4
1981	23.8	20.9	20.3	14.7
1982	22.9	19.8	19.1	13.0
1983	24.8	19.6	18.9	12.9
1984	22.3	18.6	17.5	12.0
1985	24.2	21.4	18.8	11.8
1986	22.7	17.8	14.5	6.6
1987	23.5	18.2	16.4	8.7
1988	25.6	17.8	16.6	7.4
1989	21.9	16.5	15.3	6.5

Note: The Indonesian fiscal year is from April to March. The GDP data are on a calendar year basis. The GDP series used is old series up to 1988 (appendix table 4.1). Since this provides lower estimates of GDP than the revised series, the ratios may be somewhat overstated. Thus, use of revised series would lead to NOTR to GDP ratio of 8.5 for 1988–89 as compared to 9.2 shown in the table. GDP data for 1989 are from IMF (1989.)

contrast, taxes in general are regarded as contractionary because they reduce the purchasing power of the private sector. Thus, the two types of tax revenue have quite different macroeconomic effects. Taxes on the oil sector other than the income tax, such as the VAT on the petroleum products, are included in the NOTR.

From 1978 to 1983, the NOTR to GDP ratio decreased from 7.4 to 6.0 percent, while the oil-related revenue to GDP ratio increased from 9.6 to 13 percent (table 4.1). The declining trend of the NOTR to GDP ratio during the prereform period dates back to 1973, when OPEC quadrupled oil prices. Previously, the NOTR to GDP ratio steadily increased from 1967 onward, reaching a peak of 9.2 percent in 1973 (Gillis 1989a, table 4.1). It is clear that the prereform tax structure from 1973 on was dominated by revenue from the oil sector. Between 1978 and 1983, NOTR accounted for only one-third of total tax revenue.

Among the nonoil taxes, the income tax was the most important revenue source, accounting for slightly more than two-fifths of total NOTR from 1981 and 1983 (table 4.2). Prereform sales tax (18 percent), excise taxes (17 percent), and import duties (15 percent) were the other major NOTR contributors.

Structure of Prereform Taxes

Only those taxes affected by the reform program—income, stamp duties, and sales and property taxes—are summarized here.[2] The prereform income tax consisted of separate taxes on individuals (*Pajak Pendapatan* or PPD) and on businesses (*Pajak Perseroan* or PPS), plus a special income tax

Nonoil tax revenue (NOTR)			
As a share of GDP	*As a share of total tax revenue*	*As a share of total expenditure*	*Development receipts as a share of development expenditure*[c]
7.4	43.4	33.3	37.7
6.6	34.6	27.9	34.0
5.9	29.2	24.7	25.8
5.6	27.4	23.3	24.5
6.1	31.8	26.6	27.7
6.0	31.6	24.0	46.2
5.5	31.4	25.0	36.8
7.0	37.3	29.4	30.0
7.9	54.7	33.3	42.4
7.7	46.6	32.6	65.0
9.2	55.6	36.3	81.6
8.9	57.8	40.4	68.2

a. Total domestic revenue consists of both tax and nontax revenue but excludes development receipts.
b. Other oil receipts in the budget also included here.
c. Development receipts consist of program and project aid from abroad.
Source: Bank of Indonesia (various years); IMF (various years).

Table 4.2 Central Government, Composition of Nonoil Tax Revenue

	Pretax reform period (fiscal year beginning)					
Tax category	*1978*	*1979*	*1980*	*1981*	*1982*	*1983*
Income tax[a]/amount[b]	581.0	736.0	1,045.0	1,279.0	1,606.0	1,785.0
Total tax revenue (percent)	14.2	11.3	10.5	10.8	13.4	12.8
NOTR[c] (percent)	32.9	32.7	36.1	39.4	42.1	40.6
GDP (percent)	2.42	2.14	2.14	2.19	2.56	2.42
Sales tax[d]/amount[b]	347.0	329.0	461.0	534.0	708.0	830.6
Total tax revenue (percent)	8.5	5.1	4.7	4.5	5.9	6.0
NOTR[c] (percent)	19.7	14.6	15.9	16.4	18.6	18.9
GDP (percent)	1.45	0.96	0.94	0.91	1.13	1.13
Land & building tax[e]/amount[b]	63.0	72.0	87.0	94.0	105.0	132.0
Total tax revenue (percent)	1.5	1.1	0.9	0.8	0.9	1.0
NOTR[c] (percent)	3.6	3.2	3.0	2.9	2.8	3.3
GDP (percent)	0.26	0.21	0.18	0.16	0.17	0.20
Excise taxes/amount[b]	253.0	326.0	438.0	544.0	620.0	773.2
Total tax revenue (percent)	6.2	5.0	4.4	4.6	5.2	5.6
NOTR[c] (percent)	14.3	14.5	15.2	16.7	16.3	17.6
GDP (percent)	1.05	0.95	0.90	0.93	0.99	1.05
Import duties/amount[b]	295.0	317.0	448.0	536.0	522.0	557.0
Total tax revenue (percent)	7.2	4.9	4.5	4.5	4.4	4.0
NOTR[c] (percent)	16.7	14.1	15.5	15.5	13.7	12.7
GDP (percent)	1.23	0.92	0.92	0.92	0.83	0.76
Others[f]/amount[b]	2,26.0	470.0	412.0	412.0	252.0	316.2
Total tax revenue (percent)	5.5	7.2	4.2	4.2	2.1	2.3
NOTR[c] (percent)	12.8	20.9	14.3	14.3	6.6	7.2
GDP (percent)	0.94	1.37	0.84	0.84	0.40	0.43

a. Excludes income tax on oil and gas. Includes personal and corporate income tax and withholding tax. Since new income tax was introduced in January 1984, separate receipts for income tax components are not available.
b. Amount in billions of Rp.
c. NOTR refers to nonoil tax revenue. It excludes only corporate income tax derived from oil and gas sector.

regime applied to the operations of the foreign oil companies. While, in theory, a withholding tax was levied at a rate of 2 percent on a wide range of commercial transactions, in practice it tended to act largely as a turnover tax rather than a credit against corporate tax liability (Booth and McCawley 1981, pp. 134–35). Official assessment rather than self-assessment was used to determine income tax liability.

The rate structure for individual income tax was nominally progressive, ranging from 10 to 50 percent, with seventeen steps between the lowest and the highest rates. The rates for businesses were 20, 30, and 45 percent. However, the profit levels at which the above rates applied varied according to company ownership (private or public), whether financial statements of private companies were audited or not, and, for public companies, the percentage of shares held by the public. The above rate structure provided strong incentives for splitting up companies. The tax base for the individual income tax was narrow, leaving such items as capital gains, fringe benefits, interest, and pensions either untaxed or lightly taxed. This substantially reduced the progressivity of the tax (Gillis 1989a, p. 101). For depreciation purposes, the vintage accounting method

Post-tax reform period (fiscal year beginning)					
1984	*1985*	*1986*	*1987*	*1988*	*1989*
2,121.0	2,313.0	2,271.0	2,663.4	3,949.0	5,488.0
13.9	13.0	16.2	14.1	18.4	20.6
44.3	35.0	29.7	30.3	33.2	35.6
2.44	2.45	2.35	2.33	3.06	3.15
878.0	2,326.7	2,900.0	3,390.4	4,505.0	5,837.0
5.8	13.1	20.7	18.0	21.0	21.9
18.3	35.2	37.9	38.6	37.8	37.8
1.01	2.46	3.01	2.96	3.49	3.35
157.0	167.0	190.0	275.1	424.0	590.0
1.0	0.9	1.4	1.5	2.0	2.2
3.3	2.5	2.5	3.1	3.6	3.8
0.18	0.18	0.20	0.21	0.33	0.34
873.0	943.7	1,056.0	1,105.7	1,390.0	1,477.0
5.7	5.3	7.6	5.9	6.5	5.5
18.2	14.3	13.8	12.6	11.7	9.6
1.00	1.00	1.09	0.97	1.08	0.85
530.0	607.3	960.0	938.4	1,192.0	1,587.0
3.5	3.4	6.9	5.0	5.6	5.9
11.1	9.2	12.6	10.7	10.0	10.3
0.61	0.64	0.99	0.82	0.92	0.91
229.0	258.3	269.0	406.4	448.0	447.0
1.5	1.5	1.9	2.2	2.1	1.7
4.8	3.9	3.5	4.6	3.8	2.9
0.26	0.27	0.28	0.35	0.35	0.26

d. Data are for VAT, including the luxury sales tax beginning with the fiscal year 1985. For the earlier years, the data are for sales tax existing at the time.
e. Before 1986, this tax was classified as land tax (IPEDA).
f. This is defined as a residual and includes export tax, stamp duties, and other minor taxes.
Source: Bank of Indonesia (various issues); IMF (various years).

was used. Assets were classified into ten categories, with the life of the asset ranging from five to seventy-five years. A wide variety of fiscal incentives were also provided.

Under the prereform income tax in 1983–84 the individual income tax accounted for 22.4 percent of total income tax; the corresponding amounts for corporate income tax and withholding tax were 42.4 and 35.2 percent, respectively (Asher and van Eeghen 1987, table 2.2). Nearly all sources of income were subject to withholding, including wages and salaries (*Pajak Pendapatan Buruh*, known as PPD Buruh), interest, dividends, and royalties (*Pajak Atas Bunga*, Dividend and Royalty, known as PBDR), and income from sales (*Memungut Pajak Orang*, known as MPO). The sales tax was to be paid by the collectors on behalf of the purchasers, who in turn could credit the amount against their corporate or personal income tax liability. In practice, the sales tax often acted as a turnover tax since credit was rarely claimed against the income tax (Nasution 1989, p. 33). As a result, withholding taxes under the prereform income tax system sometimes resulted in additional revenue, so withholding was more than merely a device for collecting the income tax.

In 1983–84 about half of corporate income tax was collected from state enterprises (Asher and van Eeghen 1987, table 2.2). This, however, represented a decline over the 1970 to 1980 period, during which the twenty-five largest government-owned firms (excluding the state oil enterprise PERTAMINA) accounted for nearly two-thirds of corporate income tax. Gillis reports that the effectiveness of fiscal incentives "were expensive in terms of revenue, biased in favor of capital intensive investment, and discriminatory against smaller firms. They also gave rise to intractable problems in tax administration. They were generally ineffective in achieving their central objective of attracting beneficial investments to Indonesia in general and to so-called backward regions within the country" (Gillis 1989a, p. 101).

Administrative efficiency and compliance with the prereform income tax were also low. Thus, in 1980 only 1.2 million individuals paid income tax (219,000 self-employed taxpayers and an estimated 1 million payers subject to withholding tax), out of an estimated 11 million taxable families in a population of 147 million (Asher and van Eeghen 1987, annex 1). According to the same source, corporate income tax collected in 1980 was less than 50 percent of taxes due. Moreover, tax arrears were large. There was substantial subjective interpretation of the income tax, and tax liability was often negotiated rather than determined objectively, giving rise to irregular payments and corruption. Modern management techniques and tools, for example computerized systems, were used only rarely in tax administration and in processing tax returns.

The prereform sales tax included a tax on the delivery of goods, provision of services, and the entry of goods into the customs region. The first two constituted the domestic sales tax, while the third amounted to the import sales tax. Whereas a wide range of services were covered, most sales tax revenue was derived from contractors.

The rate structure on domestic goods varied from 1 to 20 percent, with a 10 percent rate being the most common. The rate on services was 2.5 percent. Imported goods were subject to higher tax rates than domestically produced items in three ways: (1) higher nominal sales tax rates were levied on imported goods than on identical domestically produced goods; (2) the import sales tax was defined broadly as the value of imports plus import duties and charges plus a 5 percent markup applied on the sum of the first two categories; and (3) a special price, generally stated as a fraction of the selling price, was used to calculate the sales tax liability of some domestically manufactured goods. Thus, the old sales tax between domestic and imported goods was nonneutral. Since inputs were not exempt from the tax, cascading effects resulted. Moreover, sales tax exemption was used as part of a package of fiscal incentives, narrowing the tax base. The prereform sales tax operated under the Sales Tax Act of 1951. As a result of numerous amendments and regulations, it became complicated and cumbersome to administer and to pay.

Before tax reform, taxes relating to property were levied under seven separate ordinances. The IPEDA (*Iuran Pembangunan Daerah*), which may be translated into English as the "Contribution for Regional Development," and the central government net wealth tax were the most important property taxes. IPEDA was levied on the net proceeds from land classified into five sectors—forestry, estate (private and state), mining, rural (rice and nonrice), and urban. The basic nominal rate was 0.5 percent of the proceeds from land in the rural sector and 1 percent of capital value in the urban sector. Rural nonforest land and estates were subject to progressive taxes. IPEDA liability on forest land was determined at 20 percent of logging royalties, whereas tax liability for mining land was decided by negotiation. Because of uneven enforcement, ad hoc exemptions, and a 50 percent exemption for residential property, effective rates were not only much lower but also varied widely for the same type of properties. The collection cost of IPEDA was high, largely because of taxpayers whose tax liability exceeded the collection cost. Even for Jakarta, it was estimated that in the early 1970s, the collection costs of IPEDA were 20 percent of IPEDA revenue collected. It was also found that in 1972 one-third of more than 600,000 IPEDA taxpayers in the Jakarta area were assessed at less than US$1, an amount which presumably did not cover collection costs (Lerche 1980, p. 44).

Before tax reform, stamp duties were levied under the Stamp Duty Regulation Act of 1921. Under this act, the rate structure was complicated, encompassing both specified and ad valorem rates. The prereform tax system also included a limited set of excise taxes, primarily on beer and cigarettes, export duties on coffee, palm oil, rubber, timber, tin, and import duties. In 1983–84 these taxes accounted for about 35 percent of NOTRS (table 4.2).

Several weaknesses of the prereform tax system are evident from the above discussion. Among the major deficiencies were low revenue productivity, reflected in low and declining NOTRS to GDP ratio (table 4.2); uneven enforcement and compliance, in part because of complexity and ambiguities in tax laws, regulations, and procedures; and wide dispersion in effective rates within activities, sectors, and products, leading to both inefficiency in allocation and horizontal and vertical inequities between individual taxpayers. Gillis has characterized the prereform tax system as "unproductive of revenue, a source of substantial economic waste, and essentially inequitable in every important sense" (1989a, p. 80).

What was the character of the previous attempts to correct the above deficiencies before tax reform? First, the focus had been on fine-tuning certain aspects of a particular tax or a small set of taxes, rather than looking at the tax system as a whole (Booth and McCawley 1981; Asher and Booth 1983). Second, there had been an implicit acceptance of the existing tax laws and general procedures for administration and compliance although some modifications were suggested. Third, the necessity for fiscal incentives to achieve a wide variety of objectives, such as employment promotion, regional balance, and the targeting of investment in certain areas, had been generally accepted. Fourth, concerns about the effects of taxes on vertical equity received exclusive attention at the expense of horizontal equity. Greater taxation of urban real estate, better assessment and collection of income tax, and increased taxation of luxury goods were proposed to increase progressivity of the tax system (Booth and McCawley 1981, pp. 137–39). Fifth, automatic stabilization properties of various taxes and practices, such as the use of revenue targets for tax collection, were emphasized.

As will be made clear in the ensuing discussion, the underlying assumptions and thinking of these efforts differed substantially from reform programs of the early 1980s. Nevertheless, by constantly emphasizing the need to increase NOTRS and improve tax administration and compliance, the earlier discussions did help set the stage for a bolder, more comprehensive approach after years of piecemeal and ad hoc measures.

Pressures for Tax Reform

In spite of the major defects of the tax system noted above, why were no major reforms implemented during this period? It appears that one of the most important factors was the sharp increase in tax revenue from the petroleum sector brought about by the 1973 oil price increase. As noted by Gillis (1989a, p. 81) and Arndt and Hill (1988, p. 108), this large and painless source of revenue provided a considerable incentive to postpone tax reform, a well-known conclusion of the "Dutch disease" literature.

Gillis (1989a, pp. 80–81) notes three additional factors. The first is the ignorance on the part of the low-income groups of the costs of indirect taxes, primarily because they were complicated and largely hidden from ultimate consumers. It seems unlikely that even knowledgeable taxpayers could have successfully pressed for tax reform because the political system did not solicit their views. This is still the case. The implications of this for evaluating tax reform are noted later. The above suggests that reform initiatives had to come from high-income taxpayers, tax administrators, or policymakers. The second factor mentioned by Gillis is that high-income taxpayers were asked to pay little income tax, which made state enterprises, wage earners, and foreign enterprises the largest contributors. The third factor was that, given the complexity of tax design and procedures, tax administrators had wide scope for securing irregular income, thereby reducing the incentive for tax reform.

In January 1981 the process of tax reform was initiated by policymakers, or more specifically the minister of finance, and supported by influential allies in the planning agency (BAPPENAS). Three considerations appear to have been involved (Gillis 1990, pp. 4–5): (1) the key economic ministers were less optimistic than other observers and international institutions about high oil prices remaining high, (2) oil revenues used to finance government expenditure involving nontraded goods and services would most likely increase inflation, and (3) it was felt that tax reform would gain acceptance more easily if accompanied by reduced nominal tax rates.

When the minister of finance initiated reform in January 1981 (Gillis 1989a, p. 81), oil revenues were high (table 4.2) and most observers at the time anticipated that this would continue to be the case. Thus, there was no particular pressure to initiate reform to increase revenue. However, the government was concerned about the inefficiency of the tax system, and it asked the Harvard Institute of International Development to conduct technical studies and draft tax reform legislation. A team of some twenty economists (mostly from North America), six lawyers (all from the United States), and two computer specialists was assembled for the task. Except for one resident expatriate, the others made only periodic visits to Jakarta between 1981 and 1984 (Gillis 1985, p. 229).

Formulating the Reform Proposals

The tax reform legislation passed in December 1983. Thus, three years elapsed between initiation of the process and the passage of tax laws. How was this time used? The technical team kept a low profile and the reform proposals were formulated in extreme confidentiality to avoid adverse reactions. Only the minister of finance and a handful of other officials were involved in decisionmaking during this period. Harberger (1989, p. 33) has suggested that the Steering Committee, consisting of officials from the Ministry of Finance and Tax Administration and set up to liaise with the technical team, ensured that the proposals were in line with political realities and that, once adopted, the proposals would stand a good chance of being implemented. However, it was widely known that tax administrators had little to gain from tax reform, so it appears unlikely that the formal committee played a major role. Little independent information on this phase of the reform is available.

The following account, therefore, relies on two published papers on the tax reform by the director of the technical team (Gillis 1985, pp. 228–31, 1989a, pp. 81–85). Understandably, the tax reform experiences of other countries familiar to the technical team, particularly in Latin America, were taken into account in setting the parameters.

The first half of 1981 was to set broad parameters. Other than agreeing to maintain a high degree of secrecy, the following decisions were made. First, ample time (around two years) was considered essential to complete technical studies. It was felt that the quantity and quality of fiscal research were insufficient to formulate detailed proposals and it was decided that studies undertaken by the technical team, either formally or informally, would not be published. As a result, it is not possible to evaluate either the technical studies or the implications derived from them by those involved in formulating the reform proposals.

Secondly, because existing tax laws were deemed to be out of date or inappropriate, new legislation was to be drafted immediately from the policy proposals, ensuring that the rationale for policy proposals was accurately reflected in the tax legislation. This problem solving approach could only have been initiated by technocrats with autonomy in the economic sphere.

Third, issues of tax administration and compliance were integral to the tax reform proposals. In the Indonesian context, tax policy and tax administration are closely linked, and to some extent deficiencies in the latter have severely hampered the pursuit of appropriate tax policy. Hence the decision to invest in a computerized tax information system and train a new generation of tax officials to use it.

Fourth, the reform proposals were to focus on income tax (excluding income taxation of foreign oil companies), the property tax, the sales tax, and stamp duties. Excluding income tax revenue from the petroleum sector, these taxes accounted for slightly more than three-fifths of NOTRS in 1983–84 (table 4.2). Thus, excise and international trade taxes were excluded from the reform process. Excise taxes were judged to be working reasonably well, and trade taxes were considered too sensitive politically to adjust at the same time, mainly because of quantitative restrictions rather than tariffs provided substantial economic rents to powerful interest groups. Nevertheless there have been many changes in the tariff regime and customs administration.

Fifth, it was decided to present the tax reform proposals as a package in a set of draft laws to the legislature and the public, even though the technical studies and recommendations were discussed as the work was completed. The major reason for this decision seems to have been to minimize opposition, particularly from tax administrators.

In December 1983, the Indonesian Parliament approved three landmark laws concerning General Tax Provisions and Procedures (Law Number 6, 1983), Income Tax (Law Number 7, 1983), and Value Added Tax on Goods and Services and Sales Tax on Luxury Goods (Law Number 8, 1983). Laws concerning Land and Building (Law Number 12, 1985) and Stamp Duty Law (Law Number 13, 1985) were passed later, in 1985. President Suharto called the 1983 laws "national monuments," ensuring minimal open opposition. Parliamentary approval was swift, with only minor changes.

Objectives of Tax Reform

As expected, tax reform was directed at correcting deficiencies of the prereform tax system noted earlier. It is possible, however, that the objectives emerged during the process of reform rather than being preset. The main objective was to increase the NOTR to GDP ratio, thereby reducing dependence on income tax revenue from the petroleum sector. The policymakers realized that simply increasing the NOTR without curtailing government expenditure would be insufficient. The balanced budget and other characteristics of the fiscal policy of the New Order government were expected to provide discipline on the expenditure side. The initial goal was revenue neutrality in the near term. It was anticipated that base broadening and the lower nominal rates envisaged in the reform proposals would make the tax system capable of generating additional revenue quickly if the need arose. However, the need for increased revenue became evident before the submission of the proposals to parliament in late 1983 (Gillis 1989a, p. 87). Although no specific target was provided, it appears that the Repelita IV's (five-year development plan) target of a NOTR to GDP ratio of 10 percent by 1988–89 was taken as a working goal.[3]

The second objective, also directed at overcoming deficiencies in the earlier tax system, was to streamline the tax laws and improve the efficiency of transferring resources to the public sector. This was to be achieved by simplifying and clarifying tax laws and by reforming tax administration rules, procedures, and information and management systems.

The third objective was to reduce tax-induced distortions in the allocation of resources. Gillis has called this objective achieving economic neutrality (1989a, p. 91). According to him, neutrality in taxation implies "greater uniformity in tax rates and generality in not only the tax rules prescribed in tax structures but in the mechanism of tax administration" (1989b, p. 502). Given the narrow tax base, multiplicity of rates, and uneven enforcement and compliance, neutrality was sought as a workable objective to reduce tax-induced distortions in the allocation of resources and improve horizontal equity of the tax system. Thus, neutrality in the Indonesia context has a pragmatic rather than a theoretical basis. The fourth objective was to ensure that the poor did not become worse off as a result even though the reformed system did not consciously promote income redistribution.

The main differences in assumptions and objectives between the prereform period and the tax reform program can now be summarized. First, the reform program demonstrates a willingness to alter not only the nature and structures of various taxes but also laws, regulations, and administrative structure and practices. This did not exist in the prereform period. Second, the reform program emphasizes horizontal equity and economic neutrality rather than vertical equity and interventionist tax policy. Third, the program de-emphasizes automatic stabilization aspects. The common areas, as noted earlier, are the need to increase NOTR and improve tax administration and compliance.

The Reform Program

In broad terms, the reform program attempted to achieve its objectives in the following manner. The main instrument for increasing the NOTR to GDP ratio was the introduction of a VAT package. Medium-term objectives were improved tax administration and compliance, a more effective income tax, and the land and buildings tax.

To achieve the second objective, that of improving efficiency, the new tax laws are designed to be unambiguous. There are extensive definitions of taxable objects and subjects with quantified criteria in the law and the statutory bases are clearly identified. The base for each tax is defined broadly, and exemptions and exclusions are minimized. Instead of the old official assessment system, the reform program adopts self-assessment which transfers the primary responsibility for filing, determining tax liability, and paying taxes to the taxpayer. Failure to meet this responsibility results in automatic fines and penalties. Tax administration has been depersonalized, which considerably reduces contact between tax administrators and taxpayers and, thus, opportunities for renegotiating liability. Enforcement is primarily through selective auditing and internal checking. There are now provisions for modern management techniques, including computerization of tax administration and extensive training schemes.

The third objective, reducing tax-induced distortions, was to be achieved by defining the income tax base more broadly, thereby eliminating all previous fiscal incentives based on the income tax. Incentives based on import duties and certain other indirect taxes remain, however. Nominal tax rates have been reduced substantially, and a common rate structure for income tax on both individuals and companies has been established. Input taxation has been largely eliminated by the introduction of the VAT. Greater uniformity and consistency in tax administration is also expected to lead to a reduction in tax-induced distortions.

The fourth objective, that of not making the poor worse off, was to be achieved by keeping the nominal rates low; by eliminating exemptions enjoyed by high-income groups; and by using exemption levels, particularly for the income tax and the land and buildings tax, to keep the poor outside the tax net.

We now turn to a brief description of each of the taxes included in the tax reform program. The discussion covers not only the provisions in the original tax laws, but also subsequent modifications.

The Income Tax

As a result of the tax reform, there is now a single income tax called *Pajak Penghasilan* (PPH), thus ending the previous practice of applying separate laws to individuals and businesses. The new law represents a fundamental departure from the old income tax law in several additional ways. It adopts the principle of "broad-base, low nominal rates," adopts an identical rate structure for the personal and company income tax, uses a "self-assessment" system instead of the "official assessment" used earlier, and structures the income tax to reduce as much as possible the discretionary powers involved in its administration. Each taxpayer is required to register and acquire a unique taxpayer identification number. However, the individual and company income tax are not integrated—a double taxation of dividends is retained. Thus, Indonesia continues to adhere to the classical system of income taxation, which does tax the capital income originating in the corporate sector at a rate higher than the noncorporate sector.

A tax amnesty was introduced in April 1984 at the same time as the new income tax. The aim was not to generate revenue but to identify potential income taxpayers. The amnesty expired in June 1985. It covered evasion of individual and corporate income taxes, wealth tax, withholding tax, sales tax, and taxes on interest, dividends, and royalties. The penalty for taxes not paid was levied at two rates, 1 percent and 10 percent. The lower rate was applicable if the taxpayer taking advantage of the amnesty had filed an individual or corporate income tax return for 1983 and the property tax return for 1984. In other cases, the 10 percent rate was applied.

The new income tax law covers all resident and nonresident individuals and organizations, including corporations, state-owned enterprises, foundations, and associations. Individuals living in Indonesia for more than 183 days in a consecutive 12-month period and organizations which are established in Indonesia are considered residents. Nonresidents pay taxes on income received or accrued from Indonesia, while in principle residents are taxed on global income. To avoid double taxation, the income tax law (Article 24) provides tax credits for taxes paid overseas. The credits are on a per country rather than an aggregate basis. These definitions of taxable subjects are broader and more explicit than in the old system. The minister of finance, however, continues to have the authority to exempt government enterprises from income tax.

A special income tax regime continues to apply to the operations of foreign oil companies. Methods for determining their tax liabilities are spelled out in contracts with the government oil company. Gillis has noted that "the essence of taxes on oil companies was that all levies combined were intended to capture 85 percent of their net income (after deduction of all allowable costs)" (1989a, p. 100). Companies with contracts signed before January 1, 1984 continue to be subject to the former tax provisions. However, foreign oil companies with contracts signed after January 1, 1984 are now subject to the new income tax law on operations governed by such contracts (Gillis 1989a, p. 103).

The income tax rates, applicable to both individuals and companies, are 15 percent for taxable income up to Rp 10 million per year, 25 percent on the next Rp 40 million, and 35 percent on income exceeding Rp 50 million per year (the exchange rate in early 1990 was about Rp 1,800 or US$1.00). Under the old system, seventeen rates applied to individuals, ranging from 5 to 50 percent, and rates of 20, 30, and 45 percent applied to corporations.

The new law broadened the legal base for individual income considerably. Civil servants have been brought within the purview of the tax for the first time. Long-term capital gains and pensions are now subject to income tax. Under the new law, companies cannot deduct expenditures for fringe benefits, and employees do not need to report them as income. This procedure is exactly opposite to the earlier arrangement. Fringe benefits in cash, however, continue to be taxable at the individual level and deductible at the company level. Other items that were lightly taxed and are now fully taxed include honoraria, leave and educational allowances, and rental income. Interest income from time and savings deposits was exempted under the old income tax. Because of strong pressure from financial groups, this exemption, while eliminated in the law, continued in practice. However, in November 1988 interest income was made taxable, thereby closing an important loophole in the income tax base. But from 1990 onward interest on deposits of up to Rp 5 million will be exempted from the income tax. The above discussion suggests that there is now little or no scope for expanding the legal tax base for the individual or company income tax.

The deductions allowed for the taxpayer are Rp 960,000 for the individual taxpayer, an additional Rp 480,000 for a married taxpayer, an additional Rp 960,000 for a wife earning income unrelated to her husband's, and Rp 480,000 for each dependent up to a maximum of three. Thus, the maximum family tax-free income is Rp 3.84 million (or 5.21 times the 1988 income per capita), as opposed to Rp 1.08 million under the old system (or 2.32 times the 1983 income per capita). The indexing provisions of the new income tax law have so far not been applied in practice. This implies that the real income level at which tax becomes applicable has been declining over time, and that as the number of potential income taxpayers increases the tax base will expand.

The unit of taxation is the individual as far as employment income is concerned. For other income, such as rent, interest dividends, and business income, a wife's income must be aggregated with that of her husband. Joint taxation of nonemployment income is designed to minimize tax avoidance, whereas separate taxation of employment income hopefully minimizes the disincentives for women to enter and stay in the labor force. The income of children under eighteen years of age is aggregated with that of their parents. After that age, children are considered to be separate entities for income tax purposes.

The new income tax makes more extensive and systematic use of withholding as a means of collecting revenue. Among the more important withholding provisions are the following:

- Under Article 21 of the Income Tax Law, employer and other organizations (for example, a pension fund) are required to withhold applicable income tax on salary and other receipts. If the employee or the recipient has no other income, then an income tax return need not be filed. It is the obligation of the employer to correct for any under- or overwithholding of taxes. This suggests that an estimate of the number of employees having income tax withheld should be added to those actually filing the income tax to arrive at the total number of taxpayers. This point should be kept in mind when interpreting the data on taxpayer compliance.
- Under Article 22 of the law, withholding of income tax is also required when suppliers are paid out of the state budget and when

import activities are undertaken by businesses.[4] The base for imports is 10 percent of the import value if the importer has an identification number and 30 percent of the import value if such a number is not used. The base for goods and services financed from the state budget is 6 percent of the delivery price. The rate of tax is 25 percent of the base. Thus, for imports, the rate is 2.5 percent if the identification number is used and 7.5 percent if it is not. For state budget-financed goods and services, the rate is 1.5 percent. A large number of organizations, including foreign exchange banks, have been authorized to act as withholdings agents.

- Under Article 23 of the Income Tax Law, withholding at a rate of 15 percent of gross amount is required on the dividend, interest, rent, royalty, and fees for technical and management services paid to a resident taxpayer. Payment of income to nonresidents requires a 20 percent withholding rate, unless the tax treaty provisions allow for a lower rate (Article 26 of the Income Tax Law). While the withholding tax imposed under Article 23 is a credit against the domestic taxpayers ultimate tax liability, withholding tax under Article 26 presents final liability. Also, unlike Article 23, receipts from both technical and management services as well as other services are subject to withholding taxes under Article 26 (Morgan 1989, p. 75). When compared to the previous regulations, the new income tax law has effectively doubled the withholding rate on interest paid abroad by Indonesian borrowers (borrowing by the government is exempted).
- In addition, resident taxpayers are also required to make monthly advance payments of income tax equal to one-twelfth of the tax due for the preceding tax year, reduced by withholdings and collections of tax by third parties in that year. While the law provides for automatic fines and penalties for late filing, they are rarely enforced.

Since income tax on individuals and businesses is covered under one law, these provisions, whenever appropriate, apply to companies as well. The additional provisions of the new income tax law concerning companies may be summarized as follows:

- The new law abolished income tax-based incentives, including investment allowances, tax holidays, and accelerated depreciation. Indeed, in Southeast Asia, Indonesia is the only country to abolish such incentives. Incentives not based on the income tax, such as import duties and the VAT, can still be granted at the discretion of the investment agency. Companies who enjoyed fiscal incentives at the time the new law became effective can continue to enjoy them until they expire.
- Fringe benefits and charitable contributions can no longer be deducted from taxable income.
- Simplified and more unified depreciation schemes have been adopted, leading to more rapid depreciation. The vintage accounting method,

which uses historical costs, has been replaced (with one exception, when a 5 percent depreciation rate is applied) by the declining balance method, which uses the written-down value as the basis for depreciation. Under the new law, four depreciation rates (5, 10, 25, and 50 percent) apply for different assets as compared to ten rates previously. Firms were allowed to revalue their assets for depreciation purposes following the September 12, 1986 devaluation, even though the rupiah continued to depreciate against the U.S. dollar.

- Operating losses can be carried forward for five years, and under certain circumstances, for up to eight years. However, operating losses incurred before August 1983 cannot be carried over (this provision was spelled out in the circular from the Directorate General of Taxation SE-34/PJ/22/1986, August 5, 1986). This regulation is, however, contrary to the offset provisions for general loss noted above.
- When the new income tax law was passed in December 1983, firms with an annual turnover of Rp 60 million could use special calculation norms to determine their taxable income. In August 1986, this limit was doubled to Rp 120 million.[5]
- Inventory valuation now requires the first in, first out (FIFO) or the average cost method. Thus, the last in, first out (LIFO) method is not allowed under the new law.

The tax treatment of income derived from projects financed by foreign aid is as follows:[6]

- Personal income derived from these projects is fully taxable at the normal statutory rate and should be withheld by the employer. Corporate income is not taxable. For employers with income from other sources a distinction is made on the source of income, in other words, whether or not it is derived from a foreign aid project. This is cumbersome and creates an incentive to artificially shift business cost deductions to the taxable component of total income.
- Under the original regulations for the income tax, a maximum debt-to-equity ratio of three to one was established to curb the usual loophole of overstating interest expenditures and underreporting interest income. This provision was suspended shortly afterwards (March 1985) under pressure from business groups who argued that, if applied correctly, a debt-to-equity ratio should be defined more specifically and vary between sectors with different production structures. To date (September 1989), no new decree concerning debt-to-equity ratio has been issued. It should be noted that given the possibilities of financial engineering coupled with relatively weak tax administration, strict debt-to-equity ratios would be difficult to administer.
- Indonesia also levies an exit tax of Rp 250,000 on departing residents, but it is an advance payment for personal income tax and can be deducted fully from the final personal income tax liability.

The Value Added Tax Package

This package consists of the VAT and the luxury sales tax. In sharp contrast to the income tax, this package has undergone significant modifications since it was first implemented in April 1985. The original package consisted of VAT at a rate of 10 percent on domestically produced goods, imports, and contractors' services; zero rating for exports; and an additional tax at a rate of 10 and 20 percent on certain luxury goods. Petroleum products were also covered under the original package, while they had been exempt under the previous sales tax. The tax base for petroleum products, however, is the retail price, with no tax credit allowed for the VAT on petroleum products used as business inputs. Tobacco products are also subject to the excises and to the VAT. A special excise tax rate of 7.7 percent on the discounted selling price with VAT excluded is applied to tobacco products.

With the exception of petroleum products, the VAT is levied at the point of import and at the manufacturer's level for domestic products. Therefore, the ratio of tax to retail prices is not the same for all commodities, as would be the case for a consumption type of VAT that was levied at the retail level. The VAT, however, does allow tax deductions on inputs to a much greater extent than the previous sales tax. This is done through the tax credit technique. Since the luxury sales tax cannot be credited against the output tax, cascading can occur when a luxury good is used as an input for a final good that is subject to the VAT.

Under the original VAT package, unprocessed foodstuffs and activities such as farming, fishing, growing crops, and raising livestock were exempt from VAT. Certain end-use exemptions were also given (for example, three-wheeled vehicles used for public transport). Small firms, defined as those with an annual turnover of less than Rp 60 million, do not have to levy VAT on their output, but they cannot claim any deduction for input taxation paid. Thus, the government's loss of revenue is restricted only to the value added component of small firms. Small firms do, however, have the option of registering for the VAT, and can be treated as a taxable firm if it is to their advantage.

There are three ways in which VAT is also borne by the government. First, the government as a purchaser of taxable goods must pay the VAT. Although suspended until March 1987, this provision has been reimplemented, increasing both budgetary revenue and expenditure. Thus, the VAT revenue and government expenditure categories since 1987–88 cannot be compared to those of earlier years. Second, the government bears VAT on the import and delivery of certain goods and services, such as cattle and poultry feed, medical equipment, software, water, and basic materials that are not produced domestically. Obviously the VAT borne by the government cannot be credited from the output tax when these goods are used as an input. As a result, revenue is not significantly affected when the VAT is borne by the government on goods that are intermediate products. Third, the VAT on goods used in projects financed through official foreign aid and loans is borne by the government.[7] Contractors for such projects pay input tax to the relevant government agency which, in turn, pays it to the treasury. The contractor can then claim the input tax from the treasury. This procedure has been adopted to minimize revenue loss to the government.

Under the VAT provisions, input tax paid on capital goods can be credited against the output tax. However, because of the possible adverse impact on the cash flow of businesses, the collection of input tax on capital goods purchased by taxable firms has been suspended indefinitely.[8] Consequently, such input tax cannot be deducted from the output tax, and revenues are not significantly affected. The payment of VAT on imports of capital goods by selected services, such as fishing vessels, hotels, and shopping centers, can be postponed for up to five years.[9]

These services were not subject to the VAT when it was first introduced, so the postponement implied an immediate loss of revenue because it was not offset by reduced deductions. Moreover, locally produced capital goods were discriminated against. VAT on imported goods and materials used in the manufacturing of export commodities has also been suspended.[10] This provision affects the cash flow only and does not lead to a loss of government revenue, since exports are zero rated.

Extensions and Modifications to the Original Value Added Tax Package

Extending the VAT to wholesalers and services. Effective April 1, 1989, the VAT was extended to wholesalers.[11] A wholesaler is defined residually, in other words, as a trader who is not a retail trader. The regulation defines a wholesaler trader as "a trader which within its company or work delivers taxable goods to any party, except those who do this as retail trader." The wholesaler also includes those making delivery of taxable goods and services to tax collectors.[12] Authorized tax collectors pay VAT on suppliers' invoices directly to the government, not to the supplier. This is to prevent the evasion of VAT by the suppliers (effective January 1, 1989). Thus, the term wholesaler is defined pragmatically and on the basis of administrative feasibility rather than on the basis of any economic criterion.

The coverage of the VAT has also been extended to various services. VAT on domestic air transportation and telecommunication services became effective January 15, 1989. In both cases, the VAT is based on the turnover, and the tax paid, even if it is by businesses, cannot be credited against the output tax. This departure from the value added principle is only in the case of these two services. In addition, twenty-one additional services, ranging from the leasing of ships and port services to services in the delegation of intangible objects, have been subjected to the VAT effective April 1, 1989 (for a complete list of services, see the announcement of the Directorate General of Taxation, the Ministry of Finance No. PENG-139/PJ/63/1989, dated March 27, 1989). Subsequently, the VAT on services connected with the exploration and drilling of oil and gas has been postponed because of representations made by the oil industry.

Thirteen services, including medical services, are exempt from the VAT. Even so, this represents a substantial extension of the VAT, both in terms of the tax base and the potential number of taxpayers. With these extensions, only the retail sector is exempt from the VAT although it is likely that large retailers will be covered by the VAT in the future.

Enlarging the tax base and increasing the rate of the luxury sales tax.[13] Under the original tax reform program, luxury sales tax was levied at rates of 10 or 20 percent, and included only a few items. Effective January 15, 1989, a luxury sales tax was levied at rates of 10, 20, and 30 percent, and the number of items has expanded substantially to include consumer durables, such as air conditioners, refrigerators, televisions, washing machines, and video recorders (taxed at 20 percent). Among the items subjected to a 30 percent rate are alcoholic beverages and most motor vehicles. Many nonalcoholic beverages, cosmetics, and radios are taxed at 10 percent. The tax base for imported goods for both the VAT on goods and the luxury sales tax is the sum of the cost of insurance and freight (CIF), import duty, import surcharge, and other levies imposed by the customs authorities. In contrast to the earlier practice, no markup is applied.

Differing turnover limits for goods and services for exempt firms. Under the original reform program, firms with an annual turnover of less than Rp 60 million were exempt. With the extension of the VAT, the turnover limits have been redefined.[14] The Rp 60 million limit has been retained for firms dealing in taxable goods, provided they form more than 50 percent of the total turnover of a firm. For taxable services, the limit is Rp 30 million, provided they form more than 50 percent of the total turnover of a firm.

The land and building tax. As we have seen, this tax, which became effective January 1, 1986, replaced property taxes levied under seven earlier ordinances. The base of the tax is now the sales value of land and buildings instead of the annual rental value. Under the law, the government can specify between 20 to 100 percent of sales value for calculating the tax base. The property tax is payable by the user of the property.

The legislation defines taxpayers as those "having rights over land," "who possess or control buildings," or who "obtain benefits from land and buildings." Thus, the law is not very clear on whether the landlord or the tenant is liable. It appears that the tenant may be taxed if the landlord cannot be traced. Under the original tax reform legislation, land values were divided into fifty categories, ranging from Rp 33 to Rp 1.675 million per square meter, with the value in each category varying according to various criteria. Effective January 1, 1989, the corresponding values increased and now range from Rp 100 to Rp 3.1 million per square meter.[15] The formulae for determining sales values in the estates, forestry, mines, and animal husbandry, fisheries, and brackish water pond sectors have also been revised under this decree.

Under the original tax reform legislation, buildings were classified into five groups, each having four ranges of sales values. For buildings the values ranged from Rp 23,000 to Rp 700,000 per square meter. Effective January 1, 1989, the five groups were abolished, and buildings classified into twenty classes with values ranging from Rp 50,000 to Rp 1.2 million per square meter. The limit on the value of nontaxable building sales has been raised from Rp 2 million to Rp 3.5 million. No such exemption limit is provided for land. The exemption for buildings is likely to exempt virtually all rural housing and a large share of low-income housing (Gillis 1989a, p. 105).

The main advantage of classifying land and buildings is the possible economies in valuation of properties. Since it does not require valuing each and every individual land parcel or building separately it simplifies property tax administration.

The few formal exemptions from the property tax consist of land and buildings used by nonprofit education, health, national culture, and religious organizations; land used as forests, graveyards, and national parks; and properties of diplomatic representatives and international organizations. In practice, exemptions for individuals who may have large assets but little cash flow have been given on a case-by-case basis.

Although the property tax is administered centrally, regional authorities have important collection responsibilities. The revenue from the PBB *(Pajak Bumi Dan Bangungan)* is shared among various levels of government. The central government receives 19 percent, first-level governments (provinces) 16.2 percent, and second-level governments (municipalities) get the rest, 64.8 percent. The precise allocation formula is 10 percent for central government. Of the remaining 90 percent, 10 percent is allocated to the central government as the cost of collection. Of the remaining 81 percent, four-fifths is given to second-level governments and one-fifth to first-level governments. Under present regulations, the assessments of land and buildings are updated every three years. This is an ambitious target, given that the original assessment was not yet completed.

The stamp duties. The complicated 1921 Stamp Duty Law applying both specific and ad valorem rates has been replaced by a new law effective January 1, 1986.[16] It levies two specific rates of Rp 500 and Rp 1,000 on agreements of a civil nature, such as notarized and authorized land deeds, bank drafts, promissory notes, securities, and other documents mentioning amounts in excess of Rp 1 million. The rate depends on the type of document or the nominal amount mentioned in the document.

Taxes not covered under the tax reform program. Trade and excise taxes were left out because the political and bureaucratic difficulties for reforming tariffs were considered more complex than those facing tax reform (Gillis 1989a, p. 84). Excise taxes were not included because they were performing satisfactorily. As noted, most of the excise revenue is from tobacco products, particularly cigarettes. The excise base for tobacco products is, however, the retail price. For cigarettes, the rates range from 2.5 to 37.5 percent depending on the type, whether it is mechanically produced, and the level of production.[17] Items subject to the excise taxes are also subject to the VAT, and in the case of alcoholic beverages to the luxury sales tax as well. It should be noted that no separate excise tax is levied on petroleum products.

There have been significant changes in trade taxes following the introduction of the tax reform program.[18] In March 1985 the government implemented a comprehensive reform of the tariff schedule lowering the tariff ceiling from 225 to 60 percent and the number of tariff positions (ignoring specific rates) from twenty-five to eleven. The reform also raised the percentage of items below 30 percent from about 59 to 82 percent. The positive effects of the reform on the trade regime were, however, mitigated by the proliferation of import licenses. In April 1985 the government completely reorganized the customs, ports, and shipping operations. Other trade reform packages announced in October 1986, January 1987, and December 1987 removed import licensing restrictions, adjusted tariffs and surcharges, and directly reduced antiexport bias. In general, various reform packages have moved away from nontariff barriers and toward protection through tariffs. The trade regime, however, continues to be complex and the level of protection rather high, particularly in the manufacturing sector and selected agroprocessing industries. Future reforms are likely to reduce protection levels on outputs and move toward a more neutral trade regime as between export and import substitution activities.

Indonesia continues to levy export taxes (and export surcharges) on selected commodities such as iron ore, palm nuts, pepper, and certain types of wood, including sawn and processed timber, at specific and ad valorem rates depending on the type of commodity.[19] The function of export taxes, in the absence of monopoly power, is to absorb rents of inputs unique to production of the commodity.

An Evaluation of Tax Reform

Because there are no widely accepted criteria for judging the success or failure of tax reform, this preliminary evaluation is based on the extent to which objectives noted in the previous section have been realized. Tanzi (1985, pp. 227–28) raises two potential problems using such a criterion, which have been kept in mind in this evaluation. The first problem is the time horizon. As of September 1989 the new income tax had been in place for more than five and one-half years; the original VAT package for about four and one-half years (with major extensions introduced in 1989); and the land and buildings tax and stamp duties for more than three and one-half years. Gillis (1989a, p. 105) argues that this may not be enough time to evaluate the actual income distribution and resource allocation effects. Nevertheless, sufficient time has elapsed to make a preliminary evaluation of the extent to which various objectives of the tax reform program are likely to be realized. Recent extensions and modifications of various taxes have revealed the general outlines of the tax structure, which is unlikely to be altered for some time. However, we acknowledge that each reform requires a different time period to achieve its objectives. For example, an increase in tax rates or an extension of the tax base will result in higher revenue much more quickly than improvements in tax administration or resource allocation.

The second problem is that all future changes consistent with the objectives may be attributed to the team involved in formulating the original tax reform program. Because of the continuity of the government, this is not much of a problem in Indonesia. The present government has been in power for at least twenty-five years and, more relevant for present purposes, economic policy throughout this period has been dominated by like-minded technocrats. Moreover, we are not evaluating the original tax mission (Tanzi's main concern) but rather tax reform since 1983, including all modifications and extensions even if they are not consistent with the original tax reform program.

Thus, this evaluation is based on the extent to which four objectives of tax reform noted in the previous section (namely increased NOTR to GDP ratio, enhanced administrative efficiency and compliance, better resource allocation through economic neutrality, and the protection of low-income groups) have been realized by the tax reform measures introduced since 1983. Also, a brief discussion of the factors likely to influence the success or failure of tax reform in the longer term is provided.

The analysis is largely, though not exclusively, qualitative and deductive. This is unavoidable. First, the expectation that the reorganization and computerization of tax administration will lead to more detailed and timely data does not appear to have been realized. One reason is poor management of fiscal information by the authorities. In terms of published tax data, in some cases less, not more, information is available. For example, nonoil income tax revenue has not been broken down into individual, corporate, and withholding components since 1984–85 although such breakdowns were published earlier. Furthermore, the unpublished data contain gaps. Thus, a breakdown of the VAT by detailed sectors or commodities has not been available for study. A second reason is that tax reform measures are based on pragmatic considerations, not a normative framework that is consistent. Given this fact and many other factors, including noneconomic ones, that influence the outcome of tax reform measures, the use of conventional economic models is restricted.

Increasing the NOTR to GDP ratio. An evaluation of this objective is undertaken at two levels. First, the behavior of this ratio before and after-tax reform is examined at an aggregate level. Second, revenue performance of each tax is examined with a view toward identifying factors responsible for the observed performance. A brief discussion of the revenue potential of each of the taxes is provided.

Aggregate analysis. During the first year of tax reform when only the new income tax law was implemented, the NOTR to GDP ratio fell from nearly 6 percent of GDP in 1983–84 to 5.5 percent of GDP in 1984–85 (table 4.1). However, the ratio increased sharply to 7 percent of GDP in 1985–86 when the VAT package was implemented. By 1989–90, this ratio had reached 8.9 percent of GDP (table 4.1).

As noted earlier, significant extensions of the VAT, taxing interest on time and saving deposits, and increasing the tax base of the property tax became fully effective only in 1989–90. The share of NOTR in total tax revenue was 57.8 percent in 1989–90, as compared to slightly less than one-third in 1983–84 (table 4.1).

This suggests that the role of NOTR in the revenue structure has increased significantly. The increase in the NOTR to GDP ratio from 1983–84 to 1989–90 meant that the overall revenue objective has been largely met.

In order to sustain this performance, significant improvements in administrative efficiency and capability as well as continued political efforts to increase compliance levels are essential. The need for continued political efforts arises because revenue targets will require larger tax payments from professionals and businessmen who have benefited significantly from government policies over the past two decades and are important supporters of the present regime. This will require both improvement in technical capabilities of the tax administration and political effort to translate such capabilities into higher compliance and enforcement.

Disaggregative analysis. The revenue performance of each of the taxes covered by the tax reform program and of import duties is analyzed next. A major difficulty in attempting such a disaggregative analysis is the lack of detailed revenue information, such as number of taxpayers categories, and revenue streams collected by sectors, activities, and commodities for each of the various taxes. A reliance on fragmentary information makes it difficult to evaluate the data.

The income tax. At the end of the 1987–88 fiscal year, the income tax revenue to GDP ratio (2.33 percent) was lower than in 1983–84 (2.42 percent), the year immediately prior to the income tax reform (table 4.2). Until then, the revenue performance of the income tax was disappointing. An earlier study found that in 1985–86 actual revenue was only 48 percent of the theoretical potential for both the personal and corporate income taxes (Asher and van Eeghen 1987, table 3.3). In this study, the theoretical potential base is defined conservatively. Since its estimation does not assume perfect administration or coverage of the base, some leakage is incorporated. The economic recession from 1985 to 1987, coupled with administrative weaknesses, accounted for this performance. Although the introduction of self-assessment was a positive step, its effectiveness was weakened by the lack of rigorous enforcement of the tax laws; prosecutions for tax evasion have been rare and normal fines for nonfiling or for late filing are sporadically enforced.

Certain factors and measures may be characterized as favorable for revenue generation from the income tax. First, as we have seen, at the time of the income tax reform the government implemented a tax amnesty program with a view toward identifying taxpayers. This and other steps led to an impressive rise in the number of registered income taxpayers. Thus,

Table 4.3 Number of Taxpayers Registered for Personal and Corporate Income Tax

Type of taxpayer	*12/31/84*	*12/31/85*
Personal income taxpayers (PPH)[a]	330,287	536,769
Individuals withholding income tax on employee (PPH, Article 21)	12,426	15,996
Individuals withholding income tax on payment of interest, dividends, etc. (PPH, Article 23)	480	522
Corporate income tax for taxpayers (PPH)	101,370	130,016
Corporations withholding income tax for employees (PPH, Article 21)	102,376	140,507
Corporations withholding income tax on imports (PPH, Article 22)	40,396	36,951
Corporations withholding income tax on payment of interest, dividends, etc. (PPH, Article 23)	81,310	105,244

— Not available.

PPH: Income tax. Articles refer to Law of the Republic of Indonesia, number 7, 1983.

a. Excludes personal income taxpayers on whose behalf personal income tax is withheld and where the withholding tax represents the final tax liability. As of June 1986, there were approximately 1.8 million individuals in such a situation. For 1984 and 1985, there were around 900,000 individuals in such a situation.

between December 1984 and April 1988 the number of registered income taxpayers (mostly self-employed) and the number for whom income tax is withheld more than doubled. The corresponding number for corporations increased by 80 percent. But the number of actual filings is far below those registered. Although potential taxpayers have now been identified, given the way data are collected as shown in table 4.3, determining the actual number of taxpayers remains a difficult task. On the other hand, without the extensive withholding provisions, there would be even less compliance. The tax authorities are continuing to identify taxpayers by improving the third-party information network, setting up a task force (with the help of the Internal Revenue Service of the United States) to identify professional individuals, and requiring the use of the taxpayer identification number for paying utility and phone bills. Other measures to enhance the general efficiency of the tax administration are noted below.

Second, in June 1986, the Ministry of Finance created a special independent "strike force" of auditors, many of whom were trained abroad as a part of the tax reform program (Gillis 1989a, p. 108). This group of thirty officials audited two dozen companies (between June 1986 and June 1987) and reported zero or negative tax liability for 1985. The cost of the audit was Rp 250 million, but it assessed taxes and penalties of Rp 87 billion with an additional Rp 7.5 billion under dispute (Gillis 1989a, pp. 105–09). These audits provided a remarkable return on investment and demonstrated the possibility of generating additional revenue from better enforcement and compliance. There have been no additional strike force actions, but in 1987, according to the tax authorities, about 4.3 percent of income tax returns were audited. The additional revenue gain was Rp 300 billion. Capabilities to undertake more systematic and scientific auditing procedures are being acquired.

6/1/86	*12/31/86*	*9/1/87*	*4/1/88*
614,194	643,000	672,400	683,150
16,643	—	—	17,221
507	—	—	457
148,467	157,600	176,100	182,802
165,532	—	—	206,751[b]
40,062	—	—	52,708
127,505	—	—	188,554[b]

b. It is curious that, according to the official figures, the number of corporations involved in withholding exceeds corporate income taxpayers who are registered.

Source: Data up to 6/1/86 World Bank (1988a, table 2.4). Data from 12/31/86 and 9/1/87 are from World Bank (1988b, table 3.2). Data for 4/1/88 are from Ministry of Finance, Directorate General of Taxation.

Third, traditionally state enterprises contribute a significant amount of corporate income tax revenue. Although relevant quantitative estimates are not available, a recent study estimated revenue from state enterprises at 34.2 percent of total income tax revenue in 1986–87 (Asher and van Eeghen 1987, table 2.2). This provides the authorities with some flexibility in generating income tax revenue in a given year. According to the data supplied by the Directorate General of Taxation, in the last month of the fiscal year ending March of 1986–87 and 1987–88, more than one-fifth of the entire year's income tax revenue was collected. Much of this is likely to be from state enterprises.

On the other hand, the political challenges associated with the full enforcement of income tax provisions are greater in the case of state enterprises. Without more information, it is difficult to determine where the balance lies. It does seem, however, that state enterprises do provide a solid baseline revenue for the income tax.

Additional factors have adversely affected income tax revenues. First, between 1984 and 1988, many firms were still enjoying unexpired income tax incentives. Moreover, an economic slump contributed to an overall income decline. However, as these incentives expire, income tax revenue can be expected to increase. Second, tax planning opportunities still exist. The suspension of debt-to-equity ratio norms was noted earlier. Other possibilities include pension plans (contributions paid into the approved pension plans are fully tax deductible without limit, but the income from the plan is mostly exempt from tax) and leasing arrangements (Pria and Van Hien 1989, p. 17). Third, the administrative inadequacies noted earlier could allow the continuation of transfer-pricing schemes. Thus, Wakui and Van Hien (1988, p. 12) suggest that general tax planning strategies for Japanese firms investing in Indonesia could "include reasonable transfer pricing, sensibly struck licensing, royalty and technical assistance agreements, full claims for tax depreciation entitlements, efficient structuring of senior executive as well as local remuneration packages, appropriate use of permissible business structures, arrangements to ensure that tax is not incurred unnecessarily on foreign currency exchange rates, and prompt administration in respect of tax exemption certificates whenever appropriate." The authors suggest only a general outline, and that detailed tax planning arrangements must be made on a case-by-case basis. They could also allow underreporting of income of both individuals and companies.

As table 4.2 shows, the income tax revenue to GDP ratio increased sharply to 3.15 percent of GDP in 1989–90. This is substantially higher than the corresponding ratio before tax reform. Both personal and corporate income tax revenues are expected to rise significantly, but for 1989–90, some analysts expected the tax on time and saving deposit interest (which was implemented in November 1988) to contribute about one-tenth of total income tax revenues. Because disaggregate data are unavailable, it is not possible to ascertain the source of the sudden increase in income tax revenue in 1988–89. Was it the result of improved tax administration (achieved through investment in training and hardware) plus the political will to enforce the income tax law? If the answer is yes, it could signal a qualitative change in the tax structure and administration. It is also possible that certain technical changes, such as classifying a proportion of royalties from oil or dividends from public enterprises as income tax revenue, have contributed significantly to income tax revenue in 1988–89. In the absence of relevant information, it is difficult to be certain about the reasons for higher revenues.

What is the potential for increasing income tax revenue in the future? Indonesia's strategy of relying on increased nonoil exports and foreign investment requires a fiscal environment, particularly in the area of income tax, that is internationally competitive. In 1989 Indonesia's income tax rates, with the base broadly defined, are less attractive because other countries have reduced their rates, and some such as Malaysia and Singapore are expected to reduce both individual and company income tax rates even more. A comparison of the nominal rates of income tax in selected Asian economies is provided in table 4.4. Thus, neither increases in rates nor the legal tax base can be relied on to generate additional income tax revenue. Indeed, there could be pressures, largely from foreign investors, to introduce selective income tax-based incentives once again.

Foreign investment approvals have shown a sharp increase, from US$826 million in 1986 to US$4,408 million in 1988 (Sato and Aoki 1989, p. 8). However, the same authors suggest that investment by Japanese manufacturers, such as home appliance makers, have bypassed Indonesia because there are no fiscal incentives and there are mandatory localization requirements (Sato and Aoki 1989, p. 9).

Thus, better enforcement and compliance will have to generate the additional income tax revenue. Given the importance of state enterprises,

Table 4.4 Nominal Rates of Income Tax in ASEAN and Selected East Asian Economies, 1989
(percent)

Country	Individual income tax: Initial marginal rate[a] (percent)	Individual income tax: Maximum marginal rate (percent)	Individual income tax: Number of income brackets	Company income tax: Rate (percent)	Company income tax: Number of income brackets
ASEAN					
Indonesia	15	35	3	15 to 35	3
Malaysia	5[b]	40[c]	9	35[d]	1
Philippines	1	35	9	35	1
Singapore	3	33	14	32	1
Thailand	5	55	6	35[e]	1
East Asian					
Hong Kong	3[f]	25[f]	8	16.5[g]	1
Rep. of Korea	5	50	8	20 to 33[h]	2
Taiwan (China)	6	40	5	15 to 25[i]	2

a. Zero rate is ignored.
b. From the year of assessment 1985, a tax rebate of M$60 for the taxpayer and M$30 for the wife has been given to those having chargeable income exceeding M$10,000. Under the present structure, this covers the first three income brackets.
c. There is an excess profits tax of 5 percent on income exceeding M$300,000.
d. A development tax of 5 percent applicable at present will be gradually phased out starting with a reduction of 1 percent from the year of assessment 1990. This tax is in addition to 35 percent tax on company income.
e. The rate is 30 percent for companies listed on the Security Exchange of Thailand.
f. These rates apply to wage and salary income.
g. The standard rate which applies to incomes from property, interest and profits of unincorporated enterprises is 15.5 percent.
h. There is also a provision for 27 percent rate for certain categories.
i. The individual income tax rates are those proposed by the Ministry of Finance and intended to be effective in 1990.
Source: O'Reilly (1989). For East Asian economies see relevant official documents.

their improved financial performance could lead directly to an income tax revenue increase. As we have seen, both avenues will require considerable political will as local enterprises and high-income individuals will be required to pay higher income taxes.

The VAT *package.* In contrast to the income tax, the revenue performance of the VAT package has been quite impressive. The VAT revenue to GDP ratio increased from 1.01 percent of GDP in 1984–85, the year immediately prior to the implementation of the VAT package, to 2.46 percent in 1985–86, and to 2.96 percent in 1987–88. According to the figures supplied by the Directorate General of Taxation, in the last month of the 1987–88 fiscal year (ending in March), 21.7 percent of VAT was collected compared to 10.2 percent the year before. Such a sharp increase is not readily explicable. VAT revenue to GDP ratio was 3.35 in 1989–90 (table 4.2). In 1985–86 both the income tax and the VAT package generated equal revenue; however, since then the revenue importance of the latter has increased (table 4.2).

Why did the revenue from the VAT package increase so sharply during 1984–85 and 1985–86? To answer this question, it may be useful to classify the original VAT package into the following categories: (1) VAT on petroleum products; (2) VAT and luxury sales tax on alcoholic products and tobacco, which continue to be subject to excise taxes even after the introduction of the VAT package; (3) VAT on domestically manufactured goods other than those in (1) and (2); (4) VAT on construction services; and (5) luxury sales tax.

Before the VAT package was introduced, no excise or sales tax was levied on petroleum products. Alcoholic products and tobacco were subject to excises but not sales tax. Since excises could have been levied on all three items (alcoholic products, petroleum, and tobacco), the introduction of the VAT package was not necessary to generate revenue, even though its introduction did make it easier to levy or raise these taxes. This is an important point because 34.4 percent of the increment in the sales tax revenue between 1984–85 and 1985–86 was due to petroleum products, and a further 9.4 percent was due to tobacco products. If alcoholic products are added, close to half of the increase in revenue was due to the first two categories of the VAT package. However, their role has declined since 1985–86.

So far VAT returns are subject to simple audits based on monthly returns, and full auditing is not carried out. Cross-checking is a recent development. Those who do not file or who file late bear only minimal penalties. However, the reorganized Directorate of Taxation is likely to give auditing higher priority. The revenue impact of this has not been that significant because most of the VAT is collected through relatively few sources. Thus, the petroleum VAT is collected from PERTAMINA, and only twenty or so major cigarette companies account for most of production. VAT and the luxury sales tax on imports are levied at customs. As noted earlier, the number of tax collectors for the VAT and its withholding provisions have been expanded considerably. While the VAT cannot be regarded as self-checking (indeed no tax is), it does appear that the VAT is more difficult to evade than the single stage sales tax it replaced.

The recent extension of the VAT to wholesalers and to many services is, however, expected to increase the administrative task enormously. The number of wholesalers liable to VAT is around 100,000 and the number of service establishments around 500,000. This contrasts with a total of only 84,499 taxpayers registered for the VAT in October 1988. The recent extensions are likely to increase the revenue importance of the luxury sales tax.

The revenue performance of the VAT does compare favorably with that of other Asian countries. The Republic of Korea introduced a VAT up to the retail level in 1977. Its ratio of tax to GDP of 3.66 percent is, therefore, not significantly higher than the ratio of 2.96 percent for Indonesia, as the latter's VAT at that time was still at the manufacturer-importer level (table 4.5). It is the excise tax revenue which is quite low in Indonesia as compared to the other Asian countries (table 4.5). In part, this is because no excises are levied on petroleum products, whereas in other countries they form a significant proportion of the excise tax revenue. The combined VAT plus excise taxes to GDP ratio for Indonesia is greater than that of Malaysia, and below that of Korea and Thailand (table 4.5).

Table 4.5 Revenue Importance of Sales and Excise Taxes in Selected Asian Countries

Country	Year	Percentage of GDP		
		VAT[a]	Excises	VAT and excises
Indonesia	1987–88	2.96	0.97	3.93
	1989–90	3.35	0.85	4.20
Korea, Rep. of	1987	3.66	2.16	5.82
Malaysia	1987	1.35	1.62	2.97
Philippines	1986	1.30[b]	2.59	3.89
Thailand	1987	2.73	4.88	7.61

a. VAT or existing sales tax. Excludes service tax in Malaysia.
b. Philippines introduced VAT up to the retail level in 1988.
Source: Bank of Indonesia (various years); IMF (1988, Vol. XII).

With the recent extensions of the VAT package, the scope for further expanding the base or increasing the rates has been considerably reduced, though extending the VAT to large retailers does remain a viable option. Aside from an increase in the tax base brought about by economic growth, better auditing procedures and techniques and improved compliance are the major sources for increasing the revenue importance of the VAT. This would have to take place simultaneously with gearing the administrative machinery to accommodate around 600,000 additional VAT taxpayers, many of whom are likely to have relatively small turnover.

The land and building tax. This tax contributed only 0.2 percent of GDP in the year it was introduced (1986–87). This increased to 0.34 percent of GDP in 1989–90 (table 4.2). As noted, effective January 1, 1989, the base for this tax was increased. It appears that the authorities are relying on this, and on the real estate valuation and mapping exercise to increase the revenue importance of this tax. It should be noted that between 1986–87 and 1988–89, forestry (which rose from 0.5 to 1.9 percent) and mining (rose from 27.7 to 36.7 percent) were the only sectors to increase their share in revenue derived from this tax, while the share of plantation, rural, and urban sectors declined. Urban properties are widely regarded as undertaxed. It remains to be seen whether the extent of this undertaxation will be reduced in the future.

In 1988 there were about 35 million taxpayers, but the authorities estimated the potential number of taxpayers to be around 50 million. Many are, however, in rural areas where average collection is only Rp 1,000 per taxpayer. Thus, collection costs for the rural sector remain high, and a minimum tax could usefully be instituted to reduce the collection costs.

Even though import duties were not covered in the tax reform program, it is worth noting that the revenue to GDP ratio has increased somewhat since then (the move away from nontariff to tariff barriers being an important reason), even in 1989–90, when this ratio was only 0.91. This compares quite unfavorably with other Asian countries, many of whom are more outward looking. Thus, according to the data from the International Monetary Fund (IMF), in the fiscal year 1989, import revenue to GDP ratio was 1.94 in Korea, 2.83 in Malaysia, 3.86 in the Philippines, and 3.95 in Thailand. Such comparisons suggest there is substantial scope for generating additional NOTR from import duties. Given the extensive use of quantitative restrictions in Indonesia, this can be accomplished without sacrificing allocative efficiency and equity.

Improving the efficiency of tax administration. The tax reform planners recognized the need to transfer resources from the private to the public sector more efficiently to move Indonesia toward the next stage of economic development. There was also a pressing need to make the economy internationally competitive so that it could participate more actively in international trade and capital flows. Therefore, a key role was assigned to improving the capabilities of the tax administration. Many of the administrative measures implemented as a part of the tax reform program have been discussed. It should be stressed, however, that the new tax laws, regulations, and procedures represent a substantial improvement over the previous situation. Without them, it would have been immensely difficult for Indonesia to move to the next stage of economic development.

The authorities sought technical assistance from the IMF, the Internal Revenue Service of the United States, and the German Technical Team to train key personnel, modernize auditing and other procedures, and develop a new management reporting system. In addition, a large number of tax personnel have been sent abroad for training, particularly in the areas of audit procedures and computer-based tax information systems. The number of foreign and locally trained auditors now exceeds 4,000, with 1,200 specializing in VAT auditing. More scientific and objective audit selection procedures should increase additional revenue through auditing (Rp 400 billion was targeted in 1988). The tax appeal system is also functioning reasonably well.

The tax office has been reorganized along functional lines since April 1, 1989. Previously there were sixteen departments—seven direct, six special, and three regional. Now the six functional areas are audit and investigation, buildings tax, income tax, indirect taxes including VAT, legal affairs, and tax policy. In addition there is now a computer center and a center for taxpayer services, and a new management reporting system is under development. However, little detailed information is available to assess either the reorganization or the management reporting system. Gillis has argued that by 1987, in spite of the revenue gain and some reduction in tax evasion due to the introduction of the VAT, there was still little evidence of improvements in administrative practices in the tax department (1989a, p. 107). Both the recent extension of the VAT and the valuation and mapping exercises for the land and buildings tax will make even greater demands on tax administration. It does seem, however, that there has been a noticeable increase in the technical capabilities of the tax department. Nevertheless, major compliance problems remain. As in most tax reforms, wage earners and some of the large businesses, including state enterprises are easy targets and have been tackled first. However, sustaining the NOTR effort will require significant improvements in tax compliance, particularly with regard to income taxes. This in turn will depend on more transparent and objective legal and regulatory systems, and the political will to enforce tax laws uniformly and impartially.

Efficiency in resource allocation. As we have seen, tax reform planners felt that economic neutrality (greater uniformity in tax rates among activities, products, and sectors) would reduce tax induced distortions. It was recognized that neutrality was not a first or best policy but only a workable objective. To achieve this objective there was a move to a broad-based low nominal rate regime for the income tax, better enforcement of and compliance with tax laws, and the elimination of input taxation from the sales tax. The land and buildings tax was also expected to reduce tax induced distortions by reducing the dispersion of effective rates within and between types of properties.

Conventionally, the extent to which marginal effective tax rates differ among activities, assets, or sectors is measured by analyzing the effects of income tax-based incentives. These were abolished by the tax reform program, so variations in the effective rates due to this phenomenon may be lessened considerably. However, many practices (such as extensive withholding) adopted to make collection easier could also unintentionally create different tax rates for individuals with the same income. These include retaining the classical system, as well as suspending debt-to-equity norms and the possibility of "round-tripping," which refers to the different tax rate on time deposits, that is, 15 percent, and the rate of tax relief given for interest paid, that is, 35 percent (Pria and Van Hien 1989, p. 17). For both companies and individuals it pays, at the expense of the treasury, to obtain bank loans and redeposit the proceeds, thus creating a paper transaction with no economic benefits. Since the VAT does not extend to the retail level and because no input credit is provided for certain services and in the case of the luxury sales tax, nonneutralities could still arise, albeit to a lesser extent.

Significant nonneutrality, however, could arise from several additional sources. First, to the extent that enforcement and compliance differ according to various types of income (capital vs. labor), activities (wages vs. self-employment), firms (local vs. foreign, private vs. public, large vs. small, goods producing vs. service providers) and sectors (even the filing ratios for the VAT vary considerably by sectors), effective tax rates will also differ horizontally and vertically. This is one more reason for improving the efficiency of tax administration. Second, for resource allocation, what matters is the variation in marginal effective tax rates. Third, recent extensions of the VAT have decreased the availability of tax credit for input and increased the rates for many commodities. Both are likely to result in increased nonneutrality among products and type of firms.

Redistributive impact of the tax reform measures. The major thrust of the tax reform measures was to raise additional revenues and minimize deadweight loss associated with an increased tax burden. Particular attention was paid to shielding the poor from the tax system by broadening the base for the income and property taxes and eliminating tax preferences by increasing the basic exemption level. A systematic quantitative analysis of the distributional impact of the tax reform package is not yet available. Therefore, the following paragraph simply presents tentative qualitative judgments on the incidence of the tax reform measures already implemented.

Measures expected to improve the equity of the tax system include base broadening and tax rate reductions accompanied by a higher threshold for taxation for personal and corporate income, and property taxes. Such measures reduce administration costs and tax evasion. Therefore, these measures in Indonesia's context would have had positive impact on the relative position of the lower-income classes in the tax system. The VAT is now the mainstay of government revenues and, therefore, its distributional impact could have a major bearing on the equity of the overall tax system. A major portion of the new VAT falls on import trade which is subject to significant quantitative restrictions, and which in turn generates large economic rents. Gillis (1989a, p. 107) reports that the introduction of the VAT did not have any measurable impact on inflation. This suggests that much of the VAT falls on rents associated with privileged access and therefore its distributional impact would be propoor (Shah and Whalley 1990). Thus a combination of quantitative restrictions on imports and introduction of the VAT is likely to have led to the positive distributional effect. In any case, the VAT replaced inefficient and inequitable sales and excise taxes and, therefore, its differential incidence would have been equity-enhancing.

If the present practice of not indexing deductions and allowances for the personal income tax continues (even though the law does allow indexation), over time, even with more moderate inflation, many middle-income individuals will be drawn in the income tax net. Since the initial rate of 15 percent is high compared to other Asian countries (table 4.4), the costs of entering the tax net are also higher. Because of higher enforcement and compliance rates for wage earners, this could result in significant horizontal and vertical inequities.

The absence of inflation adjustment also implies incorrect taxation of interest income and profits as the tax will fall on nominal rather than real income. According to the *World Development Report 1990* (World Bank 1990), the average annual inflation rate was 8.5 percent during 1980–88. Even a moderate inflation experienced during the 1980s could have significant equity and efficiency effects. This aspect needs to be given greater attention by policymakers than has been the case so far.

Lessons

The tax system that has emerged as a result of the reform program in Indonesia represents a substantial improvement over the prereform tax system. The tax system is now more compatible with the next stage of the country's development, emphasizing the need to make the economy internationally competitive and increasing Indonesia's participation in international trade and capital flows.

Based on this analysis, seven lessons may be noted, with the first three lessons following directly from the previous discussion.

1. *It is possible to overhaul the tax system in a relatively short period of time* although improvements in tax administration take longer and are harder to achieve.

2. *It is important that the tax reforms be consistent with the country's development plan* to ensure that all national economic objectives are harmonized.

3. *Revenue generation must include a mechanism to restrain government expenditure* and ways to generate greater nontax revenue, including income from state-owned enterprises.

4. *Conditions prevailing during the initial design and implementation of the tax reform may not be those needed for the reform to succeed in the longer term.* The confidential style adopted here during the initiation, design, and legislative approval stages can implement tax reform quickly and along rational lines. But effective and sustained reform requires a fiscal system in general and a tax system in particular that is considered legitimate—an important factor in compliance at all income levels. It can be argued that in many countries effective tax reforms have been difficult to achieve because of a failure on this front.

This indicates that the tax reform program should be regarded as more than a technical exercise (though contribution of technical improvements can be quite substantial as the Indonesian experience has shown). It is a broader undertaking where the understanding of both domestic and international factors is vital. It makes a big difference when appropriate leadership and key economic policymakers strongly support tax reform and necessary institutions are strengthened to sustain the benefits of the reforms in the long term.

Thus, at some stage, greater public participation in the process of tax reform should be encouraged. It is also important to regard fiscal and related information as a public good, available to all. Publication of a detailed annual report on taxation, as in Korea, would open up the tax reform process to wider debate.

At this point, it may also be appropriate to comment on the role of equity in designing tax reform. There is considerable merit in the argument (also made in the Indonesian case) that taxation cannot be used effectively to promote income redistribution; it is important to address the equity concerns through the expenditure side of the budget or nonfiscal policy.

It should be noted that the popular perception of an equitable income distribution may not be borne out by an incidence study of the effective tax rates by income groups. The process by which such perceptions are formed is complex, but one important factor is likely to be how well and in what manner the improved tax administration procedures are used. Namely, ensuring that this improved capability is used effectively is an important challenge. This is consistent with Bird's view that "the administrative dimension should be placed at the center rather than the periphery of tax reform efforts" (Bird 1989, p. 315).

5. *The role of efficiency considerations in tax design and the best method of ensuring the elimination of tax induced distortions in the economy.* The Indonesian experience suggests that economic neutrality (essentially equalizing the tax rates that apply to different activities, sectors, and assets) as an intermediate objective for achieving economic efficiency has merit, but ensuring such a pattern of taxation is an enormously difficult task. The role of tax administration in helping to achieve economic neutrality should not be underestimated, particularly even handed enforcement and compliance across activities, assets, and sectors. A tradeoff between the ease of revenue collection and economic neutrality also needs to be explicitly recognized.

The link between economic neutrality (particularly as it can be achieved only partially) and efficiency is extraordinarily difficult to demonstrate empirically. Moreover, nurturing activities with a high potential for future comparative advantage (through research and development incentives, for example) may necessitate departures from static economic neutrality. This suggests a need for a better conceptual meaning of efficiency, a workable proxy objective, and methods to achieve it.

6. *The globalization of financial and capital markets and the increased importance of intrafirm sales in world trade is making it increasingly difficult to tax capital income.* Markets for highly skilled and professional manpower are also becoming increasingly international. Moreover, other factors, such as competition among capital importing countries to attract foreign investment and the 1986 tax reform in the United States, have also constrained the use of income tax to generate revenue. This applies to international income taxation (Bird 1988) as well as domestic income taxation. As a result, reform measures to increase tax revenue are likely to concentrate on indirect taxes, such as the VAT, excises, and import duties. To some extent, traditionally undertaxed areas such as urban properties and cost recovery programs for public and social services can mitigate the need to raise additional revenue through taxation. Nevertheless, an increased reliance on indirect taxes remains inescapable. This suggests a need for a more careful examination of their design and implications, particularly for the poor (Bird 1987). This in turn will require micro data sets to analyze expenditure patterns of households and conduct more meaningful incidence analyses. The implications for efficiency and incentives of relying heavily on indirect taxes while simultaneously increasing the tax to GDP ratio should also be borne in mind.

7. *The* VAT *is administratively feasible in developing countries like Indonesia (particularly at the preretail level).* It is capable of generating considerable additional tax revenue, and minimizing input taxation, particularly for exports. However, a VAT label does not necessarily imply uniform economic effects, as other factors come into the equation, such as detailed administration, provisions, and regulations. Nevertheless, the role of VAT has become more prominent in all tax reforms.

Conclusion

The Indonesian tax reform program represents a substantial improvement over the prereform situation. The reform has strengthened the foundations of the tax system, which is no mean achievement, particularly when compared to the tax systems of other oil-exporting countries. Its main thrust has been consistent with the requirements of the next stage in Indonesia's development. There is, however, a need to ensure that the tax environment remains internationally competitive. The key problems that remain to be solved are improving the technical capabilities of the tax administration, improving equity, and putting into practice new techniques to improve enforcement and compliance.

Appendix Table 4.1 GDP, Central Government Tax Revenue and Nonoil Tax Revenue
(billions of Rp)

Fiscal year beginning	GDP[a] Old series	GDP[a] Revised series	Total tax revenue	Nonoil tax revenue	Total expenditure
1978	24,001	—	4,074	1,765	5,299
1979	34,345	—	6,510	2,250	8,076
1980	48,914	—	9,911	2,891	11,716
1981	58,421	—	11,876	3,248	13,918
1982	62,647	—	11,983	3,813	14,356
1983	73,698	77,676	13,914	4,394	18,311
1984	87,055	89,750	15,218	4,788	19,381
1985	94,493	96,850	17,760	6,616	22,824
1986	96,489	102,546	13,984	7,646	21,891
1987	114,500	124,539	18,826	8,779	26,959
1988	129,100[b]	139,452	21,435	11,908	32,990
1989	144,700[b]	—	22,809[c]	14,909[c]	36,575

— Not available.
a. Data for calendar year
b. Estimates.
c. Budget estimates.
Source: Bank of Indonesia (various years); GDP estimates for 1988 and 1989 from The Economist Intelligence Unit (1989, p. 32).

Notes

1. Low-income economies are defined by the *World Development Report 1989* as those with a gross national product (GNP) per capita of US$480 or less in 1987. This is a reversal of the 1983 report, which classified Indonesia as lower middle-income country.

2. For a more detailed analysis of the prereform tax system see Booth and McCawley (1981), Asher and Booth (1983), Mansury (1984), Conrad (1986), Asher and van Eeghen (1987), Gillis (1985, 1989a), and World Bank (1988a, especially appendix 2).

3. According to the President's speech, the draft budget for 1989–90, the target for the NOTR at the end of Repelita V for 1993–94 was to be 15 percent of GDP.

4. See the decree of the Minister of Finance No. 382/KMK/041/1989 dated April 20, 1989 for the latest available provisions concerning Article 22 of the Income Tax Law.

5. Decree of Ministry of Finance No. 759/KMK/04/1986, August 25, 1986.

6. Presidential Decree 29/1986, July 12, 1986.

7. Ministerial Decree No. 58, July 25, 1985 and retroactively enforced as of May 6, 1985.

8. Decree of Ministry of Finance 827/KMK/04/1984, August 9, 1984. A notification by the Directorate General of Taxation, Circular No. SE-16/PJ/23/1989 has reconfirmed this suspension.

9. Presidential Decree 37/1986, August 13, 1986.

10. Ministry of Finance, Decree No. 485/KMK/01/1986, June 4, 1986.

11. Government Regulation Number 28 of 1988; and the circular of the Directorate General of Taxes No. SE-20/PJ/3/1989, May 19, 1989 provide regulations and implementing details.

12. Those authorized to act at tax collectors are enumerated in the Presidential Decree No. 56 of 1988 dated December 13, 1988.

13. The relevant official documents are Government Regulation No. 29/1988 dated December 27, 1989; Decree of the Minister of Finance No. 1335/KMK/04/1988 dated December 31, 1988 and Decree of the Minister of Finance No. 434/KMK/04/1989 dated May 2, 1989.

14. Decree of the Minister of Finance No. 303/KMK/04/1989, dated April 1, 1989.

15. Decree of Minister of Finance, No. 1324/KMK/04/1988.

16. Law number 13, 1985.

17. Decree of the Minister of Finance No. 244/KMK/00/1989 dated March 13, 1989.

18. Details of the changes in trade taxes introduced between March 1985 and December 1987 may be found in World Bank (1988b, pp. 58–59). Unless otherwise noted, the discussion in the text concerning the reform of trade taxes is from this source.

19. Details of the relevant items and corresponding tax rates may be found in the Decree of the Minister of Finance No. 210/KMK/013/1989 dated March 2, 1989.

References

Arndt, H. W., and H. Hill. 1988. "The Indonesian Economy: Structural Adjustment after the Oil Boom." In M. Ayub and C. Y. Ng, eds., *Southeast Asian Affairs 1988*. Singapore: Institute of Southeast Asian Studies.

Asher, Mukul G., and Willem van Eeghen. 1987. "Indonesia: Public Resource Management Study, Non-Oil Tax Revenue." Consultant's report. World Bank, East Asia and Pacific Region, Country Department III, Washington, D.C.

Asher, Mukul G., and Anne Booth. 1983. *Indirect Taxation in* ASEAN. Singapore: Singapore University Press.

Bank of Indonesia. Various years. *Indonesian Financial Statistics.* Jakarta.

Bird, Richard M. 1987. "A New Look at Indirect Taxation in Developing Countries." *World Development* 15(9): 1151–61.

Bird, Richard M. 1988. "Shaping a New International Tax Order." *Bulletin of International Fiscal Documentation* 7(July): 292–99.

———. 1989. "The Administrative Dimension of Tax Reform in Developing Countries." In Malcolm Gillis, ed., *Tax Reform in Developing Countries.* Durham, N.C.: Duke University Press.

Booth, Anne, and Peter McCawley. 1981. "Fiscal Policy." In Anne Booth and Peter MacCawley, eds., *The Indonesian Economy during the Soeharto Era.* Kuala Lumpur: Oxford University Press.

Conrad, Robert. 1986. "Essays on the Indonesian Tax Reform." CPD Discussion Paper No. 8608. East Asia & Pacific Region, World Bank, Washington, D.C.

The Economist Intelligence Unit. 1989. "Indonesia Country Report No.1." London.

Gillis, Malcolm. 1985. "Micro and Macroeconomics of Tax Reform, Indonesia." *Journal of Development Economics* 19(December): 221–54.

———. 1989a. "Comprehensive Tax Reform: The Indonesian Experience, 1981–1988." In Malcolm Gillis, ed., *Tax Reform in Developing Countries.* Durham, N.C.: Duke University Press.

———. 1989b. "Tax Reform: Lessons from Postwar Experience in Developing Nations." In Malcolm Gillis, ed., *Tax Reform in Developing Countries.* Durham, N.C.: Duke University Press.

———. 1990. "The Indonesian Tax Reform after Five Years." Paper presented at the Conference on Tax Policy and Economic Development among Pacific Asian Countries, January 5–7. Taipei, Taiwan.

Harberger, Arnold. 1989. "Lessons of Tax Reform from the Experiences of Uruguay, Indonesia, and Chile." In Malcolm Gillis, ed., *Tax Reform in Developing Countries.* Durham, N.C.: Duke University Press.

IMF (International Monetary Fund). Various years. *Government Finance Statistics Yearbook.* Washington, D.C.

Lerche, Dietrich. 1980. "Efficiency of Taxation in Indonesia." *Bulletin of Indonesian Economic Studies* 16(1): 34–51.

Mansury, R. 1984. "A Bird's Eye View of Indonesian Income Taxation History." *Asian Pacific Tax and Investment Bulletin* 2(12): 503–509.

Morgan, Frank B. 1989. "Withholding Taxes under Indonesia's Income Tax Law." In David O'Reilly, ed., *Income Taxation in* ASEAN *Countries*, 2nd edition. Singapore: Asian Pacific Tax and Investment Research Center.

Nasution, Anwar. 1989. "Fiscal System and Practices in Indonesia." In Mukul G. Asher, ed., *Fiscal Systems and Practices in* ASEAN: *Trends, Impact and Evaluation.* Pasia Panjang, Singapore: Institute of Southeast Asian Studies.

O'Reilly, David, ed. 1989. *Income Taxation in* ASEAN *Countries*, 2nd edition. Singapore: Asian Pacific Tax and Investment Research Center.

Pria, Eka, and Leonard Van Hien. 1989. " Indonesian Tax System: Structure and Recent Developments." *Tax Planning International Review* 16(8): 14–17.

Sato, Akiyushi, and Akira Aoki. 1989. "The Role of ASEAN States as Production Bases (Part II)." *RIM* 2(4): 7–19.

Shah, Anwar, and John Whalley. 1991. "The Redistributive Impact of Taxation in Developing Countries." In Javad Khalilzadeh-Shirazi and Anwar Shah, eds., *Tax Policy in Developing Countries*. Washington, D.C.: World Bank.

Tanzi, Vito. 1985. "A Review of Major Tax Policy Missions in Developing Countries." In Hans M. Van de Kar and Barbara L. Wolfe, eds., *The Relevance of Public Finance for Policy Making*. Detroit, Mich.: Wayne State University Press.

Wakui, Yoshiko, and Leonard Van Hien. 1988. "Japanese Investment in Indonesia." *Tax Planning International Review* 5(12):10–12.

World Bank. 1989. 1990. *World Development Report 1990: Poverty*. New York: Oxford University Press.

———. 1989. *World Development Report 1989: Economic Development*. New York: Oxford University Press.

———. 1988a. "Indonesia: Selected Issues of Public Resource Management." Report No. 7007-IND (restricted access). Washington, D.C.

———. 1988b. "Indonesia: Adjustment, Growth and Sustainable Development." Report No. 7222-IND (restricted access). Washington, D.C.

5

The Jamaican Tax Reform: Its Design and Performance

Roy Bahl

Jamaica began comprehensive tax reform by establishing the Jamaica Tax Structure Examination Project in 1983. This research, training, and administrative improvement task force lead to the enactment of structural and administrative reforms for the individual income tax in 1986, the company income tax and property tax in 1987, and the general consumption tax (value added tax) in 1991. Tanzi (1987b, p. 228) defines a "successful tax mission as one that results in reform of the tax system of a country along the lines proposed by the mission." By this criteria, the Jamaican tax reform might be viewed as one of the more successful efforts in the last two decades.

This chapter describes and evaluates the results of the four-year Jamaica tax reform project. We also include an ex post evaluation of the performance and continuing reform of the new system since its implementation. The chapter concludes with a list of lessons learned, as well as some parallels with the conventional wisdom.[1]

This work adds to the literature on tax reform in four areas. First, the Jamaican reform was sufficiently comprehensive to "shock" the tax system and still obtain a viable reform—something history and many experts warn is not likely to occur (Goode 1984; Jenkins 1989). Second, the experience in Jamaica adds to what is known about the politics and the process of tax reform, that is, how to go about involving interest groups and the general public in the design and "selling" of a comprehensive tax reform without compromising its integrity. In fact, the project remained active during public debate and implementation, and was involved in monitoring the performance of the new system. Third, we can learn something about the survival of a tax reform when political leadership changes. Edward Seaga's administration championed reform but did not gain reelection in 1989; however, the Michael Manley administration and its successor continued the work on tax reform. We examine the extent to which the proposed and implemented reforms have been either embraced or discarded and extract some principles regarding sustainable tax reform. Finally, we examine the postreform performance to assess how well the objectives of the reform were met. In the case of Jamaica, reform has been ongoing.

The Jamaica tax reform project conducted its work between 1983 and 1987. The final results are reported fully in Bahl (1991b) and summarized in Bahl (1989, 1991a). A follow-on project focused on payroll taxes, and on a review of the status of the 1987 income tax reforms (Bahl and others 1992a, 1992b, 1992c; Sjoquist and Green 1992). More recently, the tax system has been reviewed by McLure (1993), Hubbell and McHugh (1992), Garzon (1993), and Bahl and Wallace (1993).

The Economic and Political Context

Edward Seaga was elected prime minister in 1980 with a mandate to replace the direct controls that had long governed the economy with an economic growth strategy that was export driven and led by the private sector. One of the challenges he set for his administration was to find a tax package to fulfill this mandate, provide the necessary revenue, and be politically acceptable. The tax reform project was established to tackle the technical work of identifying this package. In this section we examine the work of the tax project and the economic and political factors that shaped the design of the reform program.

Macroeconomic Performance

The Jamaican economy suffered a severe and sustained contraction from 1973 through 1980 (Dawes 1982; Chernick 1978). Estimates published in the International Monetary Fund's (IMF) *International Financial Statistics* for this period show the following:

- Gross domestic product (GDP) (1980 prices) declined 18 percent.
- GDP per capita (1980 prices) declined 26 percent.
- The consumer price index (CPI) rose 307 percent.
- The local price of the U.S. dollar rose 96 percent.
- Government expenditure rose 419 percent.
- Government revenue rose 274 percent.
- Net foreign assets dropped by US$582 million.
- Estimated unemployment rose from 22 to 27 percent.

There was no economic miracle in the first half of the 1980s. The Seaga administration lived up to its mandate with new government policies to deregulate the economy and change the orientation from import substitution to export promotion. However, foreign exchange reserves were short, the treasury was almost bare, and the government had to ride out the virtual collapse of the bauxite industry. Moreover, there was considerable pressure from foreign creditors to adopt more austere economic policies. In fact, the government's new economic policies did not really take hold until the mid-1980s and any positive impacts were interrupted by a severe downturn in 1984 and 1985, when real GDP dropped by more than 6 percent.

Nevertheless the Jamaican economy did grow. Real GDP increased by 5.7 percent between 1981 and 1983, and though modest, this increase represented a reversal from the real 18 percent decline experienced between 1973 and 1980. However, economic instability also marked the first half of the 1980s. There were real GDP declines in 1984 and 1985, the Jamaican dollar was devalued in 1983 and 1984, and the rate of inflation remained over 25 percent during 1984–85 (table 5.1).

The foreign exchange shortage remained acute in the early 1980s. The Jamaican dollar was devalued by over 100 percent between 1982 and 1985. The U.S. dollar moved from an average J$2.15 in 1983 to J$3.94 in 1984 and J$5.56 in 1985. This devaluation was largely the initial response of market forces to the liberalization of a previously pegged and undervalued exchange rate. The new exchange rate system was a managed float operated through a biweekly auction. Imports, particularly of consumer goods, fell significantly in response to the devaluation, and the current account balance of payments deficit was about 13 percent of GDP in 1984. Devaluation had the expected favorable impact on exports. However, domestic real incomes, and consequently the market for domestically produced goods, grew very little. Despite rescheduling, foreign debts grew significantly relative to both export earnings and government expenditures.

Table 5.1 Selected Indicators of Economic Performance

Year	*Real growth in GDP (percent)*	*Increase in consumer CPI (percent)*	*Fiscal (deficit) surplus (percentage of GDP)*	*Exchange rate*[a]
1980	–6.2	27.3	(17.5)	1.78
1981	2.4	13.0	(16.7)	1.93
1982	0.9	6.3	(14.5)	1.99
1983	2.4	11.7	(12.4)	2.15
1984	–1.4	27.8	(12.8)	3.94
1985	–4.8	25.8	(6.1)	5.56
1986	2.0	15.1	(5.9)	5.48
1987	6.4	10.7	(2.3)	5.49
1988	1.9	8.3		5.49
1989	6.3	14.3	0.2	5.74
1990	4.2	29.8	3.7	8.10
1991	0.5	80.2	5.5	12.85
1992	1.4	40.2	5.4	22.20

— Not available.

a. Jamaican dollars per U.S. dollar.

Source: Revenue Board and Ministry of Finance data.

Inflation hovered around 30 percent in 1984. This was due to a 77 percent increase in the exchange rate, the removal of subsidies on certain foods and public utility rates, and the rapid monetary growth of past years. Because the economy was heavily dependent on petroleum, fuel price increases in 1984 affected virtually all other components of the CPI. Electricity rates increased by about 100 percent in 1984. The fiscal deficit stood at more than 10 percent of GDP in 1984, and averaged 15 percent of GDP during the first five years of the decade. All in all, the mid-1980s did not appear to be a favorable time to introduce a comprehensive tax reform.

Social Conditions

Social conditions were also not favorable for a structural tax reform in the mid-1980s, especially given the emphasis on economic efficiency and simplification. Much of the Jamaican population lives near a subsistence level of income. In addition the distribution of income is very unequal. It was estimated that at the time of the reform 40 percent of the national income was earned by the top 10 percent of the population, and that this inequality had not been significantly reduced in the previous two decades (Wasylenko 1991). The distribution of land wealth, as might be expected, was even more skewed—half of the total assessed value of land was attributable to 5 percent of the total land parcels (Holland and Follain 1990). Given the average real per capita income decrease between 1980 and 1985 of J$62, one can imagine that living standards for the poor worsened considerably in the early 1980s.

The 28 percent inflation in 1984 reflected rising housing and food prices. Deregulation and removal of subsidies also led to higher public utility prices. This explains why housing expenditures (excluding rent) grew by about 50 percent, and was by far the largest component of increase in the CPI. A food stamp program started in 1984 provided some relief to lower-income Jamaicans. The unemployment rate, though difficult to measure, appeared to be around 15 percent.

Jamaica's "brain drain" of the 1970s—educated Jamaicans migrating abroad in search of better economic opportunity—imposed a heavy cost on the economy. Between 1974 and 1980 there was a net emigration of 129,000 residents, or about 6 percent of the population. This trend continued into the 1980s at a lower rate: between 1981 and 1984 more than 30,000 persons, or 1.4 percent of the population, emigrated.

Economic Policy

The Seaga administration's economic program was outlined in "Taxation Measures 1982–83," Ministry Paper No. 9 of the Ministry of Finance and Planning. Reforms in this program fell into two classes of economic policy: general macroeconomic policy and structural economic policy. The former determined elements of public sector deficit—the growth of monetary aggregates and foreign exchange payments and receipts. The latter embraced elements that influence economic agents and their choices with respect to work effort, savings, investments, product mix, input mix, and portfolio structure.

This program counted on the controlled expansion of aggregate demand to bring order to relative price movements in commodity, labor, money, financial, and foreign exchange markets, and to bring order to the distribution of income and wealth. It implicitly promoted the proposition that economic growth and efficiency would be improved if private markets and private decisions were permitted a larger role. Accordingly, Ministry Paper No. 9 proposed to reduce public ownership of commercial enterprises, public sector control of prices (except the price of foreign exchange), and the regulation of imports, exports, and domestic investment.

The government's economic program was consistent with the strategy outlined in Ministry Paper No. 9. Import licenses and price controls were for the most part phased out. The government deficit was dramatically reduced and comprehensive income tax reform was implemented in 1986. Tax incentive policies were adjusted to favor exports and the agriculture sector, and some divestment of public enterprises was undertaken. These initiatives did not go as far as some had hoped, but the program went generally in the direction promised.

A notable exception to this economic strategy was foreign trade policy. The price of foreign exchange had not been decontrolled—except during 1983–84 when there was a controlled float of the Jamaican dollar—and foreign exchange shortages persisted. Taxation of international trade probably exacerbated the problem. In 1985 stamp duty rates were increased markedly on all imported consumer and intermediate goods. This measure raised substantial revenue, but also further protected domestic manufacturers. The government began a rollback of these increases in 1987, and instituted a program of rebates to exporters to compensate for the taxation of imported inputs. When Seaga left office, the formulation of a consistent trade policy still remained at the top of the government's list of unfinished economic reforms.

Foreign Pressure

Economic policy after 1980 was influenced by external lenders. Sometimes the conditions imposed were ignored; at other times the government had to sacrifice political, economic, and social objectives, which often imposed hardships on the population, and in some cases affected the design and implementation of subsequent comprehensive tax reform.

The government negotiated separate loan agreements with the World Bank, the IMF, and the U.S. government in 1981 and 1982. The IMF agreement provided for a target deficit level of 10 percent of GDP by fiscal year

1983–84. When the government did not meet this target, the IMF pushed for a deficit reduction program. With the unemployment rate already around 20 percent and the bauxite sector declining, substantial public employment reductions seemed out of the question. To meet the IMF target, the government turned first to tax rate increases on the most important excises—alcoholic beverages and cigarettes—and in the following year to rate increases under the import stamp duty.

These discretionary actions affected tax reform planning. It sent a message to the Jamaican public that tax reductions were infeasible, even in tough economic times. Furthermore, the import duty rate increases effectively introduced a major new indirect tax that further distorted the pattern of relative prices. On the one hand, this would be a convenient new straw man for the tax reformers to knock down. On the other hand, proposals for any general sales tax would now be harder to sell because it would shock the system even more.[2]

Another major influence was U.S. government policy. Although neither its balance of payments loans nor its project assistance carried conditions similar to the IMF or World Bank loans, U.S. foreign policy shaped tax, trade, and industrial policy. First, the U.S. government provided funding for the comprehensive tax reform project. Second, there was always the implied threat that faulty Jamaican economic policy could dampen U.S. support for the programs of the Seaga administration. Third, the U.S. tax reform of 1986 lowered the corporate tax rate to 34 percent, thereby jeopardizing the foreign tax credit position of U.S. firms investing in Jamaica and giving the tax reform program one more reason to lower the corporate rate.

The Setting for Tax Reform

All government tax reformers know that the probability of enacting a reform program depends as much on timing as it does on the quality of the program. It depends on when during the election cycle the program is introduced, whether the same political party has won both the presidency and the congress, whether other "big issues" are overshadowing tax reform, whether the president or prime minister feels secure enough to discuss taxes, and the recent history of tax reform. In Jamaica, the timing was both right and wrong.

The budgetary position in the early 1980s would make tax reform a tough sell. Successful tax reform in almost everyone's eyes meant tax reduction, an understandable reaction to the slow growth in the economy, inflation, and the income tax bracket creep. But the route to tax reduction was not clear. Budget cuts would have to accompany tax reductions, leading almost certainly to reductions in government employment. This would have taken place at a time when unemployment was high and the private sector economy was performing too poorly to absorb surplus labor.

Otherwise, tax reduction would have to come at the expense of an increase in the government's budget deficit. This was ruled out for two reasons: (1) increased domestic borrowing would have put more pressure on domestic prices, and (2) the IMF loan agreement required a reduction in the government deficit and a ceiling on domestic credit.

Government "efficiency" or cost reduction was only a slightly more promising route to budgetary balance. One possibility centered on the state enterprises, which were a known to drain the central government budget. Options included divestment and increased user charges to cover operating costs. But divestment takes time, and increased user charges on some items (for example, electricity) would have been as unpopular as increased taxes. Other deficit reduction strategies centered on removing costly government subsidies, for example, on petroleum products, or eliminating the import duty exemptions on a wide variety of producer goods. Some of these measures were eventually adopted and they proved to be as politically difficult, as expected.

The setting for a comprehensive tax reform also had some positive aspects. First, and most important, the tax system was obviously unfair. A widely held public view was that the horizontal inequities inherent in the structure, which were accentuated by the way the system was administered, went beyond tolerable limits. Piecemeal reform undertaken to fill an annual revenue gap, the approach taken for a long time, would no longer be acceptable. The public—business, labor, the press, and foreign investors—gave the Seaga administration a clear mandate for a complete overhaul of the tax system. This public dissatisfaction and the willingness of the government to think carefully about these problems contributed to the successful implementation of the income tax reform in 1986 and 1987.

A second stimulus came from foreign donors. The IMF was pressing the government to reduce the fiscal deficit and limit domestic borrowing. The Fund took its usual position of being agnostic about whether budget balance should be achieved by tax increases or expenditure reductions, but it nevertheless gave annual advice on how much tax rates would have to be increased to fill the fiscal gap. The World Bank pressed more aggressively for structural tax changes in tariffs and indirect taxation. The U.S. government did not set conditions on its aid package, but did urge changes in the tax system and financed the tax project that eventually led to reform. These external pressures motivated the government to come up with its own tax reform.

Third, the Seaga administration's political hand was strengthened in the 1984 elections when the Jamaican Labor Party (JLP) won an uncontested election. The issues underlying this political victory did not involve economic reforms, but the election meant that proposals would be reviewed by a more friendly and unified parliament. In a sense the JLP would be replacing the tax system of the opposition party.

By the time the first phase of the new reforms was put in place in 1986, the economic situation had already improved. Real GDP growth rate was positive in 1986, and over 6 percent in 1987. The decline in oil prices, a lower rate of inflation, and a good tourist season all set the stage for the individual and company income tax reforms to produce far more revenue than had been expected.

Problems with the Prereform Tax System

An analysis of the Jamaican tax system revealed four fundamental problems. First, high taxes discouraged investment in financial and human capital. Second, the tax system had badly distorted the relative prices driving economic decisions and, as a result, the economy was not performing as efficiently as it would under a system with more neutral effects on relative prices. Third, only such income and consumption as could be easily reached was taxed, thereby narrowing the effective tax base. Also, weak tax administration allowed those who could to avoid or evade taxes. A fourth overarching issue was the need for a harmonization of tax, trade, and industrial policies.

High Taxes

At the beginning of the tax project in 1983, the ratio of taxes to GDP was 23.3 percent and was thought to be too high. But complaints about high taxes can mean many things. It can signal dissatisfaction with the quality and type of public services being provided, as was the case of the U.S. tax revolt of the late 1970s (Bahl 1984). It can also mean that taxes that are high by international standards somehow make a country less attractive for investment. In the case of Jamaica, high taxes discouraged work effort and saving, and biased investment decisions leading to reduced economic growth.

International comparisons. To examine Jamaica's tax load relative to other countries we used the comparative technique originally developed by Lotz and Morss (1967), extended by Bahl (1971, 1972), and updated on a periodic basis by the Fiscal Affairs Department of the IMF (Chelliah 1971; Chelliah, Baas, and Kelly 1975; Tait, Gratz, and Eichengreen 1979; Tanzi 1987a).[3] These tax effort studies have shown Jamaica to be a country with relatively high taxes. Estimated taxable capacity increased from 16.9 to 19.5 percent of GDP during the early 1970s and then fell off to 17.8 percent during 1972–76. However, the public sector did not retrench when Jamaica's capacity to raise revenues declined; in fact, the government expenditure–GDP elasticity averaged about 2.0 over the 1974–80 period. Jamaica's actual level of taxation and tax effort increased through the 1970s, resulting in a tax effort 6.4 percent above the international norm from 1972 to 1976.[4]

An updated analysis based on a sample of fifty-two developing countries indicates that Jamaica had a predicted taxable capacity equivalent to 21.1 percent of gross national product (GNP) in 1983. The actual taxation level of 23.3 percent in 1983 placed Jamaica 10.4 percent above normal and ranked it nineteenth out of fifty-two developing countries. In the sample of Caribbean Community (CARICOM) member countries, only Dominica, Guyana, and Trinidad and Tobago showed a higher tax effort. Regression analysis shows that by international standards Jamaica's taxes were high in 1983, and that its relative level of tax effort had been increasing.

Another approach to international tax comparison is the representative tax system, which relates a country's taxable capacity to the size of its various tax bases and the "average" effective rate at which other countries tax each of these bases (Bahl 1972). Bahl, Jordan, Martinez-Vazquez, and Wallace applied this methodology to 1983 data for the same fifty-two developing countries, and show Jamaica's tax effort to be 25 percent above the international average, fourteenth highest in the sample. Only Dominica, Guyana, and St. Lucia of the CARICOM countries ranked higher. Again, this confirms Jamaica's high tax status. This approach also allows us to sort out how each tax contributes to the total tax effort. In the case of Jamaica the conclusion is clear: the high tax effort is due to high rates of personal taxes. The breakdown on the tax effort index, which is 25 percent above average, is as follows: personal taxation (personal income tax and domestic indirect taxes) is 43 percent above the international average, company income taxation and property taxation are about average, and import taxation is 10.4 percent below average.

Narrow tax bases. A recurring theme in the Jamaican tax story is that the base of virtually every tax has been significantly narrowed by exemptions, preferential rate treatment, and administrative practices. The result is that nominal (marginal) rates were set very high to satisfy revenue requirements, which explains the dissatisfaction with taxation levels. Jamaica's 23.3 percent tax share of GNP in 1983 may have been a dramatic understatement of the burden on those who actually paid taxes.

Before reform in 1986, the individual income tax base was narrowed by the exclusion of perquisites, or "allowances," from taxable income, by sixteen personal tax credits, and by the preferential tax treatment of wages earned from overtime work. More importantly, because of poor administration only the pay-as-you-earn (PAYE) sector was effectively taxed. Dividend income was not fully reached, also because of poor administration, and interest income earned from bank deposits and capital gains were not taxable. Because of these exemptions only about 40 percent of the true taxable base was actually taxed. In order to raise the necessary amount of revenue, the lowest marginal tax rate was set at 30 percent with no standard deduction. It reached 57.5 percent at the relatively low income level of J$14,000. If comprehensive income had been fully taxed, the average tax rate would have been about 11 percent. The frequently heard complaint that the income tax system discouraged work effort and investment really meant that those included in the income tax net were forced to pay very high marginal and average rates.

A similar story may be told for the five payroll taxes. Two of these, the education tax and the Human Employment and Resource Training Trust (HEART) are not contribution programs. Both share shortcomings with the individual income tax: both allow deductions for allowances and both fail to capture the self-employed in the tax net. The education tax does not allow personal allowances, and income is taxed from the first dollar earned. The other three payroll taxes, National Insurance Scheme (NIS), National Housing Trust (NHT), and Civil Service Benefit Scheme, are contribution programs, but each contains a significant tax element. The bases of these taxes are also narrowed by statutory exemptions (there is a ceiling on wages taxed under the NIS) and allowances are not taxed. Removal of the NIS ceiling and taxation of allowances alone would have permitted an (equal yield) reduction in the average rate on the three contribution programs of more than 1 percent of wages.

The base of indirect taxes also was limited by exemptions. In 1985 only about 20 percent of the value of all imported goods was subject to import taxes. As a consequence of this, the import stamp rate was over 200 percent on some items, 30 percent on capital goods, and 16 percent on raw materials. A similar story can be told about the base for domestic indirect taxes. Only about 16 percent of final consumption of services and 33 percent of domestic manufacturing output was included in the tax base. If the indirect tax system had been replaced by a value added tax (VAT) of the manufacturer-importer type in 1983, a rate of 20 to 25 percent would have been necessary to maintain revenue yield (Bird 1991b). This would have been high by world standards.

The property tax base also fell well short of its legal goal of taxing the full market value of land. The 1984 roll placed total land value at over J$5 billion in comparison with about J$2 billion estimated by the 1974 roll (Holland and Follain 1990). The property tax base had been further narrowed by derating—agricultural properties were eligible for a 75 percent reduction in assessed value, and hotels for a 25 percent reduction—which imposed a revenue cost equivalent to one-third of the 1985 yield of the property tax. The top bracket statutory property tax rate was 4.5 percent of the assessed value of land.

Allocative Effects

If the maxim of tax neutrality were followed, the tax system would raise the desired amount of revenue in such a way that the relative prices of consumption, investment, labor, and production would not be affected. The basic story is that the market, not the tax system, should guide economic decisions. In theory, the exception to this rule is that the tax system may be properly called on to compensate for market failure. As a practical matter, it is impossible to define a tax system that has no substitution effects. The modern restatement of the neutrality goal is to minimize the excess burdens associated with raising a given amount of revenue. Not every analyst or every economic planner agrees with the neutrality goal. In fact, a respectable view is that taxes should be used as levers to stimulate economic activity in desired directions (Ahmad and Stern 1991; Bird 1992, chapter 6). This interventionist approach guided taxation in the 1970s.

The tax project's major goal was to remove major distortions in relative prices, that is, to create a tax system that interfered less with the market. In fact, the final reform proposals recommended that some tax incentives be continued and that certain consumption items be exempt from sales tax to protect low-income residents. The question underlying the design of the reform, however, was how far should the government go in using the tax system to guide economic choices, correct undesirable distributional impacts, or simplify administration? The view was that the relative price distortions introduced by the tax system had gone beyond justifiable exceptions and had measurably weakened the economy's efficiency.

It is no simple matter to prove that tax-induced distortions in relative prices resulted in a significant welfare loss. The welfare loss is roughly proportional to the product of the size of the distortion in relative prices and the compensated price elasticity of demand (or substitution) for the good (or factor) in question. It turns out that the magnitude of these terms is not easy to estimate. The net change in relative prices caused by the tax code is difficult to estimate because several different provisions in the tax structure may be involved and because all may not affect relative prices in the same direction. As for the second term, there is very little evidence on the compensated price elasticities of substitution in developing countries, but what is there suggests an inelastic response to relative price changes.[5] One could have a nonneutral tax structure, then, and not suffer substantial welfare losses if the relative price distortions are small.

Labor supply. Economists have long been concerned with the effects of taxation on work effort. Theory tells us that a higher rate of tax on wages induces an individual to work less because the rewards for work are less (the substitution effect), but also that an individual will be induced to work more to make up for lost income (the income effect). There is almost no empirical evidence on this question for developing countries, but most observers would guess the price elasticity to be small (Bird 1992, chapter 7).

The impact of the tax structure on work effort in Jamaica may be of some consequence for two reasons. First, the effective rate of the combined income and payroll taxes are high, and therefore, the reduction in the net wage rate attributable to the tax system may be large. One might suppose that the labor supply response could be significant even if the price elasticity was low. Second, the compensated price elasticity of the labor supply may actually be larger than thought, because Jamaican workers have options other than to accept the tax liability. They may remain within the PAYE sector and evade or avoid taxes, or they may move from the formal to the informal sector of the economy. With an unemployment rate near 20 percent during this period, these responses would seem more realistic than choosing more leisure time or migrating abroad.

The distortive effects of the prereform system on labor supply emanated from the high tax rate imposed on marginal work effort, the mobility of labor between the formal and the self-employed sectors, the low quality of the income and payroll tax administration, and the numerous avenues left open for tax avoidance. There was widespread legal avoidance of income tax within the PAYE system in the form of nontaxable perquisites. In some cases, it also appeared that the preferential tax rate on overtime work had come to be viewed as a loophole. If labor ultimately carried the burden of both the employee and employer share of payroll taxes, and if the NIS and the NHT were contribution programs with no tax element, the average combined effective tax rate on labor would be more than 10 percent of taxable compensation.

Self-employment offered an attractive opportunity to evade taxes. About 95 percent of all income taxes were paid by PAYE workers in 1983, and only about 10 percent of those in the self-employed sector even bothered to file a return. Evasion was feasible because the probability of detection was low and enforcement of penalties was weak.

Capital-labor choice. The prereform tax system appears to have raised the price of labor relative to capital although the magnitude of the distortion is difficult to estimate because so many different taxes and subsidies were applied. It is almost as though the government had recognized this problem and attempted to correct it by adding other features to the tax system to lower the relative price of labor, for example, a preferential tax treatment of overtime earnings.

The net effect of these policies was probably to make labor more expensive, and to make the tax system more complicated and difficult to administer. Matters were further complicated by the complex pattern of tax shifting induced by these taxes and subsidies.

In the absence of a formal model to simulate the effects of the tax code on the capital-labor choices of firms, the project asked whether returns to labor were being taxed more heavily than returns to capital. The answer appeared to be yes. Before the 1986 reform, the average effective combined rate of income and payroll tax rates was on the order of 22 percent of statutory income and the combined top marginal rate was over 65 percent. The price of capital was relatively low because interest income was not taxed and dividends and capital gains were either exempt or, for administrative reasons, not fully included in the tax base. However, there were also important sectoral biases, for example, labor went relatively untaxed in the self-employed sector.

The investment decisions of firms were probably biased in favor of substituting capital for labor. To the extent that the employer's share of payroll taxes was borne by owners, the relative price of labor increased.

Moreover, capital investments were given generous allowances in computing the basis for company income taxation. The tax treatment of capital goods under the indirect tax system is not easily sorted out. Technically, capital goods were subject to an import stamp duty that had risen to 30 percent in 1986. In addition, a number of special taxes and subsidies were introduced to influence the input decisions of firms, and in many cases, these affected capital goods. More important is the incentive legislation. On the one hand, this encouraged capital intensity by providing initial capital allowances and exemptions from import taxes. On the other hand, it tied the length of the tax holiday to labor intensity, but since this was related only to CARICOM-traded goods, it was not a very powerful feature. The approval of firms for the incentive programs was said to be linked to the degree of labor intensity, but approval was a judgmental matter, and it is not clear how this effected capital-labor choices. Another preferential treatment of capital was the exemption of buildings under the property tax system.

At least one program induced firms to substitute labor for capital. The Human Employment and Training Program gives firms a tax credit (against HEART tax liability) equivalent to the wage paid to the HEART trainee. There is no hard evidence available on the impact of this program.

We could make no firm estimate of the net effects of these tax provisions on the relative prices of capital and labor. The weight of the evidence seems to side with the argument that the tax system increased the relative price of labor—a peculiar choice in a country with a highly literate but underemployed labor force. The size of the distortion may have been quite large, hence the tax system may have had a substantial effect on resource allocation, even if the elasticity of substitution was small.

Savings. If there was a strategy to use the tax system to increase the rate of private saving, it is not clear what it was. Consider the following package of effects:

- The marginal personal income tax rates were graduated and reached high levels for individuals with high marginal propensities to save.
- Four income tax credits were offered to those who would participate in specified types of savings programs.
- There were three compulsory payroll savings programs. For those who participated, the average contribution for a private sector worker was about 11.3 percent of compensation, and for a public sector worker it was 11.9 percent.
- Interest income was not taxed, but dividends were taxed under both the company and individual income taxes. In practice, however, a substantial portion of dividend income managed to escape taxation.
- Lax administration meant that a large portion of the self-employed sector completely evaded income taxes and therefore paid a marginal tax rate of zero. Since many of these were higher-income Jamaicans, the effect was to increase the rate of private saving.

- Retained earnings of companies were taxed at a lower rate than were distributions.

The tax system also affected the structure of investment although it is not clear that the effects were large. In theory, dividends were taxed at a marginal personal tax rate of 57.5 percent of income and were subject to the basic corporate rate of 35 percent. This double taxation of dividends, coupled with the tax-free status of interest income, allegedly led to the thin capitalization of Jamaican companies. There was also a bias in favor of real estate investments because the annual property tax was levied at a nominal level and because capital gains from land sales were effectively untaxed.

Commodity prices. The particularly important price distortions were the relative price of imported versus domestically produced goods, the differential tax treatment among domestically produced consumer goods, and the price of sumptuary consumption relative to all other goods.

With respect to the relative price of imports, discretionary policy had been ambivalent. Traditionally the exemption rate on imported goods was double the international average. Moreover, imports were underpriced because of the overvalued Jamaican dollar. The government responded with a system of import licensing in the early 1980s; devaluation during 1983–85; and the introduction of a new import tariff structure in 1984, which significantly increased protection of domestic industry.

The indirect tax structure was very complicated in the prereform period, with traditional goods taxed at between 15 and 30 percent, lower rates on raw materials than capital goods, and highest rates on imported consumer goods. The domestic indirect tax system mostly affected manufacturers, and the combination of the direct tax and the hidden tax on inputs led to an estimated 11 percent tax component in manufacturers' prices (Bird 1991b).

Finally, the indirect tax system relied on taxation of sumptuary goods for over four-fifths of revenues. The relative price distortions from higher tax rates on alcohol, gasoline, and tobacco may not generate substantial inefficiencies because of the low price elasticity of demand and because of the social costs resulting from drinking, driving, and smoking. On the other hand, the relative price of these goods had fallen in the early 1980s because they were effectively taxed at specific rates and because the real price of gasoline had been allowed to fall.

The indirect tax system affected relative commodity prices, but because the system was so complex it was difficult to identify these effects. A first step to getting the prices right, it would seem, is to simplify the system enough to understand its impacts on relative commodity prices.

Tax evasion and avoidance. Every income tax payer faces choices among tax evasion, tax avoidance, and fully reporting income. The potential rewards for successful evasion or avoidance under the prereform system were considerable—tax savings equal to the 57.5 percent marginal personal tax rate and the tax component of the various payroll levies. Opportunities for avoidance and evasion were certainly present. Jamaican companies awarded employees nontaxable emoluments, apparently without seeking government approval, and were able to raise both the take-home wage and reduce the firm's liability for the employer share of payroll taxes. Another vehicle for avoidance was the declaration of preferentially taxed overtime income, which was not monitored by income tax authorities. The self-employed often captured these benefits by outright evasion, taking advantage of the inability of the Income Tax Department to enforce the tax.

The costs of noncompliance to the economy were substantial. It is estimated that only one in ten self-employed Jamaicans bothered to file a tax return. This imposed a revenue cost equivalent to about 50 percent of individual income tax collections in 1983. If the prereform system had been fully complied with in 1983, it would have been possible to raise the same amount of revenue with a flat rate of about 20 percent (Alm, Bahl, and Murray 1991d).

Equity

Relative price distortions not only imposed an efficiency cost on the economy, they introduced unfairness in the system that many taxpayers found even more objectionable. The self-employed were given favored treatment by the income tax administration and paid little or no individual income tax, whereas those enrolled in the PAYE system were forced to cope with what appeared to be onerous burdens. Even within the PAYE sector, private sector workers had opportunities to avoid taxes through the receipt of untaxed allowances and low-taxed overtime earnings. On average, a public sector worker paid a higher rate of individual income tax on total compensation than did a private sector worker (Alm, Bahl, and Murray 1991a).

The price distortions in the system also compromised vertical equity. The upper-income classes gained the most from the poor administration of the income tax, and from the availability of tax preferences. Allowances tended to be concentrated in the higher-income brackets, and overtime income was claimed more heavily by salaried workers than by hourly wage earners. Jamaicans with interest and dividend income paid a lower effective tax rate and because they tended to be concentrated in the higher brackets, this reduced the overall progressivity of the system.

The indirect tax system also compromised vertical equity since taxable domestic production and imports accounted for a higher proportion of the income of lower-income families. On the other hand, the exemption of unprocessed foods and housing consumption lessened the regressivity significantly. As Bird and Miller (1986, 1991) and Wasylenko (1991) show, it is very difficult to sort out the implications of commodity tax rate differentials for the distribution of income.

Administrative Problems

The Jamaican tax system, like that of most low-income countries, was plagued by administrative problems. The tax system was complex and difficult to administer; there was a shortage of skilled staff, and assessment, collection, and recordkeeping procedures were inadequate.

Complexity. The complex system for assessment and auditing was made even more difficult by a shortage of skilled staff. Complexity also raised compliance costs for taxpayers and in so doing either wasted private sector manpower or provided additional incentives for tax evasion and avoidance.

Prior to the 1986 reform, the individual income tax included two separate rate structures and a preferential rate for overtime income, sixteen income tax credits and an even greater number of nontaxable perquisites. The forms used to establish an employee's tax credit entitlement were rarely, if ever, updated and almost never monitored by either the employer or the Income Tax Department.

The forms and instructions for year-end tax returns were long and detailed, even by comparison with other developing countries. An analysis by McLure (1984) revealed that the income tax forms did not reflect the existing law and that there were numerous errors in the instructions. Moreover, it was difficult to obtain a copy of the income tax law.

Complexity extended far beyond the income tax. There were five different payroll taxes levied on four different bases that were administered by three different government agencies. This substantially increased the burden on employers, who were required to calculate the liability for each employee, maintain appropriate records, and develop an administrative relationship with several different government agencies. There were also five different indirect taxes: the external (CARICOM) tariff, the import stamp duty, an excise tax, consumption duty, and a retail sales tax. Within this family of sales, excise, and import levies, there were over 100 rates, some with needlessly small gradations.

Staff problems. A shortage of skilled staff is a major bottleneck to improved tax administration. DeGraw (1984) reported that in 1983, a time when increased revenue mobilization was at a premium, there were 150 vacancies among the 449 positions authorized for the Income Tax Department. A disproportionately large number of these were technical positions. There were complaints about too few skilled staff throughout the tax administration service.

The reasons for the staffing problems are similar to those of other developing countries. Salaries were low, even given the job security and the prestige of a government post. In 1983 a trained accountant earning J$9,000 in the Income Tax Department could have made J$14,000 with a private sector accounting firm. The problem was more than just salary. There was no formal career development program and little opportunity for promotion. In the case of the Customs and Excise Department, entry-level personnel were recruited primarily out of secondary schools and had little background in accounting. To compound the problem, there was no training program.

Outmoded procedures. The methods used to assess and collect taxes were inadequate at the time the tax reform project began in 1983. There was no unique numbering system for either businesses or individuals, hence there was no master file of taxpayers. The system was completely manual, that is, there was little, if any, use of computers other than to print bills. This effectively ruled out the use of third-party information, cross-checking sales and income returns, and other tasks. Only about 60 percent of property tax liability was collected, and the cost of property tax administration was equivalent to about 12 percent of property tax revenues.

The income tax was essentially a PAYE levy. There was little, if any, use of presumptive assessments on hard-to-tax groups, such as self-employed professionals. The major problem was, and remains, the absence of an adequate information and recordkeeping system. The income tax file room was too small and all records were manually kept. Files were regularly misplaced or lost, and records were frequently out of date or incomplete.

Finally, there was no perceived need to monitor the performance of the tax system. There was no annual statistical volume reporting taxpayers by taxable income brackets or any attempt to develop a revenue forecasting model.

Tax, Trade, and Industrial Policy

The allocation of resources in Jamaica was distorted by the foreign trade regime and by industrial policy. These distortions were not all unwanted. Some policies were designed expressly to favor one industry or sector, others to discourage the consumption of imported goods, and still others to protect certain domestic production activities from foreign competition. In other words, taxation was not the only instrument of economic policy in the hands of government, and in the mid-1980s, it probably was not even the most important. Clearly, the design of a comprehensive tax reform—especially one that sets out to correct distortions in relative prices—must take the goals and impacts of trade and industrial policy into account.

The problem is how to do this. Is it good advice to stay with the basic taxation maxims of horizontal equity and neutrality, even though these might run counter to foreign trade policy? Alternatively, should tax policy play more of a supporting role and focus on reinforcing the allocative impacts of other government policies? Or is it possible to design tax reforms that can be relatively neutral in their effects on the allocation of resources and at the same time support the government's goals of conserving foreign exchange, encouraging export development, and stimulating investment (Shoup 1991)?

Government policies for trade and industrial growth had the objectives of stimulating investment and stabilizing the nation's external balance to ensure competitiveness in export markets and allocate enough foreign exchange to support the demands for local industrial growth and necessary imported goods. Many different instruments were used to support these policies in the early 1980s: multiple exchange rates, devaluation, import licensing, tax incentives, protective tariffs, import duty exemptions, preferential tax rates for certain commodities, and special capital depreciation allowances. Sometimes the effects of these policies were reinforcing, but at other times they were offsetting. Hence, the net impacts were not always consistent with the Seaga administration's stated strategy to support export-driven growth through the private sector. To complicate matters, the government's approach to trade and industrial policy was continuously changing in the early 1980s—in part to accommodate pressures from external creditors.

The policies of the 1970s and early 1980s were interventionist in spirit. The policy mix was designed precisely to affect economic choices and, therefore, stimulate certain production and consumption activities and discourage others. Horizontal inequities and relative price effects were at the very heart of this strategy. This left open the possibility that a more neutral tax program would push the government to an even greater use of targeted, direct controls to reestablish preferences that the tax reform may have taken away.

By 1984, the Jamaican dollar had become considerably overvalued (Whalley 1984). The policy of a fixed exchange rate effectively taxed exporters by forcing them to sell foreign exchange earnings at a low price and to buy imported inputs at world market prices. Not surprisingly, this resulted in a foreign exchange shortage and an active, illegal foreign currency market. The situation worsened after 1983 with the collapse of the Jamaican bauxite industry (a major source of foreign exchange) and the heavy drain on foreign exchange reserves for debt repayment and oil purchases. The government responded with an extensive system of import licenses, and finally with a devaluation.

Beginning in 1984, the policy instrument used to shape trade policy was the stamp duty on inward customs warrants, essentially a surtax on the value of imported goods levied independently from the common external tariff. During 1984–85, the import stamp tax rates were increased dramatically as an emergency revenue measure. Collections nearly tripled in one year, and in 1985–86, the import stamp duty accounted for over 13 percent of total taxes. Revenues were derived principally from a 16 percent tax on raw materials, a 30 percent tax on capital goods, and a 40 percent tax on consumer goods. Although successful as a revenue measure, raising the stamp duties may have harmed the Jamaican economy in other ways: it was protectionist and, because it was so complicated, it appeared to be arbitrary in its application.

Objectives of the Reform

It is tempting to claim that comprehensive tax reform can and should satisfy all of the criteria for a "good" tax system. Indeed, many tax reform studies are unable to resist this temptation and design a system with multiple or even conflicting objectives. In fact, there are important decisions to be made about exactly which objectives of tax reform are the most important and which can be given up.

The primary objectives of Jamaican tax reform were simplification and neutrality. The goal was to put in place a system that could be efficiently administered and "get the prices right." To be sure, there were important constraints: political resistance to taking back tax preferences, the need for progressivity in the system, the requirement of adequate revenue, and the goals of trade and industrial policy. Still, the primary thrust was to restructure the tax system to lessen the distortive effect on relative prices and, therefore, on economic decisions.

Both of these objectives pointed toward a reform package that would broaden the tax base and flatten the tax rate. A broader tax base can generate the same amount of revenue as the present system, but at lower marginal rates, which can reduce some of the harmful efficiency effects. With fewer exemptions and special features, taxes could be more easily assessed and collected. This would minimize the time required to police those already in the system and allow tax officials to expand the base and bring those who are hard to tax into the net. Simplification also makes the tax system more understandable and reduces compliance costs.

The other choice for an overall objective would have been to stay with the interventionist spirit of the prereform system. The project rejected this approach for two reasons. First, this was clearly out of step with the Seaga administration's economic program, which promised a market-oriented growth strategy. Second, even if one believed that manipulation of tax rates and bases was the best route to Jamaica's economic and social development, there were serious doubts about the ability of the tax authorities to implement a finely tuned system.

What about equity in this comprehensive tax reform? The view of the project was that equity should not be the primary objective in the design of a comprehensive tax reform in a developing country. The history of Jamaica's tax system is a case in point. The steep progressivity of the individual income tax rate structure was designed to increase the tax system's vertical equity. What it did instead was increase the incentives for evasion and avoidance. Because the income tax administration was too weak to enforce the system properly, the loopholes and noncompliance grew. Eventually individual income tax burdens became quite regressive.

Another problem with taking vertical equity as a primary reform objective is that efficiency costs may be imposed. One example is the tradeoff between what are usually viewed as special "equity" features of a tax—high marginal income tax rates on the rich and higher taxation of luxury goods—and the disincentives to saving and investment that such measures bring. Finally, there is the tradeoff between introducing selective tax treatments to enhance vertical equity and defining a tax base that is broad enough to provide adequate revenues. It would be unthinkable to prohibit the taxation of alcoholic beverages, cigarettes, or petroleum consumption on grounds of improving the overall equity of the tax system.

Equity was not ignored in the design of tax reforms, and the following constraints were included in developing reform programs. First, the overall system should not be made more regressive. Since Wasylenko's (1991) analysis showed the system to be proportional over the first eight deciles and regressive at the top end, a program of broad-based, flatter tax rates would not appear to compromise this objective. Second, there should be no increase in the tax burdens on very low income households. A low-income household survey by Miller and Stone (1987) identified the consumption patterns of low-income families and, therefore, the necessities that should be excluded from the base of new general sales tax (Bird and Miller 1986, 1991).

Horizontal equity was an important objective of the reform. Getting the prices right and equal treatment of equally situated individuals and businesses are very closely linked objectives. Horizontal inequities not only induce uneconomic behavior by firms and workers, they undermine confidence in the tax system and encourage noncompliance. There is probably no better rationalization for shirking one's taxes than pointing to the perceived unfairness in the tax system.

It is important to distinguish structural tax reforms from revenue-raising programs. The objective of reform was to design a revenue-neutral system. In truth, "one period" revenue neutrality is about the best that can be expected. One might design a system to yield the same revenue as the present system in the first year of the reform, but it is unlikely that the revenue-income elasticity of the restructured system will be the same. As the reform program unfolded, the government agreed to some reduction in revenue compared to the prereform system, but remained silent on the elasticity issue.

Nor was there any clear directive to change the mix of taxes away from its relatively heavy reliance on income taxation. In 1984 about half of government revenues were raised from direct taxes (table 5.2), a relatively high share for a low-income country (Tanzi 1987a).

The Individual Income Tax

Prior to the 1986 reform, the individual income tax base, in theory, included all sources of income except bank deposit interest. In practice, however, there was no tax on capital gains and most self-employed income was outside the tax net. There were two rate structures, depending on whether income was

Table 5.2 The Structure of Taxes
(percent of total taxes)

Tax	*1984*	*1985*	*1991*	*1992*	*1993*
Customs duty	8.5	8.5	10.1	13.3	13.8
General consumption tax	—	—	—	9.9	21.3
Special consumption tax	—	—	—	12.9	10.2
Consumption and excise duty	28.3	25.8	23.2	5.8	2.3
Stamp duty	7.0	9.1	11.2	9.0	6.2
Retail sales tax	1.7	1.8	2.6	1.4	—
Betting and gambling taxes	1.4	1.3	0.6	0.6	0.5
Income tax	46.6	44.7	46.8	42.2	41.4
Land and property tax	2.0	1.2	0.8	0.6	0.5
Motor vehicle licenses	1.1	2.2	0.6	0.6	0.3
Education tax	0.4	2.0	3.8	3.7	3.5
Total[a,b]	1,504.7	2,022.6	7,923.0	11,481.0	19,050.4
Exhibits					
Total indirect share	49.9	51.0	48.0	52.9	54.3
Total indirect share	49.9	51.0	48.0	52.9	54.3
Total direct share	50.1	49.0	52.0	47.1	45.7
Taxes as a percent of GDP	—	—	23.8	22.9	25.7

— Not available.
Note: Years are fiscal years.
a. Includes other indirect taxes.
b. In J$ millions.
Source: Revenue Board and Ministry of Finance data.

above or below J$7,000. The top marginal rate was 57.5 percent. There was no standard deduction, but taxpayers could qualify for up to sixteen tax credits for purposes that ranged from personal allowances to stimulation of saving to employment of helpers in the home. Because the credits were not indexed to inflation, their value substantially eroded during the early 1980s. The income tax administration did relatively little monitoring of the credit system.

The tax base was further narrowed by the practice of permitting employers to grant nontaxable perquisites (allowances) to employees. These perquisites were negotiated between employee and employer (including government ministries) and did not have to be reported to the income tax commissioner. A sample survey taken for the project showed allowances to average 15 percent of taxable income, and frequently over 30 percent for taxpayers with incomes above J$18,000. Perhaps as important, there was a general perception that allowances were even greater—some prominent Jamaican analysts argued, from anecdotal evidence, that the allowance–taxable wage ratio averaged 40 percent.

Analysis of the Tax System

The general direction for reform was to broaden the tax base, reduce top-end marginal rates, and protect the real income position of low-income families. This had to be done within a constraint of revenue neutrality and the almost certain opposition of interest groups who had long since come to expect (and rely on) some of these tax preferences.

Theoretical analysis of the income tax became an indispensable part of the blueprint for reform, and the project actually began with an analysis of the relative merits of alternative forms of income taxation (Break 1991; Alm, Bahl, and Murray 1991a, 1991b). However, it became clear at an early stage that any proposed reform would rise or fall on the empirical evidence. The reform would surely bring many winners and losers as lower rates replaced exemptions, and hard estimates of the revenue consequences of such a sweeping change would be crucial. The prime minister, who ultimately would have to champion the program, wanted to see the numbers at every turn of the work.

Gathering adequate data for the analysis turned out to be a major undertaking. There were no statistics of income tax, records were not computerized, complete and up-to-date information could not easily be obtained from the manual filing system, and there was no complete master file of either taxpayers or firms. To complicate matters further, neither nontaxable allowances nor overtime income were reported to the income tax authorities. There was no hard information on the rate of filing by the self-employed, but there were some records on audit activity.

The analysis of reform options required estimating the number of taxpayers, taxable incomes, nontaxable perquisites, and tax credits—all by income class. This was done by drawing a large random sample of taxpayers and manually recording data on taxable income, tax credits, tax liability, and so forth, from the files on each employer and each individual. The prime minister organized a special survey of employers, which yielded data on the value of nontaxable allowances by income bracket for about 60,000 workers. This was supplemented with a sample survey of a large number of self-employed individuals to determine the extent of evasion by nonreporting.

This analysis reached five general conclusions about the performance and failings of the existing system.

1. *The income tax base had been narrowed dramatically by tax credits, allowances, and various forms of evasion.* More than half of potential individual income tax liability was not included in the tax net in 1983. According to rough (and arguably conservative) estimates, full taxation of allowances and unreported income would have doubled individual income tax revenues. To give some idea of the opportunity cost, this amount would have fully covered the government deficit in 1983.

2. *The income tax system was not as progressive as its legal rate structure suggested.* When measured against statutory income, effective tax rates showed a progressive pattern, but when tax liability was measured against a more comprehensive definition of income—including allowances and unreported income—the progressivity disappeared. The progressivity of the statutory rate structure was all but negated by evasion and avoidance. Movement away from a nominally progressive income tax rate structure, therefore, would probably not compromise income distribution goals.

3. *The system contained substantial horizontal inequities.* Differential tax treatment of individuals in the same income bracket depended on an individual's

ability to conceal income and to receive a larger share of income in allowances. For example, the average tax rate for individuals in the highest income classes ranged from 50 percent for PAYE employees who complied with the tax law to zero for nonfilers, with an estimated average of less than 10 percent.

4. *Inflation raised effective tax rates via bracket creep.* Simultaneously, inflation had an offsetting effect on the vertical equity of the tax because the value of credits declined in real terms and because the increasing tax rate stimulated evasion and avoidance. The three main avenues for escaping the high rates of individual income tax—evasion, allowances, and overtime—were all concentrated in the upper-income brackets.

5. *Marginal income tax rates were high enough to affect work effort, investment, saving, and compliance choices.* Given the distortion in relative prices, this seemed a plausible argument, but there was little hard evidence. On the other hand, there is some empirical support for the argument that higher marginal tax rates are associated with higher rates of evasion and avoidance, even after taking account of the level of penalty rates and the probability of detection (Alm, Bahl, and Murray 1990).

Tax Evasion

Tax evasion, tax avoidance, or reporting income are the choices facing every taxpayer. The 57.5 percent marginal tax rate and the tax component of the various payroll levies were powerful incentives for evasion and avoidance. Opportunities for avoidance were certainly present, and the self-employed often captured these benefits by outright evasion, taking advantage of the Income Tax Department's inability to fully enforce the tax (Alm, Bahl, and Murray 1991c).

To roughly assess the revenue loss from avoidance and evasion, the working population of six professions was estimated: accountants, architects, attorneys, physicians, optometrists, and veterinarians. Based on a random sample of this group, it was determined that only one out of five paid income taxes between 1981 and 1983, and 60 percent did not have an income tax reference number. The revenue loss was about one-half of the total income tax collections for 1983 (Revenue Board 1985). This analysis was extended to nine other self-employed occupations, with similar results. Even with liberal allowances for late filing, we concluded that less than 20 percent of self-employed persons filed a return. Based on this sample, the estimated revenue costs from evasion were on the order of 50 percent of income tax collections (Alm, Bahl, and Murray 1991d).

The Reform Program

The key elements of the 1986 reform program were as follows:[6]

- The credit system was replaced by a standard deduction of J$8,580.
- A flat tax rate of 33.3 percent replaced the progressive rate structure.

- Allowances were included in taxable income with some exceptions.
- The preferential treatment of overtime income was eliminated.
- Interest income was made taxable above a threshold level.

Two important principles drove this proposal. First, as a package of reforms, the individual pieces would make little sense if viewed in isolation. For example, it would not have been politically possible to eliminate the overtime preference without reducing the marginal tax rates. Second, this program would not work without major improvements in administration. It was unrealistic to expect that an announced change in the tax status of allowances and lower marginal tax rates would automatically broaden the income tax base.

Projected revenue and tax burden impacts. One can estimate the structural impacts of the reform program by using historical data. At the time this proposal was being evaluated, the hypothetical question was: What would have happened had these reforms been implemented in 1983?[7] The results showed that the proposed system would have reduced average taxation from 14.5 to 9.8 percent of taxable income for those who actually paid income taxes in 1983 (Revenue Board 1985). Enactment of the full program would have led to a revenue loss equivalent to 26 percent of revenues. The distribution of tax burdens would have been more progressive because the impact of the interest tax, the taxation of allowances, and the relatively high standard deduction of J$8,580 would have offset the effects of the lower nominal rates on higher-income taxpayers. The estimated revenue-income elasticity of the reformed system would not have been significantly less than that under the prereform system. This is because the new system is not really a flat rate tax but a two-rate tax—0 and 33.3 percent—and income growth bumps a substantial number of workers into the taxpaying range. The standard deduction of J$8,580 was not indexed, hence average tax rates for all taxpayers would rise as income increased.

The project made out-year projections of the impact of the proposed reform, and compared these with projections for the prereform system. The results suggested that in 1987 the flat rate tax would yield only 7 percent less than the prereform system. The progressivity would be greater under the new system: it was estimated that those in the over J$50,000 income class would face an effective rate of 32.5 percent under the new system in 1987—about twice the effective rate they would have paid under the prereform system. This increased progressivity was primarily due to the tax on interest. The effective tax rate on earnings in all income brackets was projected to drop. Those who would emphasize the potential economic impacts of lower marginal rates on higher-income taxpayers applauded this reduced taxation of earned income, and those who look to the tax system to reduce disparities in the distribution of income were happy with the increase in the average rate of taxation in the top brackets.

Allocative effects. Would there be significant investment, saving, and work effort responses to this reform package? Even if the price elasticities of work effort, saving, and income tax compliance were very small (as the evidence suggests), the impact could be substantial because marginal tax rates were reduced so dramatically. At the time the reform program was designed, the project could only speculate on the size of any effects. It seemed plausible to argue that the after-tax return to investors and to increased work effort could be significantly increased. The rewards of outright evasion are lessened, and the upgraded system of enforcement had a better chance for success than a system with higher marginal rates.

The prospective impact on saving is more complicated taxes to assess. Since interest would be brought into the tax base and one-third of the gross return on savings accounts would be taxed away, there would be a reduction in the demand for commercial bank savings deposits relative to equity investments. Moreover, the reform program also removed other preferential tax treatments of investments by eliminating the income tax credits for the purchase of life insurance premiums and unit trust shares. Two responses were possible: (1) all investments would be put on an equal footing, thereby improving the relative attractiveness of purchasing stocks, or (2) the interest tax would encourage avoidance via capital flight, a shift to consumption, or a shift in investment to the more lightly taxed real estate sector. The project relied on a variety of arguments to suggest that the latter effect would not dominate, but in truth, there was no hard evidence.

Results of the Reform

After seven years of experience with the reformed system permit some evaluation of its success might center around the following questions:

- Has it proven to be a sustainable reform?
- Have revenues grown as expected?
- Has administration improved?
- How have economic choices been affected?
- Have equity goals been served?

Sustainability. The first question is whether the individual income tax reform of 1986 was sustainable. Did it give way to government pressure to recover lost revenues or restore special preferences to target groups? Jamaica certainly presents an acid test of sustainability, because the new reform had to survive a political campaign and a change of administration. Although the income tax was not a major issue in the campaign, the winning People's National Party made a number of statements during the campaign about its dissatisfaction with the taxation of interest.

In fact, there have been few structural changes in the individual income tax since the 1986 reform and for the most part, these were consistent with the recommendations of the project. In the last year of his administration, Prime Minister Seaga announced an increase in the standard deduction from J$8,580 to J$10,400 effective January 1, 1989.

The project originally had recommended indexing the standard deduction, and the 1989 increase was approximately equal to the general rate of inflation during the period since the reform was enacted. The relief provided through this increase in the exemption was substantial—an estimated 32,000 individuals were dropped from the taxpaying population in the first year. The estimated cost of this program in fiscal year 1989–90 was J$106 million, or about 7 percent of estimated collections (excluding interest) for that year (Bahl and others 1992a, 1992c).

The Manley administration was active in reforming the income tax, and with two minor exceptions, these reforms remained consistent with objectives laid down for the 1986 reform. The standard deduction was increased to J$14,352 in 1992. This increase was less than the rate of inflation during the project. In 1992 the Manley administration also reduced the tax rate to 25 percent. These changes strengthened the reform program by emphasizing allocative effects and setting the stage for a higher rate of the GCT.

The two minor exceptions are a tax-free allowance for productivity worker increases, and a tax-free status for certain bonus payments and gratuities. Since there is little hope for effective monitoring of such provisions, there is ample opportunity for avoidance. Both became loopholes in the income tax system that are only available to some taxpayers.

Perhaps the real test of sustainability of the income tax reform was changes that did not happen. The flat tax structure was retained, the standard deduction was retained, and global base (including interest income) was retained. In fact, the Manley administration passed on the option to abolish the tax on interest income in favor of increasing the standard deduction and lowering the tax rate. Credits were not reintroduced and allowances (fringe benefits) were not expanded. The Jamaican flat tax, therefore, would seem to have passed a fairly stern sustainability test.

Revenue performance. Revenues from the individual income tax in the postreform period increased beyond the project's and the government's expectations. After the reform took effect in 1987, the individual income tax share of total GDP ratcheted up from its average level of about 4.6 percent in the 1980–86 period (table 5.3). In the first four years under the reformed system, individual income tax revenues averaged 6.3 percent of GDP. This was in spite of an increase in the standard deduction in 1989. This rapid growth enabled the government to further increase the standard deduction and lower the tax rate in 1992, bringing the individual income tax closer to its historical level of 5 percent of GDP.

There are several reasons for the revenue growth in the postreform period. The Jamaican economy grew and the income-elastic income tax responded. The tax on interest income was more elastic than had been expected. The amount collected from the withholding tax on interest has been equivalent to over 40 percent or more of the amount collected on earned income since 1987.

The conclusion we draw from these trends is that the revenue-increasing impacts of base broadening and simplification significantly

Table 5.3 Individual Income Tax Revenues
(millions of J$)

Fiscal year	*Pay-as-you-earn sector*	*Individual other*	*Total on earned income*	*Interest*	*Total*	*Percent of GDP*
1980	155	16	171	—	171	3.6
1981	206	15	221	—	221	4.1
1982	270	23	293	—	293	5.0
1983	347	25	372	—	372	5.3
1984	411	33	444	—	444	4.7
1985	512	35	547	—	547	4.9
1986	583	44	627	—	627	4.7
1987	640	40	680	256	436	6.0[b]
1988	766	59	825	348	1,173	6.6
1989	941	69	1,010	429	1,439	6.7[b]
1990	1,141	75	1,216	651	1,867	6.1
1991	1,489	196	1,685	604	2,289	5.2
1992	2,390	—	—	—	3,411	4.7[b]
1993	3,408[a]	—	—	—	—	—

— Not available.
a. Preliminary.
b. Denotes years when discretionary changes took place.
Source: Revenue Board and Ministry of Finance data.

outweighed the revenue-reducing effects of rate reduction and a standard deduction. The reformed system gives the Jamaican government an income tax with an elasticity slightly above unity, that is, it grows (and declines) at a slightly greater rate than GDP.

Allocative effects: Compensation adjustments. One of the main features of the reform was a broadening of the tax base—the replacement of tax credits and allowances with a standard deduction. However, the broad-base concept did not clear parliament untouched and important loopholes remained in three areas. The first is specifically outlined in the legislation. Housing allowances, one of the more prominent abuses of the prereform system, are permitted as nontaxable income if certain criteria are met. The other two important loopholes are provided at the discretion of the income tax commissioner. A nontaxable allowance for an automobile provided by an employer is permitted. The amount is calculated according to a schedule that relates the allowance to the engine displacement in cubic centimeters of the automobile.[8] The third, also given at the discretion of the commissioner, is a uniform allowance. This provision was originally intended for occupationally required uniforms, for example, policemen, but has been extended to cover business suits in some cases.

The first two allowances benefit primarily the higher-income classes, while the latter is a more general form of income tax relief. As experience with the new system has grown, clever tax avoiders have devised new schemes to beat the system.[9] There has been no significant move to expand the list of allowances, probably due to the reduced tax rate. On the other hand, there has not been a significant push to tighten enforcement on these perquisites. The income tax department, for the most part, accepts the declared amount of allowances.

The magnitude of distortions created by these loopholes is not known. Income tax reform should have resulted in a significant adjustment in the compensation package for PAYE employees—away from allowances and toward wages and salaries. However, as pointed out above, the remaining loopholes could dampen the propensity to convert allowances to wages. Unfortunately, there are no readily available data that will enable us to test these hypotheses.

To study the initial compensation adjustments, the project drew out a 5 percent sample of PAYE firms and government agencies and then sampled 20 percent of the employees. An inspection team headed by a senior auditor visited each firm/agency and recorded the wage and allowance particulars for each sampled employee. The data were taken for November 1985 before the reform was enacted, and for May 1986, after most firms had converted to the new system. The results of this sample survey are revealing (Wasylenko and Riddle 1987).

- Before reform, allowances were equivalent to an average of about 22 percent of total compensation. However, the tax reform led to a short-run base expansion of only about 7 percent because some allowances remained untaxed.
- An initial adjustment to the reform was for allowances to migrate to other categories, particularly uniforms.
- During the adjustment period, average wages increased by 19.9 percent, average allowances by 17.4 percent, and inflation by 11.6 percent.

One might discount these results on the grounds that it was too soon to measure the impact of the reform in May 1986. The other possibility is that these data do tell the true story—that allowances will not be brought fully into the base until the loopholes are closed. It does seem clear that base broadening alone does not explain the substantial revenue increase during 1986 and 1987. When the microsimulation analysis was redone with a 1988 sample, the results showed that nearly 60 percent of all allowances were untaxed, and that bringing all allowances into the base would increase revenues by about 13 percent (Bahl and others 1992a).

Administrative dimensions. One tenet of the reform program was that a simpler income tax structure would make administrative improvements possible, and that the reduced rates would remove some of the incentives to evade and avoid taxes. Since nonfiling by the self-employed was estimated to be the primary source of evasion, one test of the reform is whether there has been an increase in the share of total collections from non-PAYE taxpayers. The data in table 5.3 indicate that this has not been the case. For most of the 1980s, collections outside the PAYE sector were equivalent to 7 percent of PAYE collections. The rewards for evasion may have fallen and the avenues for avoidance may have been closed down after 1986, but the self-employed do not appear to have been drawn into the tax net. As of 1991, only 3.8 percent of the estimated 70,000 self-employed workers filed a tax return. This points to enforcement as a continuing bottleneck to a more revenue-productive income tax. From this information, one can only conclude that there is no strong evidence that a major enforcement push on the self-employed has accompanied the structural reform.

Payroll Taxes

Five payroll tax programs use wages as the base for the tax or contribution (Alm and Wasylenko 1991). These include the education tax, the Human Employment and Resource Training Trust Fund, the National Housing Trust, the National Insurance Scheme, and the Civil Service Family Benefits Scheme (CSFBS). The latter three have a tax element, but can be viewed as contribution programs because individuals are entitled to benefits in some proportion to their contribution. The education tax and HEART are surcharges levied on the individual income tax. In total, these payroll tax contributions generated sizable revenues, equivalent to roughly half of individual income tax collections in the mid-1980s.

The Programs

The education tax was established to advance educational goals; however, collections from the tax go into the general fund. The base of the tax is total earnings, that is, there is no ceiling above which income is not taxed, and there is no floor. Until 1989, the employee and the employer were each taxed at the rate of 1 percent of wages, and self-employed persons were taxed at the rate of 1 percent. Education tax revenues were equal to 6.5 percent of individual income tax revenues in fiscal year 1989. In 1989 the rates were increased to 2 percent for individuals and 3 percent for employers, and revenues increased to 3.5 percent of total tax collections (see table 5.2).

The HEART Fund was established in 1982 by the Human Employment and Resource Training Act to develop employee training schemes. Private sector employers whose monthly payroll exceeded J$7,222 were required to pay a 3 percent tax on total gross emoluments of employees. Unlike the education tax, HEART payments are deposited in an account earmarked for use by the Trust and do not go into the general fund. During 1984–85, revenues from the HEART tax were about 4 percent of individual income tax revenues. By law, compensation in the form of allowances should be included in the base for both the education tax and HEART; in practice allowances are not taxed.

The National Housing Trust was established to improve the existing stock of housing. To accomplish this, the Trust imposes a contributory rate on the wages of workers, and then uses these contributions to finance a variety of housing benefit programs. For an employed individual, the legal tax base is gross emoluments; the employee pays 2 percent and the employer 3 percent. The self-employed pay 3 percent of gross earnings, and domestic workers pay 2 percent of gross earnings. Allowances are in principle subject to the contribution, but in practice are excluded. An individual is exempt if annual wages are less than the minimum wage of J$3,120 per year. An employee's contributions entitle him to a variety of benefits, all of which are related to the amount of his contributions. Employee, but not employer, contributions are vested with the employee. In 1988 NHT revenues were equivalent to 20 percent of income tax revenues (excluding revenues from the tax on interest).

The NIS is a funded social security system. Contributors are entitled to a variety of benefits, which are based on past contributions. In 1987 total contributions were just over J$82 million, and the NIS Trust Fund generated income of J$124 million. The contribution rate for PAYE and self-employed workers is 5 percent of weekly gross earnings between J$12 and J$150 (split equally between the employee and the employer in the case of PAYE workers).

The CSFBS is a forced insurance scheme for some government employees. All persons in pensionable offices must contribute. In practice, coverage is limited and less than 25 percent of government workers participated in 1985. A contributor must pay in 4 percent of total salary. It was not possible to obtain detailed data for contributions and earnings under this program. It appears that revenues have grown erratically, and that contributions were only about J$2.2 million in 1983.

Problems and Proposed Reforms

There were (and are) two major problems with structural reform (Alm and Wasylenko 1991; Bahl and others 1992b). The first is the narrow base on which the payroll taxes are levied and consequently the high nominal rates of tax that must be imposed to reach the revenue target. The second problem is that the administration is fragmented and there is little integration or even communication among the five programs. There are five separate recordkeeping systems, each with its own audit program, and (except for the education tax) each is responsible for assessment and monitoring collection efficiency. Compliance with the education tax is monitored by the Revenue Board, but in the past, only two people have been assigned exclusively to this task. NHT and NIS officials have the authority to audit company records and to obtain income tax information, but their compliance staffs focus primarily on the internal consistency of the records. The monitoring division of the HEART Trust Fund looks mainly at the training capacity of participants. And for all of these programs, almost no attention is given to bringing the self-employed into the payroll tax net.

The project made concrete recommendations for a reform of the payroll tax system to accompany reform of the individual income tax. The general thrust was that payroll tax reform should concentrate on simplifying the system, broadening the tax base, and lowering rates, as well as overhauling the administration of these five taxes. As a first step, it was recommended that the education tax be abolished as a separate payroll levy. To protect revenues, if necessary, this would have required an estimated increase in the individual income tax rate from 33.3 to 35 percent (in 1986).

HEART is a more difficult case because one might argue the benefits principle as a justification for financing worker training with an employer tax on private sector payrolls. Alternatively, it might be argued that the benefits of such a program are economywide, which makes it a better candidate for general fund financing than for earmarking. Moreover, the burden of employer payroll taxes may be shifted onto employees, and this would not place the burden for financing HEART where the government wants it. In general, the inclusion of these levies in the general income tax would improve the horizontal equity of the system because the income tax base is more comprehensive than the payroll tax base. It would also improve vertical equity because interest income would be taxed and the standard deduction would not.

The project urged the government to consolidate the administration of the two largest contribution programs, NIS and NHT. Centralized assessment, audit collection, and recordkeeping could lead to substantial reductions in administrative costs and in compliance costs. This consolidation, along with a simplification of the rate and base structure of the two taxes, would make enforcement easier and give officials more time to concentrate on bringing the self-employed within the payroll tax net.

If the base of the payroll taxes could be broadened, the rates could be lowered. Elimination of the ceiling on NIS contributions and taxing allowance income were seen as the most likely routes. With only these base-broadening measures, the combined tax rate on payrolls for the four remaining programs could have been reduced from 11.4 to 10.4 percent of compensation in 1986. Elimination of the education tax would have further reduced this average rate to 8.8 percent. With a stronger enforcement program to increase contributions from the self-employed, the rates could have been dropped even further.

The Postreform Period

Payroll taxes had not been restructured at the time of the income tax reform in 1986, so the rate of payroll tax contribution was frozen in absolute amount at 1985 levels. This had a number of consequences:

- There was a revenue loss because the income tax base grew dramatically during the immediate postreform period.

- The value of benefits to CSFBS, NHT, and NIS enrollees was compromised.
- The overall regressivity of the tax system increased because payroll taxes do not allow a standard deduction. Even those families whose income tax liability was wiped out by the introduction of the standard deduction in 1986, or its increases in 1989 and 1992, had no reduction in their payroll tax liability.

The failure to reform the payroll tax system compromised the spirit of the income tax reform. Payroll taxes are perceived as income taxes in the mind of the Jamaican worker, who reads the amount of deductions on his pay slip every week. However, payroll taxes have separate rate, base, and administrative structures, and consequently create different equity, efficiency, and elasticity impacts. In some cases they offset rather than reinforce the goals of the income tax reform. Most important in this regard is that the payroll tax system makes the taxation of income more complicated, increases the tax burden on the lowest-income workers, and provides a substantial incentive to tax evasion.

Why did the Seaga administration not move to reform the payroll taxes at the time of the 1986 and 1987 income tax reforms, or even later? In the case of the CSFBS, NHT, and NIS contribution programs, part of the answer was that much administrative preparation work needed to be done, and that the benefits side of the program needed a thorough review. Then there were the political issues associated with merging the programs and bringing them under a more uniform and centralized scrutiny.

The education tax and HEART are a different story. Both are pure revenue raisers, and to abolish them would have meant looking elsewhere to make up the lost income. The administration felt that raising the income tax rate above 33.3 percent would have compromised its political acceptability, even though the trade for elimination of the education and HEART taxes would have significantly lowered the overall regressivity of the system. An increase in consumption duty rates was out of the question, and the general consumption tax (GCT) was not introduced until 1991. Finally, the Seaga administration had by now been associated with taxation for nearly four years, and with an election approaching its attention had shifted to the expenditure side of the budget.

The Manley administration was active in the payroll tax area. In 1989 the rate of education tax was increased from the previous levels of 1 percent each on employers and employees, to 3 percent on employers and 2 percent on employees. Revenue tripled between 1988 and 1989, as shown in table 5.2. However, the education tax remains an income tax in disguise. As noted above, it is more regressive than the general income tax because it allows no standard deduction. Moreover, because it increases the rate of tax on payrolls, it may discourage the growth of labor-intensive activities. The fact that the government has not reformed payroll taxes is one of the major failures of the tax reform project.

Company Income Taxation

The company income tax has been a reliable, growth-responsive source of revenue. In the years prior to the reform, however, the structure of this tax has come under scrutiny because of preferential treatment, the absence of any mechanism to adjust taxable profits for inflation, and the separate treatment of a company and its shareholders.

Rate and Base Structure

Before the 1987 reform, the company income tax was levied at a basic rate of 35 percent on chargeable income. In addition, there was an additional company profits tax of 10 percent levied on the same base. Companies were required to withhold tax of 37.5 percent of the value of dividends paid, but could credit these withholdings against ACPT liability. Wozny (1991) estimates that companies that distributed 40 percent of their pretax profits would recover all of the ACPT they had paid on these profits. ACPT credits could be carried forward indefinitely.

The tax base was defined in much the same way as that in other developing countries, with at least the same degree of complexity. Jamaican law permitted deductions for capital allowances rather than book depreciation. Enterprises may claim a prescribed initial allowance[10] and an annual deduction computed on a declining balance basis against historical cost. Inventories are valued using the first in, first out (FIFO) method. Losses can be carried forward for five years but there is no provision for loss carrybacks. Capital gains on the sale of shares listed on the Jamaican Stock Exchange are not taxed.

There are many exceptions to this basic treatment. Financial institutions are taxed under a separate and very complicated regime, as is the case in most countries (Martinez-Vazquez 1991; Brannon 1991). Separate incentive legislation provides a different rate and base of tax for incentive companies (Thirsk 1991), and preferential treatment is given in the taxation of public enterprises (Davies and Grant 1991). Some resident shareholders are given special relief on the taxation of dividends, and dividends paid to nonresidents may be subject to a special withholding tax rate. Before the reform, the company income tax was very complicated and very difficult to administer. Reform removed some but not all of these complications.

The company tax declined from about 20 percent of taxes to 16 percent over the 1980–85 period, even though there was a substantial reduction in the payment from bauxite companies (see table 5.4). The revenue yield from the company tax was unstable through the first half of the 1980s, but the general trend was downward because administrative improvements and collection campaigns could not offset the loss in revenues from the

Table 5.4 Company Income Tax Revenues
(millions of J$)

Fiscal year	*Revenues*	*Percent of total taxes*	*Percent of GDP*
1980	158	19.0	3.3
1981	210	22.5	4.0
1982	242	21.2	4.1
1983	293	20.0	4.2
1984	236	17.0	2.5
1985	364	15.7	3.3
1986	547	19.5	4.1
1987	597	12.4	3.8
1988	488	11.5	2.8
1989	623	12.8	2.9
1990	689	13.3	2.2
1991	1,280	11.2	2.9
1992	1,635	14.2	2.3

Source: Revenue Board and Ministry of Finance data.

bauxite companies and the downturn in the economy. Tanzi's (1987a) comparative analysis for the 1980s shows that Jamaica relies less on the company income tax than do other countries at a similar income level.

Problems and Reform Needs

The view at the outset of the tax reform project in 1983 was that restructuring of the company tax was essential. The rate and base structure were not totally compatible with the economic policies of the new administration, and revisions of other taxes would change the way the company tax fits into the total system. In particular, the tax structure was biased in favor of certain types of investment decisions (for example, debt vs. equity) and certain types of firms (for example, incentive firms and some public enterprises). The major problem was its complexity: this imposed a high compliance cost on payees and a high administration cost on the government.

Complexity and administration. The company tax was not easy to administer because of its special rate and base structure. These problems were magnified by a shortage of skilled staff and outmoded operating procedures. Such administrative difficulties not only raised costs, but also led to arbitrariness in assessing the tax base, and inevitably to some unfairness in the way different firms were treated.

Two good examples of how a complicated structure can compromise administration relate to capital consumption allowances and inventory valuation. The system of capital allowances was quite complex (and was not changed by the 1987 reform). There are numerous schedules for different asset types, special allowances for different industries, and incentive laws that provided special treatment to both favored industries and favored types of assets. Income tax officers spent too much time classifying taxpayers at the cost of too little time on book audits, with the result that monitoring was lax. Compliance costs, in one form or another, are also raised by such a complicated system. Large enterprises make use of accounting firms to assist them in compliance, but smaller enterprises can less easily take advantage of the available compliance options. This introduced an unintended but potentially important nonneutrality into the system.

The other example has to do with valuing inventories. The law required that inventory be valued at the lower of cost or market value, and most firms used the FIFO method for determining the cost of their sales. However, some large firms had shifted to the last in, first out (LIFO) method, which had neither been sanctioned in the courts nor approved by the commissioner. Others avail themselves of even more advantageous approaches, such as writing off stocks that are over a certain age and excluding the proceeds of their sales from chargeable income. These practices went unchallenged because the Income Tax Department lacked an effective audit branch.

Inflation. Brisk inflation during the prereform period in concert with the present tax structure drove up real company tax rates, influenced investment choices, and provided additional incentives for tax avoidance and evasion. The law contained no provisions for infiation adjustments, except for the crude 20 to 90 percent approximation of an initial first-year allowance.

Under inflationary conditions this approach led to understated capital consumption (to a differential degree for assets of different lives) and FIFO accounting understated the cost of goods sold. Both practices caused profits to be overstated and dampened the rate of investment. Wozny (1991) demonstrated that the effective tax rate on an equity-financed capital investment in a basic industry increased from 42 to 60 percent when the inflation rate was 10 percent higher.[11]

Because the effects of inflation may also work in the direction of overstating profits, firms were given an incentive to adjust their financing structures. Inflation caused a decline in the real value of corporate debt, which resulted in untaxed gains that varied among companies according to the degree to which they issued debt. The deductibility of interest expenses under the previous system allowed a firm to compensate for the fact that capital allowances were not indexed, by substituting debt-for-equity financing of its capital assets. In the example of the capital investment presented above, the effective tax rate would actually have been lower with a 10 percent higher inflation rate if 80 percent of the investment had been debt financed.

Finally, the availability of three important avenues of tax avoidance—the preferential tax treatment of incentive activities, interest income, and capital gains—encouraged enterprises to undertake tax arbitrage, or transactions whose sole purpose was to achieve a reduction in tax liability. Avoidance techniques observed were revaluation and sale of assets with leaseback arrangements, revaluation and sale of assets with a distribution of the (nontaxable) proceeds to shareholders, and the leasing of capital equipment by incentive firms to affiliated nonincentive firms.

Debt-equity choice. Like most countries in the world, Jamaica taxes distributed and undistributed corporate profits under the company tax and dividend income under the individual income tax. The method by which the prereform system provided relief from double taxation of dividends (the ACPT) was complex and only partial. Interest expenses were deductible and, before the 1986 individual income tax reform, interest income received by individuals was exempt. This tax structure was widely criticized on grounds that it biased investment decisions in favor of debt and against equity investments. This, it was argued, led to thin capitalization of corporations, inhibited the development of the domestic capital market, and created horizontal inequities, that is, investors paid different amounts of tax depending on their portfolio composition.

The 1984–85 tax system was in fact horizontally inequitable and did favor debt-financed investment but this had nothing to do with a lack of dividend relief. It was due to the fact that borrowers were able to deduct nominal interest payments from their gross book income, whereas true economic income would have been computed by deducting only payments of real interest (Wozny 1991). The tax penalty on dividends that existed under the 1984–85 tax system was due to the overly favorable treatment of retained earnings, not to an overtaxation of distributed earnings.

Perhaps because the impact of the system had not been fully understood, there had long been a movement to remove the company tax bias between debt and equity finance. The focus in these proposals was on removing the double taxation of income. The Private Sector Organization of Jamaica (PSOJ) called for integration in the 1970s, but predictably did not propose that the undertaxation of capital gains be corrected. In 1979 the IMF studied the company tax and concluded not only that the existing system was effective in inducing corporate retention but that this was beneficial. The government agreed with this conclusion and it did not change the basic structure of tax, but it did lower the ACPT from 15 to 10 percent. The possibility of indexing depreciation or interest deductions was not seriously considered.

Even with respect to the narrow question of the double taxation of dividends, it is not clear how much economic loss resulted from distortions introduced by the system. One could take the position that these price effects either were not significant or that they were offset by some other distortion. With respect to the latter, consider that the bias in favor of debt was to some extent offset by the absence of a capital gains tax on securities traded on the Jamaican exchange. Moreover, all dividend recipients were not being subjected to double taxation. Less than 10 percent of the self-employed—a large proportion of those expected to face marginal tax rates in excess of the withholding rate of 37.5 percent—even filed a return.

The integration issue was raised again in connection with the comprehensive tax reform project. The Seaga economic program called for elimination of those features of the tax system that discouraged investment and called for the development of a domestic capital market. Moreover, the 1986 individual income tax reform forced reconsideration of the issue because the company tax rate and the withholding rate of 37.5 percent on dividends were now above the maximum individual rate of 33.3 percent.

The Reform Program

The broad objectives of the company tax reform called for by the project and stated by the prime minister (Revenue Board 1985) were to simplify the present rate and base structure, remove the differential tax treatment of debt and equity finance, and provide effective investment incentives. The Tax Reform Committee took this charge to mean developing a simpler system and eradicating disincentives to investment and biases against equity finance.

There were several important constraints. An initial challenge was to assure revenue neutrality. This requirement was later relaxed, but it was clear that any proposal that carried too great a revenue loss would have no chance. Second, the new system would have to work within the existing administrative capabilities of the income tax department. Administrative improvements would come with a simpler, more rational system and with a better training program for the tax administration service, but these improvements would not be available immediately. The administrative constraint ruled out reforms such as inflation indexing of capital allowances. Third, the reformed system of taxing companies and dividends would have to fit the new individual income tax structure. This almost certainly meant a general rate reduction in the company tax. Fourth, the reform would have to be sensitive to the politics of detaxing the business sector. This ruled out disallowing deductibility of interest expenses or exempting dividend income from individual income tax.

Proposed changes. The most important component of the proposed reform was to reduce the tax rate from 45 percent (including ACPT) to 33.3 percent. The project and the Tax Reform Committee further recommended that dividend distributions to residents be exempted from individual income tax. A strong argument in favor of this proposal was that the system would be greatly simplified, and thus easier to assess and monitor. There would be a 33.3 percent tax on companies and no withholding on the personal tax liability of their shareholders.

More important, this reform program would reduce the tax incentive to employ debt. Along with the rate reduction and the proposed elimination of the transfer tax on capital gains arising from the transfer of corporate shares, it would all but eliminate the tax disincentive to distribute earnings. Both distributed and retained corporate income would be taxed at the same rate as any other income. In other words, full tax integration would be achieved for resident Jamaicans. The new flat rate individual income tax would permit this without a complicated imputation and credit mechanism. The strongest argument in favor of either integration or dividend relief is that higher rates of corporate profit distribution would improve the allocation of capital by subjecting investment decisions to the test of the market.

It was also proposed that the withholding tax on nonresidents be retained, and that branch and subsidiary firms be placed on a comparable tax basis. The magnitude of the basic rate reduction meant that the overall tax borne by foreign investors would be lower than it had been under the existing system and lower than the taxes levied by Jamaica's closest competitors in the region. Most foreign investors would receive a real tax benefit from the elimination of the withholding tax (it would not simply have resulted in an offsetting increase in their home country tax liabilities), but the line between investment attraction and revenue sacrifice had to be drawn somewhere.[12] It was decided that the greatest efficiency gains would be achieved by lowering the basic corporate rate.

To reduce the bias against risk taking a three-year carry-back of operating losses and an unlimited carry-forward was proposed. This proposal to consolidate returns was rejected because of its bias in favor of larger established enterprises, the administrative complexities involved, and the implied revenue loss.

The proposed reform was not compatible with existing industrial policies. A general rate reduction is not targeted, that is, the lower rate is available to all firms, and not just to those favored by industrial policy. This is inconsistent with the approach taken under the incentive legislation, where only approved firms are eligible for tax subsidies. The spirit of the targeting approach would have been to invest an amount equivalent to the company tax rate reductions in tax holidays for an expanded list of approved firms.

Despite its view that a targeted industrial policy was ill advised, the project held that scrapping incentive programs would be ill advised. Most competitor countries give comparable subsidy packages, and withdrawal by Jamaica could be viewed as a less hospitable business climate. Jamaica's political climate was considered risky by some investors, and its economy had only recently shown signs of reversing a long-term decline. It was not considered a good time to take actions that might shake investor confidence.

A middle ground articulated by Thirsk (1991) called for restructuring the incentives program; namely, to adjust other fiscal policies to strengthen rather than offset the attraction of Jamaican incentives. Among the possibilities were the following:

- Tax incentives could target investors whose home countries have negotiated tax treaties containing tax sparing clauses. In countries without such clauses, the United States, for example, tax incentives may represent little more than a transfer to the foreign treasury.

- Tax incentives could be replaced with expenditure subsidies for infrastructure development. This could be especially effective for investors whose home countries deny tax sparing.
- Criteria for choosing eligible firms could be redesigned to remove the bias in favor of capital-intensive activities.
- Income tax returns of all incentive firms could be audited and the effectiveness of the program continuously monitored.

Adopted changes. The tax authorities adopted the recommendations to reduce the company tax rate to 33.3 percent, abolish the ACPT, and retain the withholding tax on dividend payments to nonresidents. Branch and subsidiary firms were given equal tax treatment, subject to administrative problems of valuing branch profits, and an unlimited loss carry-forward was adopted.

They rejected the proposal to exempt dividends from personal tax liability and decided instead on a separate entity approach whereby company profits and dividends would each be taxed at 33.3 percent, the latter under a withholding system. In doing so the government passed by the opportunity to fully (and simply) integrate the income tax.

Why would the government forgo the opportunity to integrate the income tax? One reason given was that the government was in a crucial stage of its negotiations with the IMF and was under pressure to minimize the revenue cost of the reform package. The full program would have cost an estimated J$98 million in fiscal year 1987, an amount equivalent to less than 20 percent of company tax revenues, and about 2.6 percent of total tax revenues. This would not have been a big revenue loss. A more likely explanation is political; namely, that the JLP did not want to be perceived as the party of the "big man." The government was still being criticized for the taxation of interest income introduced along with steep cuts in the marginal personal income tax rates the year before. There also was the problem of explaining the difficult concept of integration to the public such that they fully understood that the exemption of dividend income and the taxation of interest income represented equivalent treatment.

Wozny (1991) modeled the economic impact of the reform, with specific concern for the integration issue. Corporate income would bear a lower overall tax burden than it had under the prereform system but, because the tax burdens on other forms of income would have been reduced by a greater degree, corporate-source income would still be relatively disadvantaged, especially when distributed. The end result of this discrimination would be a lower supply of funds for equity investments, compared to what would have existed if the full integration proposal had been adopted. The 1987 tax system would also discourage the distribution of earnings to resident shareholders to a greater degree than the prereform system. Wozny (1991) estimated that this reform would lead to a reduction in the payout rates of widely held companies from about 0.32 to between 0.23 and 0.26.

The lowering of the corporate tax rate from 45 to 33.3 percent in 1987 increased the posttax return on corporate investment, and stimulated growth in the sector and the demand for corporate equities. However, the imposition of a higher tax penalty on dividends impeded the flow of investible funds out of established, widely held companies and into the hands of investors who presumably would have found the highest returns available for these funds. Even though both the corporate and personal tax rates are lower than they were before 1986, the retention of full double taxation of distributed corporate income is inconsistent with the government's long-range economic strategy, which calls for a reallocation of resources out of the low-return import substitution sector into the higher-return export promotion sector.

Revenue performance. The short-run revenue loss associated with the rate reduction was realized in fiscal year 1988, during the first full year of the reform. Revenues actually increased 3 percent in nominal terms, which actually translated into a reduction in the effective rate of taxation from 3.8 percent of GDP in 1987 to 2.8 percent of GDP in 1988 (table 5.4). The effective rate has remained below 3 percent since that time. The company income tax after reform is as cyclical as before, but appears to play less of a role as a revenue raiser. Apparently, the aggregate effect of the income tax reform was to shift the onus of payment from companies to individuals.

Sustainability. The company tax reform and the objectives of a lower tax rate, or simplification of the tax structure, appears to be sustainable in that neither the Seaga nor the Manley administrations proposed changes. Implicitly, this suggests some degree of support for an industrial policy that provides a generally lower tax rate for all companies, rather than lower rates for some companies at the expense of others. There was one bit of backsliding in a ministry paper issued by the Seaga administration in May 1988 in a proposal to provide a tax credit to any firm that expanded its equity base from its own profits through the issuance of bonus shares. This tax credit would have been difficult to monitor and its revenue cost would have been paid by other taxpayers in the form of a higher effective tax rate. In general, this proposal, which was eventually rejected, went against the spirit of the reform program in that it narrowed the tax base, implied an increase in the nominal rate for nonbenefiting firms, and complicated the tax structure.

Indirect Taxes

The history of changes in the structure of indirect taxes in Jamaica has been one of piecemeal adjustments to cover annual revenue shortfalls. As a result, underlying problems within the system have persisted or even worsened. The conclusion reached by virtually all who have studied the system was that it should be replaced with a general sales tax (Due 1991a). The project and the Tax Reform Committee reached the same conclusion (Bird 1991a).

The prereform indirect tax system included five separate taxes (Due 1991b). From a revenue standpoint, the most important was the consumption duty levied on the value of imported and domestically produced goods and collected at the import and the manufacturing stage. Excise taxes on alcoholic beverages, cigarettes, and petroleum accounted for significant revenues, but the retail sales tax yielded only a small amount. Domestic revenue from these taxes represented about 40 percent of all indirect tax revenues, 20 percent of all tax revenue, and 5 percent of GDP in 1983. Two other taxes levied on the import base were the customs duty and stamp duty on inward customs warrants. The customs duty proper was a relatively small revenue source by international standards (less than 10 percent of revenues during the prereform period), primarily because of Jamaica's membership in CARICOM. However, with significant rate increases beginning in 1984, the stamp duty grew to become a major fiscal instrument.

Problems

The Jamaican economy simply outgrew its indirect tax system. The same laws, regulations, and structure designed forty years earlier for the duty on rum and a few other items were now unable to also cover manufacturing activities. Cnossen (1991) described the situation well: "As Jamaica's economy has grown more complex, the administration of its indirect tax system, which is largely based on production checks, has become more cumbersome, impeding the free functioning of business and trade. The inherently fragmented nature of the present indirect system's coverage, its multirate structure, and its complexity may have undesirable economic effects. Its distributional effects are largely undeterminable."

Complexity. Administrative problems with the indirect tax system were in part due to its complexity. The five taxes were levied under separate legislative acts, administered by different divisions within the Customs and Excise Department, had different licensing and return requirements and separate recordkeeping systems. Taxable bases were not the same, and the rate schedules were a mixture of ad valorem and specific rates with many fine gradations. Because of this complexity tax officials spent far too much time classifying commodities for purposes of selecting the proper rate. Furthermore, the base was not clearly defined in either the law or the regulations, so tax officials often had to make a notional assessment of the taxable value of an object. The result was that the tax administration service, already understaffed, had much less time to assure compliance.

Efficiency. Jamaica's system of indirect taxation did not foster tax neutrality. It distorted the relative prices of consumer goods from what they would have been in the absence of taxation, it gave enterprises an incentive to alter their methods of doing business, and it offered inefficient protection to domestic producers. All of these concerns could be traced to a single underlying problem: the tax base was too narrow. Thus the need for revenue forced high effective tax rates on commodities where assessment and collection were relatively easy. Less than 20 percent of the final consumption of services and less than one-third of gross manufacturing sales were taxed.[13] Domestic value added coverage was thin because the consumption duty was essentially a manufacturer's sales tax and did not reach the distributive sector, small firms, or most of the service sector. Excluding the traditional excise taxes, the average effective rate of indirect taxation on those commodities actually in the base was 3.6 percent in 1983.

Perhaps more of a concern were the distortions potentially introduced by the consumption duty. Because the taxes are levied at the manufacturer and import stage, differential wholesale and retail margins are not recognized. As a consequence, the final tax burden on consumers varied by commodity in unintended ways. Using an input-output table for 1983, Bird (1991c) estimated that the average (pyramided) effective tax rate on inputs was equivalent to 2.4 percent of the gross value of manufacturing output, compared to an average tax rate of 7.8 percent on total manufacturing output. Since the rate of import taxation on raw materials and capital goods was increased substantially after 1983, it is reasonable to expect that the proportion of hidden tax increased.

To what extent does the Jamaican indirect tax system protect domestic producers from foreign competition? The system is not neutral in its treatment of internationally traded and domestically produced goods. Although a large proportion of imported goods previously entered the country tax free, the stamp duty on imported goods was levied at a high rate. Bird (1991c) estimates that imports were taxed at a rate 19 percent higher than that for domestic production in 1983–84. Moreover, consumer durables and capital goods were taxed at significantly higher rates than were other imports. With the shift in revenue reliance from consumption duty (which does tax imported and domestically produced goods at the same rate) to the import stamp duty, the rate of protection increased. To the extent that the tax incentive program favored domestic producers with lower rates for raw materials or outright exemption for intermediate goods, it accentuated this protection.

Inelasticity. The revenue-income elasticity of indirect taxes in Jamaica was lower than in most countries. One reason is because the tax base excluded much of the rapidly growing service sector and about 80 percent of all imports. Second, the tax rate structure had not fully shifted from a specific basis to an ad valorem basis, and so was not as automatically responsive to income and price level growth as would otherwise have been the case. Bird (1991c) estimated that over the 1978–84 period, the buoyancy of all indirect taxes approximated unity, while for the consumption duty it was 0.78. Were it not for the discretionary rate and base increases for import stamp duties and traditional excises, indirect revenue growth would not have kept pace with GNP.

Administration. The indirect tax system was beset by serious administration problems. As noted above, some of these problems were traceable to the complexity of the system and could be addressed by nothing short of a restructuring of the tax. Beyond this, however, there were important shortcomings in the areas of personnel, recordkeeping, and procedures that would compromise the effective operation of even the best-designed general sales tax.

The major problem was the shortage of qualified staff. Under the prereform system, most of the inspectors lacked sufficient training to audit effectively. The inspection program was also burdened by antiquated operating procedures and the virtual absence of an information system. The ratio of inspectors to accounts was acceptable, but the frequent visits to enterprises were not true audits. Due (1991b) reported that "there is no system of priorities for inspection nor guidelines for the inspectors, no system for them to report their findings, and little supervision." Even in the case of the traditional excises, where administration is relatively more manageable and physical methods of control are used, there was evidence that procedures were inadequate and that qualified staff were in short supply. For example, Cnossen (1991) reported that "consumption duty supervision of the largest beer factory in Jamaica is exercised by only one junior officer, largely on the basis of the brewing book."

Revenue Performance

After 1985 the reliance on indirect taxes increased. By 1988, indirect taxes as a percent of GDP had climbed to over 13 percent, and indirect taxes accounted for 56 percent of all taxes (table 5.2). This represented substantial increases in the import stamp duty and in the consumption duty—in the latter case, particularly on alcoholic beverages, fuels, tobacco, and utilities. As the tariff reform began to take hold after 1988, the revenue from the import stamp duty began to fall. As table 5.5 shows, the split between the share of indirect taxes coming from imports and local goods has not changed markedly since the mid-1980s. The share of indirect taxes in GDP remains above that of the prereform period.

Proposed Reform

The reforms proposed in 1986 and 1987 were to make the indirect tax system more neutral with respect to economic choices, less arbitrary in the way it treats similarly situated individuals and firms, more closely tied to the performance of the economy and less to annual discretionary actions, and more administratively efficient. The major constraints in designing such a reform program were revenue neutrality and protection against increased tax burdens on low-income Jamaicans.

The project proposed a GCT to replace the existing domestic indirect taxes. The proposed GCT had the VAT feature of allowing credit for taxes

Table 5.5 Distribution of Indirect Taxes
(percent)

	Taxes on imports			Taxes on local products	
Fiscal year	Customs	Stamp	Consumption duty/GCT	Consumption duty, GCT, SCT, excise taxes	Stamp
1985	18.9	12.3	9.8	50.9	8.0
1986	17.1	26.4	6.7	41.4	8.2
1987	13.0	21.9	5.7	46.1	14.2
1988	16.5	22.1	7.1	44.1	10.2
1989	23.0	17.0	7.9	41.8	10.2
1990	24.2	13.0	8.9	43.1	13.1
1991	25.7	15.1	9.3	36.7	13.2
1992	29.3	7.0	7.2	43.6	12.9
1993	26.8	1.7	15.6	45.6	10.3

Source: Revenue Board and Ministry of Finance data.

paid on inputs and was to be levied on importers, manufacturers, and large distributors. It was broad based,[14] and had a simple rate structure—most goods would be covered under a single general rate and some would be subject to a single luxury rate. The project also supported the possibility of bringing the import stamp duty into the GCT, but a higher GCT tax rate would be necessary, at least temporarily, to protect revenues (Bird 1991b).

The introduction of the GCT with its VAT feature would provide exporters (who would be zero rated) with an automatic rebate for taxes paid on inputs. This was preferred to giving rebates to exporters based on the estimated import content of their exported output, or on the value of exports. The proposed basic rate structure of the GCT—a single basic rate and (possibly) a luxury rate—was consistent with the goals of giving equal tax treatment to imported and domestically produced goods, and discouraging nonproductive uses of foreign exchange. Finally, the proposed GCT fit the government's policy of broadening the indirect tax base.

It was proposed that taxes on alcoholic beverages, cigarettes, and petroleum products remain unchanged in order to protect revenue and minimize disruptions associated with the administrative transition and the potential short-run revenue losses. Accordingly, only about 30 percent of collections from the consumption duty, retail sales tax, and excise duty would initially come under the GCT. The project recommended these taxes be included eventually.

Despite the recommendations of the project and the Tax Reform Committee, the GCT implementation was delayed for a number of reasons: the proper administrative machinery was not in place, inspectors had to be trained, firms registered, and a recordkeeping system designed. An organizational structure for assessment was needed, collections and appeals had to be decided on and put in place, and a taxpayer awareness program had to be completed. These are important issues and the world is full of instances where tax reforms have not succeeded, precisely because an administrative infrastructure was not in place before the tax became effective. Some observers feared that the GCT would be inflationary—or at least jolt prices with a onetime increase—and affect the market prices paid by lower-income families. This fear was compounded by the reasonable (and with hindsight, well-founded) expectation that a devaluation was probably not far off. With an election around the corner, the administration was especially concerned about any public policy that would potentially increase the price of consumer goods.

Unfavorable revenue consequences was another important reason why the implementation of the GCT was delayed. The project, the Tax Reform Committee, and the Seaga administration believed that there would be a transitional revenue loss because of administrative difficulties, and the government felt that it would be politically dangerous to set a high rate (recommended by the project) to provide an effective margin of safety. Moreover, the World Bank, as part of their lending program, had outlined a tariff reform that would significantly reduce revenues from customs and the import stamp duty over a four-year period.

Three other important considerations slowed the implementation of the GCT. The proposed export rebate scheme described above may have pleased exporters less than the previous drawback system. This kept pressure off the government to find a way to simplify the indirect tax system and to credit producers for taxes already paid on inputs. Another issue was the relaxed external pressure to complete the tax reform. Once the World Bank reached agreement with the government on the tariff reform and the export rebate, it withdrew its condition that the GCT be implemented in the following year. Finally, the Seaga administration had grown weary of treating tax policy as its political platform, and was ready to move on to more popular expenditure programs.

Actual Reforms

Despite these reasons for delay, there seems never to have been any question that there eventually would be a GCT. By the end of 1989, the Manley administration had announced its intention to enact the GCT, and the tax became effective October 1991.

The GCT was introduced with a general rate of 10 percent. The rate was lower than the project estimated as a revenue-neutral rate because the (former) stamp duty was merged with the GCT.

No special rate was set for the taxation of services. The taxation of items at a single rate under the special consumption tax was not enacted. Although administratively the taxation of these items is listed under a heading of special consumption taxes, the rate structure for these items was not altered from the former excises taxes. The final law allowed for the following categories of exemptions and zero ratings: agriculture inputs, books and newspapers for education, expenditures by diplomatic organizations, export goods and related services, foodstuffs, government (bauxite treated like the government), health supplies, miscellaneous, and religious organizations.

The government opted to allow a transitional stock-in-trade credit scheme, with credit given as an allowance for preexisting indirect taxes paid by establishments on material inputs held as inventory by firms at the onset of the GCT. The law became effective October 22, 1991. In 1993 the tax rate was increased to 12.5 percent.

Impacts

It is too early to tell whether the long-term yield of the GCT will meet expectations, but early indications are of substantial revenue productivity. It was expected that transition would have reduced the revenue productivity of the GCT for at least a five years by comparison with the consumption duty. These transition issues include the choice of a 10 percent rate, the adaptation of the administration, and the transition credits. However, as may be seen from table 5.2, the revenue productivity of the GCT has been strong, and by 1993, it accounted for 31 percent of revenues (including the special consumption tax) as compared with 34 percent for the combined consumption duty, excise, and stamp duty in 1991.[15]

Property Tax

The political sensitivity to property tax policy is far out of proportion with the amount of revenue raised. Its problems are not primarily structural.[16] In fact, the tax system itself is one that, properly implemented, could serve as a model for other countries. However, for a number of years, the government has been unwilling to levy the tax according to the intention of the law, with the result that it yields negligible revenue and distributes its burden unfairly.

The System

The base of the property tax is the unimproved market value of land, that is, the value of land as it would be if there were no structures on the site.[17] The valuation roll is supposedly updated every three to five years, though in practice updating is much less frequent. There are three important classes of preferential treatment: agricultural properties, hotel properties, and low-value properties. In effect, Jamaica has a classified property tax with lower rate schedules for agricultural and hotel properties.

About 60 percent of Jamaica's 550,000 parcels have land values less than J$2,000 (about US$500 in 1984) and are subject to a nominal property tax charge of J$5 per year. The remaining properties are taxed according to a progressive rate structure in terms of land value, rising to 4.5 percent on a site value of J$50,000.

Revenue from the property tax has fallen over the past decade from a little over 5 percent of total revenues to about 2 percent (table 5.2). This decline has occurred principally because of the failure of government to revalue on a regular basis. The property tax is a central government levy, and local governments have no influence over the rate or base.

Problems

There were two problems with the property tax. The first grew out of the failure of the government to bring in a new valuation roll between 1974 and 1986, and the second arose from inadequacies in the day-to-day administration of the tax. More debatable as problems are whether the progressive rate structure discouraged compliance and whether the improved value of property (land and structures) would be a more appropriate base.

Before the 1986 reform, the tax base was the 1974 value of sites. Because this base was fixed for over a decade, the yield from the property tax had fallen to a negligible revenue position. This lack of buoyancy created a number of problems:

- The low revenue yield put that much more pressure on other taxes to carry the revenue load.
- Landowners were undertaxed relative to labor and owners of capital. This discriminatory practice fit neither the economic nor the social policy of the government.
- Landowners were taxed according to the 1974 values. Because the pattern of land values had changed dramatically since 1974, distribution of tax burdens had become unfair.

The problems with the property tax were not limited to the delay in bringing in a new roll. There had been a policy of derating certain properties, most notably agricultural and hotel properties. Such tax incentives defeated the very purpose of a site value property tax, namely, to stimulate owners to use their land well. While the incentive benefits of derating are not at all clear, the revenue cost of derating is substantial—about 30 percent of revenues collected.

Finally, there were problems with the administration of the property tax. Typically, one thinks of four areas of property tax administration: identification of parcels, recordkeeping, valuation, and collections. Identification and valuation are less of a problem, and a 1984 valuation roll is now in place. Though constructing the new roll was difficult and showed the shortage of valuation expertise in the public sector, the more serious administrative problems had to do with recordkeeping, information flow, and collections. The following situation existed at the time the tax project made its recommendations:

- The collection rate for the property tax was only about 60 percent.
- The collection cost was about 12 percent of revenues collected.
- There were no reliable records on derated properties, that is, no list of the number that received preferential treatment or the nature of the preference.

- About one-third of the sites on the roll did not have any land use coding.
- Total property tax arrears were equivalent to over two years' revenue yield.

Reform Options and Recommendations

The project and the Tax Reform Committee considered three structural options for reforming the property tax: (1) expanding the revenue importance of the property tax, (2) changing the base from land to capital value, and (3) changing the rate structure to be flat rather than progressive.

Revenue importance. The value of taxable land increased from about J$2 billion in 1974 to over J$5 billion in 1984. Application of the existing rate structure to the new valuation roll would have moved the revenue importance of the property tax to the position it held in the late 1970s, when it accounted for about 5 percent of total revenues. If this increase in property tax was to be accomplished at the expense of an equal yield reduction in the income tax, the economy could benefit from efficiency and equity gains. The conclusion from Follain and Miyake's (1991, p. 654) general equilibrium analysis of such a tradeoff demonstrated the gains to labor and the losses to landowners from such a switch.

Land versus capital value. The major structural reform considered was a shift from a land value tax (LVT) to a capital value tax (CVT). The popular appeal of such a change is apparent—large buildings and high value residences indicate more taxable capacity. To tax these structures would generate revenue and would be equitable, since these properties are in the hands of higher-income individuals and businesses. Another consideration is that a developing economy may not want the development intensity that a LVT brings. On the other side are arguments that land value taxes do not impose an excess burden, that they capture the windfall gains of urban development from landowners, and that they are an effective way to tax those who accumulate wealth in the form of land. Of all the arguments in favor of the LVT the most persuasive is that it is administratively easier to implement because structures do not have to be valued. The Follain and Miyake (1991) analysis confirms the land development intensity advantages for the LVT, but also points out that "to oversimplify, the CVT generates a process with smaller buildings that house more labor and machines. The LVT encourages the construction of larger plants which house fewer workers and fewer machines."

Rate structure. Property tax rate structures are often progressive in developing countries to take account of the fact that the benefits from public investments and services are skewed toward those with higher property values (Bahl and Linn 1992, chapter 4). Moreover, a progressive rate permits lighter taxation of a larger number of holdings and concentrates the revenue collection on fewer properties where collection is administratively easier. Jamaican officials also felt that the tax supported the objective of forcing large landholdings into active use. Finally, a progressive rate can offset distortions in favor of holding land when the income tax base does not include capital gains.

These are strong arguments, but not totally persuasive. If a progressive rate is meant to force large landholdings into active use, then the base of the tax should be the aggregation of all properties owned by an individual, rather than "parcel progressivity." However, aggregation provides an incentive for taxpayers to use costly methods to avoid tax payment, and burdens an already inefficient enforcement system. This problem would be avoided under a flat rate tax. Moreover, a lower, single rate would avoid some of the shock of switching to the new tax roll and lessen political opposition to the reform. Finally, the flat rate would be more in keeping with the spirit of the proposed reform of the income and indirect tax systems—to broaden the base and to levy lower, flatter rate taxes.

Recommendations. The major recommendations of the project were as follows:

- The land value base should be maintained.
- The revenue from the property tax should be three times the present level. This would restore the property tax to the revenue level it would have achieved if its base had been kept up to date.
- The 1984 valuation roll should be adopted, and a flat rate should be applied to all land with value in excess of J$6,000. This would keep about 60 percent of properties off the roll. To avoid tax shock, this new system should be phased in gradually by even increments over a three-year period.
- Agricultural and hotel derating should be eliminated or substantially cut back.
- The administrative system should be completely overhauled, with particular emphasis on (1) improving collection procedures and recordkeeping, (2) developing a sales data bank that would permit assessment-sales ratio studies, (3) indexing land values to update the valuation roll between general revaluations, and (4) monitoring relief and derating in a more systematic way.

The Tax Reform Committee generally accepted these recommendations in proposing a tripling of the revenues to be accompanied by an equal amount reduction in the income tax, bringing in the 1984 roll, the retention of the land value basis, and a three-year period to phase in the reform. The committee was silent on the issue of derating, and held to the present progressive rate structure.

Adopted reforms. The government followed the recommendations to bring in the new land value base and to roll back the tax rates to minimize the amount of tax shock. The proposal to switch to a capital value base was dropped. The derating program for agricultural and hotel properties was continued. On the subject of future revaluation, the government committed itself to work on an indexing system, but little progress has yet been made.

The reform program enacted in 1986 has been a mixed success. Revenue growth did increase significantly after 1987, thanks to significant administrative efforts. However, the government froze the rates at the 1986 level, probably guaranteeing that the property tax would remain at 2 percent of revenues or less. Property taxes were estimated at about 1.2 percent of total taxes in 1990, and had shrunk to 0.5 percent by 1993.

In all countries, property taxation draws criticism from the public that is far out of proportion to the tax burden involved. This was also the case in Jamaica. One possible reason for the opposition is that it is a national tax and therefore the financing link to specific local services cannot be easily seen. Another is that the government was unsuccessful in convincing voters that one price of the popular income tax reduction was an increase in the property tax.

There have been no significant changes in the structure of the property tax since 1987. A new valuation roll was due to be put in place by the end of 1991, but public opposition was strong enough to delay this action.

Conclusions

Although the reform is not yet complete, we might ask whether it has begun to address and resolve three problems: high taxes, a tax system that was affecting economic choices in inefficient ways, and weak tax administration.

Are Taxes Still Too High?

At the beginning of the tax reform project, the ratio of taxes to GDP was 23.9 percent, a high ratio by comparison to other developing countries. In 1993 the ratio is estimated to be 25.7 percent. Does this increase suggest that the tax reform missed one of its goals? In part, the answer is yes and raises some questions. Why has the tax ratio risen? Why was a supposedly "revenue-neutral" individual income tax reform not neutral? In fact, revenue neutrality was achieved in two respects. First, the initial-year yield of the reformed system did not exceed the projections of the yield of the prereform system, and first-year revenue neutrality was the mandate given to the project. Second, there is a sense in which the long-run performance has been revenue neutral. A backcasting study done by the project indicated that, with some improvements in enforcement, revenues from the prereform system could also have grown to these levels.

It should also be noted that the increase in the overall tax ratio overstates the degree to which the reform itself stimulated revenue growth. Changes in the structure of taxes on consumption of locally produced and imported goods were not part of the reform program until 1991, but rose from 9.2 percent of GDP in 1985 to 11.1 percent in 1990.

The other half of the high-taxes story is that the tax bases were too narrow. Consequently, nominal tax rates had to be very high to meet revenue targets and the income and consumption in the net were subject to exorbitant rates. The strategy of the reform was to broaden the bases and flatten out (lower) the rates. Here the reform gets good marks.

- Interest income and a substantial portion of fringe benefits were brought into the individual income tax base; and the rate was lowered first to 33.3 percent and later to 25 percent.
- The proportion of imports taxed has increased dramatically, and the top tariff rate was lowered to 30 percent.
- The 1984 roll of property value (and under present plans a 1991 roll) replaced the 1974 roll, thereby expanding the base of taxed property values.
- The company income tax rate was reduced to 33.3 percent.

Other proposals to broaden the tax base have not yet been acted on. The payroll tax regime has not been changed. Transfer taxes have not been amended to reach capital gains. Broadening the income tax base through greater taxation of the self-employed has not yet happened to any great extent, and certain allowances remain nontaxable by law or practice. Finally, the GCT did significantly broaden the tax base by including sales down to the level of large distributors, but it provided for an extensive list of exemptions and zero-rated goods.

This said, however, the tax base clearly is broader than in the prereform period. Taxable individual income is a significantly greater share of GDP, as is company income and taxable property value. Some of this growth is attributable to the stronger performance of the economy rather than to the tax reform per se, but it is doubtful that the prereform system would have captured this growth as well as the new system has.

Allocative Effects

The tax reform set out to remove some of the more obvious price distortions. In many ways it has been successful in cleaning up the tax structure and moving the system toward neutrality.

- The individual income tax was converted to a flat rate structure and previously nontaxable fringe benefits were brought into the base.
- The preferential treatment of overtime earnings was eliminated.
- The company income tax rate was lowered and the ACPT was eliminated.
- An indefinite loss carry-forward was introduced.
- Interest income was brought into the individual income tax base.
- A greater share of imports was brought into the tax base.

In other cases, there was no move toward neutrality. Dividends remain as before, taxed under both the company and individual income tax. Capital gains remain untaxed. The heavy employee and employer tax on payrolls has not been changed. The tax treatment of exports does not appear to provide the intended relief for taxes paid on inputs, since it involves an export rebate based on the value of exports rather than the tax content of inputs. Depreciation rates still do not reflect economic depreciation. Finally, the system of tax incentives, awarded to companies on a case-by-case basis, remains in place. A reasonable statement may be that the reform has gone far and in the right direction, but that many important distortions persist.

Economic Performance

The bottom line in a tax reform that stresses neutrality is whether the economy somehow performed better than in the prereform period. But it is no easy matter to separate the effects of tax reform from everything else—a foreign exchange crisis, a change in administration, a devastating hurricane, and the recession and recovery of the U.S. economy.

The introduction of the tax reform program in 1986 did coincide with a stronger growth in the economy. There have been positive rates of GDP growth in every year although performance in 1991 and 1992 has been sluggish and characterized by high inflation in the aftermath of a significant devaluation of the Jamaican dollar (table 5.1). Still, the economy has been stronger after 1986 than before. On the basis of available evidence, no one could argue the extent to which these changes are due to the individual and company tax reform, but many, including the former prime minister, believed that the economy could not have performed as favorably under the old regime.[18]

To the extent that tax reform can be said to have stimulated economic growth, three areas offer support for the hypothesis. First, by increasing the revenue flow and controlling expenditures,[19] it reduced the fiscal deficit, relieved some of the pressure from the IMF and the World Bank, and perhaps increased the overall level of confidence in the government. The "confidence factor" is less directly measurable, but it is important. The stronger performance of the Jamaican stock market after 1986 gives some evidence of confidence in the government's fiscal policies.

A second possible effect is an increased work effort and an increased propensity to invest and take risks. These changes would be a partial response to the lower marginal tax rates. However, there is no way to measure or even estimate this with available data, although the small increase in the number of taxpayers suggests no significant movement from the informal to the formal sector. Third, one might argue that there have been some portfolio adjustments to the reform. The 33.3 percent tax on deposit interest on accounts over J$2,000 may have stimulated activity in the stock market, direct investment, and perhaps real estate. The aggregate effect on the savings rate is unclear.

Equity

The vertical equity objectives of the reform were (1) not to increase the burden of taxation on low-income Jamaicans, and (2) not to increase the overall regressivity of the system. Both objectives were achieved.

Wasylenko (1991) estimated the distribution of tax burdens in the prereform period and found it to be progressive (table 5.6). In a later study, on 1990–91 data, Sjoquist and Green (1992) estimated that the reform had led to slightly more progressivity, and had lowered the tax burden for the lowest 40 percent of income earners.[20] There were three main contributing factors to this result: (1) the inclusion of interest income in the tax base, (2) the increase in the standard deduction, and (3) the slight progressivity of the GCT.

No estimates of the burden are available for a more recent period. One can speculate, however, that the overall progressivity of the system may have declined. The individual income tax rate was reduced to 25 percent (reducing progressivity), the standard deduction was increased (lowering the tax burden on the poor), and the GCT rate was increased (raising the tax burden on the poor).

Administration

A major problem to be addressed by the tax reform project was weak tax administration.[21] Procedures were outmoded, staff was inadequate to do the job of administering a modern tax system, and the recordkeeping system was primitive. The strategy of the project was to simplify the tax system, put in place a modern training system, offer technical assistance

Table 5.6 Distribution of Tax Burdens before and after Reform

	Prereform, 1984		*Postreform, 1990–91*	
Income group[a]	*Income in J$*	*Tax (percentage of income)*	*Income in J$*	*Tax (percentage of income)*
1	0–1,814	22.14	0–6,212	20.30
2	1,815–2,987	22.88	6,213–10,287	19.33
3	2,988–4,314	23.39	10,288–15,555	19.75
4	4,315–6,258	28.85	15,556–21,220	27.79
5	6,259–8,279	31.48	21,221–28,280	31.56
6	8,280–10,999	33.68	28,281–36,268	34.67
7	11,000–14,574	35.54	36,269–45,901	38.90
8	14,575–19,403	37.26	45,902–59,089	37.04
9	19,404–29,702	35.30	59,090–81,137	30.42
10	Over 29,702	31.75	81,138–103,941	26.84
11	n.a.	n.a.	Over 103,941	34.76

n.a. Not applicable.

a. Income groups are deciles for 1984; for 1990–91, groups 1–9 are deciles and groups 10 and 11 are quintiles.

Source: Prereform data from Wasylenko (1991) and Bahl (1991b, chapter 28); postreform data from Sjoquist and Green (1992).

in procedures, and help in the establishment of a computer center for the tax service.

The success of the project in improving tax administration has been uneven, but this is clearly the weak link in the work of the project. Subsequent evaluations by Hubbell and McHugh (1992), Garzon (1993), and Bahl and Wallace (1993) have confirmed this.

Staff and training. The numbers of trained staff remain inadequate (Bahl and Wallace 1993). The government's "Comprehensive Training Plan, 1992–95" (Revenue Board 1991) indicates a need to train an additional 120 revenue agents, and to train substantial numbers of staff in the areas of electronic data processing and general administrative duties.

Each department claims a significant staffing problem.

- The income tax department has an authorized staff of 458, but has filled only 274 of these positions with full-time personnel.
- The total number of GCT employees is 260. The GCT commissioner indicates that the office is in need of 50 more field officers (revenue agents) to determine whether the 9,000 enterprises that claim no liability in fact do owe tax.
- The inland revenue commissioner has called for more, highly trained staff, and particularly for additional revenue agents.
- The stamp and transfer tax office is grossly understaffed, by any standards. For example, there are only six officers to handle all verification of declared land transfer values, and none of the six are trained.

Registration. The failure to bring all taxpayers into the net remains a crucial problem. Not surprisingly, departmental officials do not have good estimates of the numbers of persons and companies who do not pay taxes. But even illustrative data seem to indicate that a substantial fraction of potential taxpayers are outside the tax base.

There are about 80,000 registered self-employed taxpayers, but only about 25 percent filed returns in 1992, a performance roughly comparable to that in 1986 (Alm, Bahl, and Murray 1991c). The comparable filing rates are 50 percent for companies filing on behalf of PAYE taxpayers, and 20 percent for companies filing for company tax. By any account, this is a substantial compliance gap, and it has not closed significantly.

As of 1993 there were about 26,000 enterprises registered for GCT—17,000 are taxpayers and 9,000 claim to be below the threshold. The large enterprise group (J$1 million turnover) has a compliance rate of 83 to 87 percent, and accounts for approximately 90 percent of the tax collected. The smaller enterprises have a compliance rate in the low 60 percent range.

Underreporting. There is significant underreporting of income. This is not unusual for a developing country, and tax administrations usually attempt to limit this with an aggressive audit program. In Jamaica the effectiveness of audits is compromised by inadequate staffing, inadequate access to information, and a penalty structure that does not provide a sufficient disincentive to evade taxes.

A bank secrecy law makes it impossible to properly audit the income tax on interest income. The Bank of Jamaica can do an audit of banks, but cannot examine the records of individual taxpayers; hence a "proper" audit is impossible. There are also restrictions on the sharing of information from income tax returns. The income tax department can obtain data from the Inland Revenue, customs, GCT, and Motor Vehicles Department, but cannot communicate these data to other departments without permission from the minister of finance. These laws and restrictions have seriously compromised audit effectiveness.

A major form of tax avoidance is the claiming of nontaxable allowances. Few adjustments are made by the Income Tax Department to the reported amounts. The auditors do not have detailed information on nontaxable allowances.

Assessment procedures and recordkeeping. There are inadequate and inefficient procedures used to assess and collect taxes. This raises administrative costs and slows the efforts to move toward full compliance. Basic problems are as follows:

- The Income Tax Department does not check the arithmetic of each return. Inland Revenue does a rough manual check at the time of collections.
- Inland Revenue manually posts *all* payments in ledgers.
- The GCT office does not do an annual report of activities when they monitor the efficiency of their operations.
- The Stamp and Transfer office keeps only hard copy of all transactions, but files these in numerical order each day.
- Inland Revenue accepts payment of taxes from businesses and individuals without verifying the taxpayer identification number.
- Hard copy returns are filed by the Income Tax Department by date received, and are kept for six years.

One of the major recordkeeping problems with the income tax has to do with the PAYE portion of the individual income tax. Each company files a tax list with summary information about each employee. Information on allowances is not reported, nor is information about payroll tax collections. In order to get full information about the tax payments by individuals, the tax officials must visit each company. For this reason, relatively little is known about the taxpaying status of individual PAYE employees.

Complexity and structural issues. The 1986 reform simplified the income tax system and probably reduced both compliance and administrative costs. But some loopholes were left open, and this has continued to compromise administrative efforts by failing to reduce the complexity of the income tax system and by continuing to give taxpayers an incentive to avoid income taxes. Major problems remain with nontaxable allowances in the areas of housing, uniforms, and automobiles (Bahl and others 1992a).

A second type of complexity in the current income tax structure is an exemption program for approved productivity schemes and for income received from gratuities. The productivity incentives are very hard to administer and are not well regulated in the law. The idea is to encourage higher wages in certain industries such as agriculture, tourism, and bauxite by exempting a portion of income from tax. In the tourism industry, gratuities may be exempted up to 10 percent of total sales receipts. The companies may deduct whatever is paid to the employees. The law is currently written so that the exempt portion is exempt from individual income tax, education tax, and NIS. These programs greatly complicate administration.

Finally, the recent rate reductions under the individual income tax have led to an eight-point spread between the top rates of the individual and company tax rates. This opens the door for tax avoidance by devising schemes to transfer income from company to individual status. Such schemes complicate the job of the tax administration and raise the cost of compliance for the taxpayer.

Information. A major problem with the entire tax administration system is an inadequate flow of information to support the assessment, collection, and audit process. The problems begin with the absence of a unique taxpayer identification number that is universally used. The current system permits multiple identification numbers. As a result of inconsistent identification numbers, an inadequate electronic data processing system, and secrecy laws, there are no integrated records for individual taxpayers or businesses showing their history of payment, tax liability, or financial summaries. In some cases, the law denies this information to the tax officials.

The process for moving taxpayer information is also flawed. In the past, the information was directed from each department to Fiscal Services, where theoretically it could be accessed by the various departments. However, access has not been adequate and the system remains manually operated. Finally, there has been inadequate training, and hardware and software development to allow each department to move beyond a manual approach to data entry and records storage.

On the positive side, there is now a master file of PAYE employers and self-employed filers although these computerized lists show only the names of the filers and contain no other taxpayer information. Another improvement in procedures has been the development of a registration, assessment, collection, and recordkeeping system for the GCT. The project assisted the Revenue Board in developing its own computer center and data processing facility. A professional staff is in place and other donors have been assisting in software development. Unfortunately, none of the major taxes were chosen for early emphasis. The land valuation roll is now on the system, as is certain customs valuation information, but neither the income tax nor the consumption duty files have been brought into the computerized recordkeeping system. Rather, the income tax files are still kept manually although there has been some improvement in file room procedures and organization.

Lessons

The Jamaican tax reform presented a rare opportunity to do a comprehensive study of a tax system in a developing country. It was a big project in every sense, that is, importance to the government, resources committed, public interest, and willingness to implement a sweeping structural and administrative reform. The mandate was open at the outset in that it called for a restructuring of the entire tax system to fit a new economic program of a newly elected administration, and the project had substantial latitude in defining the scope of the work. The way in which the Jamaica study was carried out also makes it different. The prime minister was very much a part of the study and met with the team on numerous occasions about the specifics of the work. Although the project team was composed of foreigners, its members developed a very close working relationship with their Jamaican counterparts, and since the project continued for four years there was time to get to know the country. Work continued through the implementation stage (and with follow-on analyses in 1992 and 1993), so there was some opportunity for the project team to be involved in the selling and implementation, and even to observe the first results of the new system. Finally, because resources to support the project were adequate it was possible to assemble a team of experts with extensive knowledge of systems in other countries.

The Jamaica tax reform project has served up nine lessons about successful tax reform that may be transferrable to other settings.[22]

1. *Tax reform and the economic setting.* The best time to do a comprehensive reform of the tax structure is when the economy is performing poorly. There is a sense of urgency and tax policy is one area where the government can take aggressive action. At such times, it is easier to focus the attention of policymakers on structural problems of the entire tax system and to think through the ways in which the tax system may be retarding economic growth. Inefficiencies that are so visible when the economy is not going well tend to become invisible in periods of economic growth. Consequently, when the economy is growing, the attention of tax reformers shifts to piecemeal adjustments that are "popular" or that appear to improve vertical equity, and to administrative improvements. The attention of politicians shifts to the expenditure side of the budget during periods of economic growth, and this shift accelerates as elections approach.

2. *How much can a tax system be shocked*? The Jamaican experience provides some strong arguments in favor of shocking the system with a comprehensive reform. First, the prereform system had gotten so far away from desired government tax policy that incremental reform could not possibly repair it. Only a complete overhaul would suffice. Second, the best time to inflict the painful parts of a tax reform is when it can be done simultaneously with measures that provide taxpayer relief. For example, removal of allowances and tax credits under the Jamaican individual in-

come tax could never have been accomplished if a dramatic rate reduction and a high standard deduction had not been introduced at the same time. Third, if the primary objective of the system is to rectify distortions in relative prices, then large changes are called for (because the tax-price elasticities of saving, investment, work effort, and evasion are low).

3. *The role of equity considerations*. Vertical equity cannot be the driving force behind a comprehensive tax reform program in a developing country. In part this is because developing countries cannot successfully implement progressive tax systems and in part it is because the costs of vertical equity are very high. Consider the case of the individual income tax, where the issue most often arises. It is one thing to recite the rhetoric linking progressivity in nominal rates to vertical equity, but quite another to show that such a linkage actually exists. The problem is with administration. The individual income tax had the look of a progressive tax with a steeply graduated nominal rate structure, but in fact the tax was regressive because of the extent of evasion and avoidance at the top end. Doing away with the progessivity in Jamaica's nominal rate structure had very little effect on the distribution of income.

Probably more important is the goal of horizontal equity, which the Jamaicans equated with fairness in taxation. The prereform system was riddled with many such inequities: private sector workers received more income in nontaxable perquisites than public sector workers, self-employed workers paid lower taxes than those in the PAYE sector, those in certain industries had access to the preferential "overtime" tax rate while others did not, only some types of businesses could engage in arbitrage to avoid income taxes, and so on. Such unequal treatment undermined confidence in the tax system. A primary goal of the Jamaican study was to find a way to eliminate these horizontal inequities and the distortions in economic choices which they promoted.

4. *The power of data*. Empirical estimates of the impact of proposed tax structure changes on revenue yield and on tax burdens were key elements in selling the reform package. The quality of the underlying data was not without problems, but the fact that data were available gave a basis for removing much of the guesswork in evaluating the options and lifted the debate to a much higher level. It provided a reasonable basis for guessing at the differential impacts of alternative reform programs and simulating the impacts of alternative specifications of the rate and base.

5. *First policy, then administration*. A first principle for successful tax reform is to get the policy right and then deal with the administrative problems. The consumers and sponsors of a reform often cannot see beneath a plethora of administrative problems to the real issue, which may well be a badly structured tax. Or, how many times do we question the sense of creating a new tax structure when the old one cannot be properly administered? Too often the call for technical assistance in tax administration from the internal revenue service or from one of the international agencies is premature.

There are three good reasons for giving policy reform priority over administrative reform. First, administrative improvements can often generate a quick revenue impact. Second, the true, underlying problem may be the tax structure. It may be so complicated as to be beyond the capacity of the government to properly administer, or it may be so unfair that payment of taxes will be resisted no matter how much the administration improves. Third, if the reform goes no further than administration, the government will not go through the exercise of questioning whether the tax system is affecting the economy in ways that reinforce government objectives.

6. *Monitoring*. The results of a tax reform should be monitored in the first years after implementation. While it is essential that the reform study generate the best possible forecasts of revenue yield, tax burden impacts, and economic effects, it is also essential that the tax planners know the actual outcome and be ready to adjust the new system as needed. It is especially important for monitoring to begin immediately after the reform is implemented and before new avenues of avoidance become entrenched. Taxpayers (and tax evaders) are far more adept at finding loopholes in new legislation than tax reformers are at closing all the avenues for tax avoidance. The more dramatic the structural reform and administrative shock, the more likely such loopholes will appear and go undetected. The continuing reform of the Jamaican tax system has been greatly helped by the evaluations, analyses, and technical assistance since the close of the tax project in 1987.

7. *Tax reform or fiscal reform*? Is it better to do a comprehensive fiscal reform—which also includes consideration of the expenditure side of the budget—than a comprehensive tax reform? The former is a more difficult job, requires more resources and time, and probably raises more controversial issues. However, it allows the government to get a better picture of the overall implications of the tax reform under consideration. With hindsight, a fiscal reform would have been the better route because expenditure policy was a key element of Jamaica's taxation policies.

8. *Neutrality in taxation and economic policy*. Perhaps the most difficult part of designing a comprehensive tax reform is matching the tax policy design with the set of economic policies already in place. It is easy to recite the maxims of a "good tax" and to come to the proper conclusion that it should be neutral in its effects on economic choices. In the case of Jamaica, this matched the stated goal of the government to rely more on market forces to guide economic growth. However, there is almost always another set of policies in place that raise questions about whether neutrality is an appropriate objective of the tax reform. It is particularly difficult to define an efficient tax policy when existing trade and industrial policy has led to a set of preexisting distortions.

9. *Implementation*. The Jamaican experience suggests five rules about how to successfully implement a tax reform.

Rule 1. The government must see the project as its own and not that of a donor or even that of a technical assistance research team.

Rule 2. The technical assistance team should have the right mix of skills and experience, and, above all, have expert credentials. Nothing short of well-known tax policy experts with extensive policy experience would have satisfied the Jamaicans.

Rule 3. Tax reform should not be hurried. It takes time to get the technical proposals properly in place and include public debate. The Jamaican press and public interest groups were all involved in the debate, at a surprisingly technical level, for a full six months before the income tax reforms were implemented. By the time the law was enacted, a very major change in the system was not seen by the public as a tax shock.

Rule 4. Timing is important. Elected government officials are not willing to be associated indefinitely with tax reform, even good tax reform. Such programs carry unfavorable connotations for most citizens and politicians, and the zeal for even so noble a goal as getting the prices right wanes as time goes by and election time approaches.

Rule 5. Implementation requires a great deal of attention. The project did have two income tax administration experts and a customs expert resident in country to work out administrative procedures and assist with training, and a sales tax administration expert to do the same for the GCT office. On the other hand, probably too little attention was given to carefully drafting the new legislation and implementing regulations.

Notes

1. Sections that refer to the design of the reform program and the estimated impacts are revised and abridged versions of two earlier papers by Bahl (1989, 1991a).

2. An agreement with the World Bank led to a trade liberalization program beginning in 1987. This, in effect, rescinded the stamp duty rate increases enacted in 1985 and 1986. The new program flattened the duty rate structure and eliminated most import exemptions. Hence it moved the import stamp tax structure back in the direction of the proposed general consumption tax (GCT). This would have made the introduction easier, except that the Bank and the government agreed to postpone implementation of the GCT in favor of a program of export rebates. At the time the project ended its work in 1987, the indirect tax system still had not been reformed.

3. For a good critique of this method see Bird (1976).

4. For this analysis is see Bahl (1991a).

5. For a good review of this literature see Ghandi (1987) and Skinner (1989).

6. The proposed income tax reform program went through a number of iterations before reaching the structure adopted in January 1986. The initial analysis and evaluation of the reform options was completed by the summer of 1985. The government's policy paper released in June (Revenue Board 1985) outlined the general format for the reform and provided a menu of alternative rate structures. The Tax Reform Committee accepted these in principle but recommended an even further broadening of the base and lower tax rates. The prime minister and the cabinet

generally accepted these proposals, but made some modifications in the proposed treatment of fringe benefits, the tax rate level, and the income exemption level. The revenue neutrality constraint was relaxed somewhat at the last minute: the prime minister instructed the Tax Reform Committee that a first-year revenue loss could be accepted, but it should not exceed J$40-J$60 million (about 9 percent of projected 1986 individual income tax collections).

7. At the time this work was done, the most recent available data were for 1983.

8. In 1991 the government began taxing the use value of automobiles provided by an employer, and treating only automobile allowances as an allowable deduction.

9. One recent example is the creation of company internal savings programs that pay an above-market interest rate but apply no interest withholding tax. This provides the firm with a ready source of capital and provides the employee with a higher after-tax return at the cost of less revenue to the government and less horizontal equity in the tax system.

10. Industrial buildings and machinery are given an initial allowance of 20 percent, but other asset investments receive a lower percentage according to a complicated schedule.

11. He defined the effective tax rate as the ratio of the present value of the tax payments (individual and corporate income) to the present value of the economic income arising from the investment. Economic income is measured as the difference between revenues and economic depreciation.

12. The international implications of the company tax reform are described in Oldman, Rosenbloom, and Youngman (1991).

13. The latter excludes food, petroleum products, cigarettes, and alcoholic beverages.

14. The traditional excises would continue to be levied under a separate structure.

15. The 1991 number is overestimated for purposes of this comparison because it includes stamp duty on domestic transactions.

16. The property tax is described in detail in Holland and Follain (1990, chapter 25).

17. For a good discussion of the site value base as applied in Jamaica see Oldman and Teachout (1979).

18. Prime Minister Seaga as quoted in "The Daily Gleaner," October 10, 1987.

19. The government deficit reduction was accomplished by a substitution of external for domestic borrowing, tax increases, and an expenditure reduction program. About 4,000 positions (4 percent of the civil service positions) were cut after 1984.

20. Sjoquist and Green (1992) included an estimate of the GCT burden in their computations, even though the tax was not introduced until late 1991.

21. This section draws heavily on Bahl and Wallace (1993).

22. These lessons are reported in more detail in Bahl (1990).

References

Ahmad, Ehtisham, and Nicholas Stern. 1991. *The Theory and Practice of Tax Reform in Developing Countries*. Cambridge, U.K.: Cambridge University Press.

Alm James, Roy Bahl, and Matthew Murray. 1990. "Tax Structure and Tax Compliance." *The Review of Economics and Statistics* 77(4): 603–13.

———. 1991a. "An Evaluation of the Structure of the Jamaican Individual Income Tax." In Roy Bahl, ed., *The Jamaican Tax Reform*. Cambridge, Mass.: Lincoln Institute of Land Policy.

———. 1991b. "A Program for Reform." In Roy Bahl, ed., *The Jamaican Tax Reform*. Cambridge, Mass.: Lincoln Institute of Land Policy.

———. 1991c. "Income Tax Evasion." In Roy Bahl, ed., *The Jamaican Tax Reform*. Cambridge, Mass.: Lincoln Institute of Land Policy.

———. 1991d. "Tax Base Erosion in Developing Countries." *Economic Development and Cultural Change* 39(4): 849–72.

Alm, James, and Michael Wasylenko. 1991. "Payroll Taxes." In Roy Bahl, ed., *The Jamaican Tax Reform*. Cambridge, Mass.: Lincoln Institute of Land Policy.

Bahl, Roy. 1971. "A Regression Approach to Tax Effort and Tax Ratio Analysis." *International Monetary Fund Staff Papers* 18(3): 512–612.

———. 1972. "A Representative Tax System Approach to Measuring Tax Effort in Developing Countries." *International Monetary Fund Staff Papers* (March): 87–124.

———. 1984. *Financing State and Local Government in the 1980s*. New York: Oxford University Press.

———. 1989. "The Political Economy of the Jamaican Tax Reform." In Malcolm Gillis, ed., *Tax Reform in Developing Countries*. Durham, N.C.: Duke University Press.

———. 1990. "Jamaica: Tax Reform: Evaluation and Lessons." *International Bureau of Fiscal Documentation* 44(January): 37–43.

———. 1991a. "The Economics and Politics of the Jamaican Tax Reform." In Roy Bahl, ed., *The Jamaican Tax Reform*. Cambridge, Mass.: Lincoln Institute of Land Policy.

———, ed. 1991b. *The Jamaican Tax Reform*. Cambridge, Mass.: Lincoln Institute of Land Policy.

Bahl, Roy, and Johannes F. Linn. 1992. *Urban Public Finance in Developing Countries*. New York: Oxford University Press.

Bahl, Roy, and Sally Wallace. 1993. "Tax Administration Reform Project: Economic Evaluation." Georgia State University, Policy Research Center, College of Business Administration, Atlanta.

Bahl, Roy, Robert E. Moore, David L. Sjoquist, and Richard Hawkins. 1992a. "Analysis of Jamaican Income Taxes and Allowances." Jamaica Tax Review Project, Report No. 1. Georgia State University, Policy Research Center, College of Business Administration, Atlanta.

———. 1992b. *Analysis of Jamaican Payroll Taxes*. Jamaica Tax Review Project, Report No. 2. Georgia State University, Policy Research Center, College of Business Administration, Atlanta.

______. 1992c. *The Jamaican Tax Reform: Review, Assessment and Recommendations for Next Steps*. Jamaica Tax Review Project, Report No. 4. Georgia State University, Policy Research Center, College of Business Administration, Atlanta.

Bahl, Roy, Jorge Martinez-Vazquez, Michael Jordan, and Sally Wallace. 1993. "Intercountry Comparisons of Fiscal Performance." Guatemala Fiscal Administration Project, Technical Note No. 7. Georgia State University, Policy Research Center, College of Business Administration, Atlanta.

Bird, Richard. 1976. "Assessing Tax Performance in Developing Countries: A Critical Review of the Literature." *Finanzarchiv* 34(2): 244–65.

———. 1991a. "An Introductory Overview." In Roy Bahl, ed., *The Jamaican Tax Reform*. Cambridge, Mass.: Lincoln Institute of Land Policy.

———. 1991b. "Choosing a Rate Structure." In Roy Bahl, ed., *The Jamaican Tax Reform*. Cambridge, Mass.: Lincoln Institute of Land Policy.

———. 1991c. "Sources of Indirect Tax Revenue in Jamaica." In Roy Bahl, ed., *The Jamaican Tax Reform*. Cambridge, Mass.: Lincoln Institute of Land Policy.

———. 1992. *Tax Policy and Economic Development*. Baltimore, Md.: Johns Hopkins University Press.

Bird, Richard, and Barbara Miller. 1986. "The Incidence of Indirect Taxes on Low-Income Households in Jamaica." Jamaica Tax Structure Examination Project, Staff Paper No. 26. Syracuse University, The Maxwell School, Metropolitan Studies Program, Syracuse, N.Y.

———. 1991. "The Incidence of Indirect Taxes on Low-Income Households in Jamaica." In Roy Bahl, ed., *The Jamaican Tax Reform*. Cambridge, Mass.: Lincoln Institute of Land Policy.

Brannon, Gerald. 1991. "The Taxation of Financial Institutions in Jamaica." In Roy Bahl, ed., *The Jamaican Tax Reform*. Cambridge, Mass.: Lincoln Institute of Land Policy.

Break, George. 1991. "The Jamaican Income Tax System: A Framework for Policy Formation." In Roy Bahl, ed., *The Jamaican Tax Reform*. Cambridge, Mass.: Lincoln Institute of Land Policy.

Chelliah, Raja J. 1971. "Trends in Taxation in Developing Countries." *International Monetary Fund Staff Papers* 18(3, November): 570–612.

Chelliah, Raja J., Hessel Baas, and Margaret Kelly. 1975. "Tax Ratios and Tax Effort in Developing Countries, 1969–1971." *International Monetary Fund Staff Papers* 22(1, March): 187–205.

Chernick, Sidney, ed. 1978. *The Commonwealth Caribbean: The Integration Experience*. Baltimore, Md.: Johns Hopkins University Press.

Cnossen, Sijbren. 1991. "The Extended Excise Tax System." In Roy Bahl, ed., *The Jamaican Tax Reform*. Cambridge, Mass.: Lincoln Institute of Land Policy.

Davies, David, and Lauria Grant. 1991. "The Taxation of Jamaican Public Enterprises." In Roy Bahl, ed., *The Jamaican Tax Reform*. Cambridge, Mass.: Lincoln Institute of Land Policy.

Dawes, Hugh N. 1982. *Public Finance and Economic Development: Spotlight on Jamaica*. Lanham, Md.: University Press of America.

DeGraw, Sandra. 1984. "Current Administrative Procedures of the Income Tax Department of Jamaica and Some Recommended Changes." Jamaica Tax Structure Examination Project, Staff Paper No. 4. Syracuse University, The Maxwell School, Metropolitan Studies Program, Syracuse, N.Y.

Due, John. 1991a. "A General Consumption Tax." In Roy Bahl, ed., *The Jamaican Tax Reform*. Cambridge, Mass.: Lincoln Institute of Land Policy.

———. 1991b. "Jamaica's Indirect Tax Structure." In Roy Bahl, ed., *The Jamaican Tax Reform*. Cambridge, Mass.: Lincoln Institute of Land Policy.

Follain, James, and Emi Miyake. 1991. "Land Versus Property Taxation: A General Equilibrium Analysis." In Roy Bahl, ed., *The Jamaican Tax Reform*. Cambridge, Mass.: Lincoln Institute of Land Policy.

Ghandi, Ved P., ed. 1987. *Supply Side Tax Policy: Its Relevance to Developing Countries*. Washington, D.C.: International Monetary Fund.

Garzon, Hernando. 1993. "Jamaica: Overview of the Tax Administration and Potential for Improvement." World Bank, Washington, D.C.

Goode, Richard. 1984. *Government Finance in Developing Countries*. Washington, D.C.: Brookings Institution.

Holland, Daniel, and James Follain. 1990. "The Property Tax in Jamaica." In Roy Bahl, ed., *The Jamaican Tax Reform*. Cambridge, Mass.: Lincoln Institute of Land Policy.

Hubbell, Kenneth L., and Richard McHugh. 1992. "Evaluation of USAID/Revenue Board Assistance Project." Paper prepared for the Revenue Board Assistance Project, U.S. Agency for International Development, Washington, D.C.

Jenkins, Glenn P. 1989. "Tax Changes before Tax Policies: Sri Lanka, 1977–1988." In Malcolm Gillis, ed., *Tax Reform in Developing Countries*. Durham, N.C.: Duke University Press.

Lotz, Joergen, and Elliot Morss. 1967. "Measuring Tax Effort in Developing Countries." *International Monetary Fund Staff Papers* 14(3, November): 478–99.

Martinez-Vazquez, Jorge. 1991. "The Taxation of Financial Institutions in Jamaica." In Roy Bahl, ed., *The Jamaican Tax Reform*. Cambridge, Mass.: Lincoln Institute of Land Policy.

McLure, Charles, Jr. 1984. "Defects in Forms and Instructions: Jamaican Individual Income Tax." Jamaica Tax Structure Examination Project, Staff Paper No. 1. Syracuse University, The Maxwell School, Metropolitan Studies Program, Syracuse, N.Y.

———. 1993. "A Review of Tax Policy in Jamaica." World Bank, Washington, D.C.

Miller, Barbara, and Carl Stone. 1987. "The Low-Income Household Expenditure Survey: Description and Analysis." Jamaica Tax Structure Examination Project, Staff Paper No. 25. Syracuse University, The Maxwell School, Metropolitan Studies Program, Syracuse, N.Y.

Ministry of Finance. 1983. "Taxation Measures 1982–83." Ministry Paper No. 9. Kingston.

———. 1993. "New Tax Measures." Ministry Paper No. 20. International Monetary Fund, Washington, D.C.

Oldman, Oliver, and Mary Miles Teachout. 1979. "Some Administrative Aspects of Site Value Taxation: Defining 'Land' and 'Value': Designing a Review Process." In Roy Bahl, ed., *The Taxation of Urban Property in Less Developed Countries*. Madison, Wisc.: University of Wisconsin Press.

Oldman, Oliver, David Rosenbloom, and Joan Youngman. 1991. "International Aspects of Revisions to the Jamaican Company Tax." In Roy Bahl, ed., *The Jamaican Tax Reform*. Cambridge, Mass.: Lincoln Institute of Land Policy.

Revenue Board. 1985. "Comprehensive Tax Reform, Parts I and II." Kingston.

———. 1991. "Comprehensive Training Plan (1992–95)." Kingston.

Shoup, Carl. 1991. "Integrating Tax Policy, Industrial Policy and Trade Policy in Jamaica." In Roy Bahl, ed., *The Jamaican Tax Reform*. Cambridge, Mass.: Lincoln Institute of Land Policy.

Sjoquist, David L., and David Green. 1992. *Distribution of Tax Burdens*. Jamaica Tax Review Project, Report No. 3. Georgia State University, Policy Research Center, College of Business Administration, Atlanta.

Skinner, Jonathan. 1989. "Do Taxes Matter? A Review of the Effect of Taxation on Economic Behavior and Output." Policy, Planning and Research Working Paper No. 48. World Bank, Office of the Vice President, Development Economics, Washington, D.C.

Tait, Alan, Wilfred Gratz, and Barry Eichengreen. 1979. "International Comparisons of Taxation for Selected Developing Countries, 1972–1976." *International Monetary Fund Staff Papers* 26(1, March): 123–56.

Tanzi, Vito. 1987a. "Quantitative Characteristics of the Tax Systems of Developing Countries." In David Newbery and Nicholas Stern, eds., *The Theory of Taxation for Developing Countries*. New York: Oxford University Press.

———. 1987b. "A Review of Major Tax Policy Missions in Developing Countries." *Tax Policy Missions in Developing Countries*. Detroit, Mich.: Wayne State University Press.

Thirsk, Wayne. 1991. "Jamaican Tax Incentives." In Roy Bahl, ed., *The Jamaican Tax Reform*. Cambridge, Mass.: Lincoln Institute of Land Policy.

Wasylenko, Michael. 1991. "Tax Burden Before and After Tax Reform." In Roy Bahl, ed., *The Jamaican Tax Reform*. Cambridge, Mass.: Lincoln Institute of Land Policy.

Wasylenko, Michael, and Bruce Riddle. 1987. "Payroll Tax Reform in Jamaica." Jamaica Tax Structure Examination Project, Staff Paper No. 35. Syracuse University, The Maxwell School, Metropolitan Studies Program, Syracuse, N.Y.

Whalley, John. 1984. "Tax Reform and the Foreign Trade Regime in Jamaica." Jamaica Tax Structure Examination Project, Staff Paper No. 7. Syracuse University, The Maxwell School, Metropolitan Studies Program, Syracuse, N.Y.

Wozny, James. 1991. "The Taxation of Corporate Source Income in Jamaica." In Roy Bahl, ed., *The Jamaican Tax Reform*. Cambridge, Mass.: Lincoln Institute of Land Policy.

6

Tax Policy and Tax Reforms in Korea

Kwang Choi

The Republic of Korea has sustained remarkable economic growth for over three decades. The average annual rate of growth during 1962–92 was nearly 9 percent. As a result of such rapid and sustained growth Korea is recognized as one of the most successful newly industrialized economies, with the promise of making a successful transition to a mature industrial economy by the turn of the century.

Economic progress was not smooth, however. The government encountered new and difficult problems along the way and to overcome them it undertook many structural adjustments. Some of these adjustments were highly successful; others were not.

Because taxation is a powerful tool of economic policy, national tax reform involves broad issues of economic policy, as well as specific problems of tax structure design and administration. In the process of crafting tax policies to meet the needs of a rapidly changing economy, society, and polity, tax revenues in Korea increased about three times faster than gross national product (GNP). Although the long-term growth trend has been upward, the growth in the overall tax burden has been erratic.

The evolution of the tax system in Korea reflects the economic development needs as laid out in the objectives of the Five-Year Economic Development Plan. The primary objective of Korean tax policy during the industrialization process was to mobilize resources for capital formation in the public sector without discouraging investment in the private sector. In addition to promoting savings and investment, most tax reforms emphasized efforts to simplify the tax system, eliminate corruption and tax evasion, and obtain a more equitable distribution of the tax burden among sectors of industry and classes of income.

This chapter examines the effects of tax policies and tax reforms on resource allocation and income distribution, and infers some lessons that may be of use in the design of tax policy in other developing countries.

An Overview of Fiscal and Tax Policy

Through hard work and single-minded effort Korea has achieved impressive economic growth for the last few decades, transforming itself from a poor agricultural economy with a typical dual structure into a fairly sophisticated industrial economy.

Economic development during 1961–92 can be divided into four periods. From 1962 to 1971, Korea, for the first time in its modern history, implemented an outward-looking development strategy with great success. From 1972 to 1978, the government promoted the accelerated development of heavy and chemical industries, a process that greatly interfered with the operation of the market mechanism. The country achieved high growth, but at the cost of high inflation and severe structural imbalances in the economy. From 1979 to 1986, government policy reforms were aimed at undoing the structural distortions caused during the previous period. As a result, by 1986, the country regained not only price stability and high growth momentum, but also much of its international competitiveness.

Table 6.1 Economic Structure of Korea

	1962	*1967*
GNP		
Billions of dollars (current)	2.3	4.3
Billions of Korean won (1975 constant)	11,215.8	13,864.2
Population (millions)	26.5	30.1
Per capita income		
Dollar (current)	87.0	142.0
Thousands of Korean won (1985 constant)	423.0	568.0
Gross investment (percentage of GNDI)	9.9	17.0
Gross savings (percentage of GNDI)	11.0	15.4
Export (millions of current dollars)	54.8	320.2
Export (percentage of GNP)	2.4	7.4
Import (millions of dollars, current)	421.8	996.2
Import (percentage of GNP)	18.3	23.3
Industrial structure (percent)		
Agriculture, forestry, and fishing	37.0	30.6
Mining and manufacturing	16.4	21.0
Other	46.6	48.4
Heavy industry in manufacturing (percent)	28.6	34.7
Heavy industrial goods in exports (percent)	—	8.7
Employment structure (percent)		
Agriculture, forestry, and fishing	63.1[a]	55.2
Mining and manufacturing	8.7	12.8
Other	28.2	32.0

— Not available.

Note: GNDI is gross national disposable income.

a. 1963 data.

From 1987 to 1992, the economy underwent considerable structural changes due to the democratization process at home and the end of the Cold War abroad. Certain serious adverse consequences of the structural changes of this period still affect the economy today.

In the early 1960s economic conditions were similar to any resource-poor, low-income, developing country today. As table 6.1 shows, GNP per capita in 1962 was a meager $87 and the level of domestic savings was negligible. The population was growing at nearly 3 percent a year and there was widespread unemployment and underemployment. There were no significant exports and the country had been running a chronic balance of payments deficit. In 1962 total exports amounted to $55 million or about a quarter of its imports.

Fiscal policy played a supporting role in the Korean industrialization process by contributing to the overall savings rate and by minimizing the tax disincentives of investment. Although the focus of fiscal policy changed sharply over the past three decades, there continued to be a relatively small public sector, comparatively low taxes, liberal use

1972	1977	1982	1987	1991
10.7	36.8	71.3	128.9	281.7
28,504.7	46,135.4	59,322.2	99,611.6	141,623.2
33.5	36.4	39.2	41.6	43.3
319.0	1,012.0	1,824.0	3,110.0	6,518.0
850.0	1,269.0	1,516.0	2,403.0	3,274.0
17.7	28.4	25.2	32.3	36.1
17.2	27.6	24.2	36.2	36.3
1,624.1	10,046.5	21,853.4	47,280.9	71,870.1
15.0	27.2	30.7	36.7	25.5
2,522.0	10,810.5	24,250.8	41,019.8	81,524.9
23.6	29.4	34.0	31.8	28.9
26.8	22.4	14.7	10.5	8.0
23.5	28.9	30.4	33.0	28.6
49.7	48.7	54.9	56.5	63.4
36.1	49.2	452.8	57.0	65.1
21.3	32.8	50.8	52.9	57.7
50.6	41.8	32.1	21.9	16.7
14.2	22.4	21.9	28.1	26.9
35.2	35.8	46.0	50.0	56.4
4.5	3.8	4.4	3.1	2.3
7.5	5.8	6.0	3.8	2.6

b. 1965 data.
c. 1976 data.
d. 1980 data.
Source: Economic Planning Board (1980, various years); Bank of Korea (various years a, various years b).

Table 6.2 Fiscal Indicators
(percentage of GNP)

Year	*Budget expenditure*	*Total tax burden*	*Government consumption*	*Government saving*
1960	17.1	10.3	14.5	–2.1
1961	19.4	9.7	13.6	–1.8
1962	24.9	10.6	14.0	–1.6
1963	14.3	8.6	10.9	–0.4
1964	10.5	7.1	8.5	0.4
1965	11.6	8.6	9.3	1.7
1966	13.6	10.7	10.0	2.7
1967	14.1	12.0	10.2	4.1
1968	15.9	13.9	10.4	6.1
1969	17.2	14.6	10.3	5.9
1970	15.4	14.3	9.5	5.2
1971	15.1	14.4	9.8	4.5
1972	16.6	12.5	10.2	1.9
1973	12.2	12.1	8.4	3.0
1974	13.3	13.4	9.7	2.0
1975	15.1	15.3	11.1	2.4
1976	15.4	16.6	11.0	4.3
1977	15.4	16.6	10.8	4.5
1978	14.7	17.1	10.4	5.3
1979	16.4	17.4	9.9	6.5
1980	17.6	17.9	11.5	5.2
1981	17.4	18.0	11.6	5.3
1982	17.6	18.2	11.5	5.7
1983	16.5	18.5	10.7	6.9
1984	15.8	17.7	10.0	6.5
1985	15.9	17.3	10.1	6.3
1986	15.2	17.0	10.1	6.2
1987	14.9	17.5	9.9	6.8
1988	14.3	17.9	9.8	8.2
1989	15.3	18.5	10.5	8.3
1990	16.0	19.4	10.7	8.8
1991	15.1	19.6	10.8	8.2

Source: National Statistical Office (various years).

of tax incentives for saving and investment, heavy reliance on indirect taxes, increased public savings, and relatively little emphasis on spending for redistributive social services. Fiscal planners applied the logic of supply-side economics much earlier than their counterparts in the United States and the United Kingdom.

Table 6.2 shows four indicators of the capacity of the government to influence the economy: (1) share of the budgetary expenditures of the central government in GNP, (2) total tax burden measured as a ratio of total national and local tax revenue to GNP, (3) government consumption as a proportion of GNP, and (4) the government saving rate.

Despite active involvement by the government in activities of the private sector, the size of government—whether measured as budgetary expenditure, a percentage of GNP, or the overall tax burden—is still low compared with other industrial and developing countries. The total bud-

Table 6.3 Measures of Tax Revenue Growth
(percent)

Period	*Average tax rate*	*Marginal tax rate*	*Income elasticity*
First Plan Period (1962–66)	9.3	12.3	1.27
Second Plan Period (1967–71)	14.5	18.4	1.28
Third Plan Period (1972–76)	15.6	16.8	1.04
Fourth Plan Period (1977–81)	17.6	18.4	1.04
Fifth Plan Period (1982–86)	17.7	16.2	0.91
Sixth Plan Period (1987–90)[a]	18.3	21.6	1.18

a. Four-year average.
Source: Bahl, Kim, and Park (1986); National Statistical Office (1991).

getary spending by the central government fluctuated widely during 1957–91 and showed no consistent trend. The share of government consumption fluctuated less than the share of general budgetary expenditures. Because of a concerted effort by the government to raise revenue, the share of total national and local tax revenue as a percentage of GNP, or the tax burden, increased from 6.7 percent in the mid-1950s to 19.6 percent in 1991. There has been an almost uninterrupted increase in government savings since 1964, when the government sector moved from a position of net spending to one of net saving.

The expenditure policy has been basically restrained with the ratio of the central government expenditure to GNP remaining at 20 to 23 percent throughout the 1962–92 period. The ratio of the government's fiscal investment averaged around 27 during 1962–66, but gradually declined to around 15 during 1987–92, indicating that the government's role in capital formation was important but declining.

An important characteristic of the central government expenditure pattern is the large share of defense expenditures and small share of social development expenditures throughout the industrialization period. In 1980 defense expenditures accounted for 35.6 percent of total government outlays and 6.3 percent of GNP. The shares of defense expenditures in government outlays and GNP declined to 25.4 percent and 3.9 percent, respectively, in 1991. Social development expenditures remained smaller than defense and economic development expenditures. However, expenditures for social development increased gradually from 13.3 percent of total expenditure in 1971 to 24.5 percent in 1991.

The upward trend in the tax burden has not been smooth, partly because of fluctuations in economic activity and partly because of revenue loss from extensive tax incentives and major tax reforms that reduced the tax burden. Table 6.3 summarizes the tax efforts during each of the five-year plan periods. Since no adjustments have been made for discretionary changes, estimates of the income elasticity do not reflect exclusively the automatic or built-in elasticity. The tax system as a whole shows an observed income elasticity of greater than unity except for the fifth plan period (1982–86). The relatively high elasticity during the initial period of development can be attributed to a combination of economic growth, inflation, and discretionary actions. Low elasticity during the fifth plan period is due to a drop in excise taxes and a substantial cut in the high marginal rates for both personal and corporate income taxes.

Any national tax effort has three dimensions: level, structure, and administration. The structure is the centerpiece, as it determines the ease with which any given level can be achieved and the efficiency with which taxes will be administered. It also determines the allocating, redistributing, and stabilizing functions of tax policy.

According to table 6.4, the structure of the national tax system depended heavily on domestic indirect taxes on goods and services and

Table 6.4 Structure of National Taxes

	Tax as a percentage of GNP					*Tax as a percentage of total national taxes*				
Tax	*1970*	*1975*	*1980*	*1985*	*1990*	*1970*	*1975*	*1980*	*1985*	*1990*
Income, profit, capital gains	4.7	3.4	4.1	4.4	6.2	35.0	24.3	25.5	28.7	37.5
Social security contributions	0.1	0.1	0.2	0.3	0.8	0.8	1.0	1.2	1.7	5.1
Property	0.3	0.5	—	−0.1	0.4	2.5	3.9	0.6	0.7	2.4
Goods and services	6.3	7.1	8.4	7.6	6.3	46.5	51.1	52.4	49.0	38.4
International transactions	1.9	2.0	2.8	2.5	2.1	13.8	14.4	17.2	16.2	13.0
Other taxes	0.2	0.7	0.5	0.6	0.5	1.3	5.5	3.0	3.8	3.6
Total	13.5	13.5	16.0	15.5	16.5	100.0	100.0	100.0	100.0	100.0

— Not available.
Source: Ministry of Finance (various years).

Table 6.5 Characteristics of the Tax Structure
(percent)

Year	*Direct taxes to total taxes*[a]	*Indirect taxes to total taxes*[a]	*Income taxes to national income*	*Indirect taxes to GNP*
1970	33.9	66.1	5.9	9.4
1971	36.9	63.1	6.2	9.1
1972	32.4	67.6	4.9	9.2
1973	29.8	70.2	4.4	8.5
1974	29.9	70.1	4.9	9.4
1975	25.8	74.2	4.6	11.4
1976	29.6	70.4	4.8	11.7
1977	28.2	71.8	4.2	11.9
1978	28.0	72.0	4.3	12.3
1979	27.9	72.1	4.5	12.6
1980	25.2	74.8	4.0	13.4
1981	25.7	74.3	4.3	13.5
1982	26.0	74.0	4.7	13.9
1983	24.5	75.5	4.6	14.6
1984	24.8	75.2	4.3	14.0
1985	27.8	75.2	5.0	12.7
1986	27.3	72.7	4.9	12.6
1987	29.8	70.2	5.6	12.3
1988	33.3	66.7	6.1	12.1
1989	36.5	63.5	7.0	11.9
1990	35.9	64.1	4.9	12.4
1991	35.7	64.3	7.2	12.1

a. The classification of direct and indirect taxes is based on national income accounts.
b. General sales tax in Korea before July 1977 (a business turnover tax) was replaced by the VAT.

import duties, which accounted for 52.4 and 17.2 percent, respectively, of total tax revenue of the central government in 1980. Taxes on income and profits in that year accounted for only 25.5 percent of the total national tax revenue. Although taxes on income have gained more importance in recent years, income taxes do not occupy the central position in the revenue structure.

Taxes on wealth at the central government level, such as the inheritance and gift tax, assets revaluation tax, and the securities transaction tax, are hardly significant in terms of their revenue yield. Collections from the above taxes in 1990 comprise only 2.4 percent of the central government's total tax revenue. Wealth taxes at the local government level, such as the land acquisition tax, property tax, registration tax, city planning tax, fire services facilities tax, and automobile tax, are major fiscal resources for local governments accounting for 49.6 percent of their total tax revenue. Revenue from wealth taxes as a percentage of total tax revenues at both levels of government is about 9.9 percent, which is quite low by international standards.

One important characteristic of the tax structure is that it relies heavily on indirect taxes. The first two columns of table 6.5 show that more than 70 percent of total tax revenues, national and local, came from indirect taxes until the mid-1980s and then decreased to 64.3 percent in 1991. Many people have mistakenly believed that the overall tax burden and share of indirect taxes in the total tax revenue increased with the introduction of the value added tax (VAT). The VAT led to neither an increase in the overall tax burden nor a heavier reliance on indirect taxation for government

National taxes on goods and services to private consumption	*General sales taxes to taxes on goods and services*[b]	*National taxes to total taxes*	*Local taxes to total taxes*	*Local taxes to GNP*
8.3	18.1	91.7	8.3	1.2
8.4	17.4	91.9	8.1	1.2
7.4	21.2	91.9	8.9	1.1
7.6	10.8	88.6	11.4	1.4
8.4	21.4	89.4	10.4	1.4
9.9	27.6	89.8	10.6	1.6
10.8	26.3	90.5	9.5	1.6
11.6	34.3	88.6	11.4	1.9
11.8	47.9	89.2	10.6	1.9
12.0	46.3	88.8	11.2	1.9
12.5	47.6	88.3	11.7	2.1
12.5	47.3	88.9	11.1	2.0
12.9	47.7	88.2	11.8	2.2
14.0	49.0	87.8	12.2	2.4
13.8	48.0	87.8	12.2	2.3
12.4	48.9	87.8	12.2	2.1
12.8	48.8	88.3	11.7	2.0
12.5	49.8	88.2	11.8	2.1
12.7	50.0	86.3	13.7	2.5
10.7	63.7	81.1	18.9	3.5
11.8	64.0	80.8	19.2	3.7
10.5	71.9	79.1	20.9	3.9

Source: Bank of Korea (1993a); Ministry of Finance (various years); National Statistical Office (1993).

revenue. As table 6.5 shows, the ratio of indirect taxes to GNP rose from 1974 to 1983 and there was no significant difference in the tax burden ratio before and after introduction of the VAT. Table 6.5 also shows that the share of indirect taxes in total taxes exhibited only a minor increase after the introduction.

The most important change in the tax structure in connection with the introduction of the VAT was the increased reliance on the general consumption tax rather than the selective excise taxes. It must, however, be pointed out that the revenue yield of the VAT relative to that of excise taxes is still lower than other countries with a VAT.

While the ratio of total tax revenues to GNP has increased markedly over the three-decade period, the composition of total tax revenues has not changed much. During the 1970s and 1980s national government revenues accounted for more than four-fifths of the total tax revenues and local government for the remaining, less than one-fifth. In contrast to the national tax burden, the local tax burden remained more or less unchanged until the mid-1970s, but it increased somewhat to over 2 percent of GNP in the 1980s and 3.9 percent in 1991.

Background on Tax Policies and Tax Reforms

Since the establishment of the government in 1948, there have been nine major tax reforms. In the years 1956, 1961, 1967, 1971, 1974, 1976, and 1982 the changes were substantial enough to be labeled "comprehensive." Although the standard litany of equity and efficiency objectives was cited along with improved tax treatment, the primary concerns differed. In the reforms of 1956, 1961, and 1971 the concern was the poor budgetary performance of the government sector. In the reforms of 1967, 1974, 1976, and 1982 government budgetary performance was relatively strong, and the focus was on structural changes or improvements.

Since the characteristics of the tax system are derived from its social and economic functions, the development of the tax system can be divided into four distinct periods, each with a different emphasis on developmental policies and strategies: (1) Before 1961 the focus was on import substitution and post–Korean War rehabilitation, (2) 1961 to 1972 was the development-supporting period, (3) 1973 to 1979 was the heavy industry promotion period, and (4) from 1980 to the present has been the structural adjustment and liberalization period.

Independence to Post–Korean War Reconstruction Period

After Korea regained its independence after World War II, the short-lived ecstasy of liberation in 1945 was followed by the unwanted division of the country and finally by the Korean War, which left the country in ruins, the economy decimated, and the national psyche traumatized. The termination of the U.S. Military Government concurred with the establishment of the Government of Korea and a drastic reform of the tax system to reduce cumulative budget deficits, activate depressed business sectors, and increase taxes on property and unearned income as a means of redistribution. In 1949 the government set up the Tax System Reform Committee to make a proposal for increasing tax revenue and to review the equity of tax burden distributions. It was the Tax System Reform Committee that prepared new democratic and modern tax laws. According to the recommendations of the committee, sixteen new laws, which include the Personal Income Tax Law, the Corporation Tax Law, the Commodity Tax Law, and the Liquor Tax Law, were passed and three taxes were abolished. The new tax system reduced the tax burden of landowners, whose holdings were decreased by the Land Reform, and increased the tax burden of the wealthy urban class.

The Korean War, which broke out on June 25, 1950, less than two years after the establishment of a new independent government, abruptly changed the tax system. To finance enormous defense expenditures the government introduced the Special Measure Law for Taxation, the Temporary Tax Increment Law, and the Temporary Land Income Tax Law. In December 1950 the Temporary Tax Increment Law drastically increased various tax rates. Under the Special Measure Law for Taxation, which was passed in January 1951 and superseded all existing tax laws, the personal income tax became a schedular tax with progressive rates applying to wages and salaries, real estate income, and business income. At the same time taxes on business income were made payable in quarterly installments on the basis of annual estimates of current-year liabilities.

During the early post–Korean War years more attention was given to the mechanics of spending rather than to the means of mobilizing revenues. This was because the depressed state of the economy and the impoverished condition of the people limited the potential for mobilizing financial resources from within Korea, and the prospects for substantial foreign aid seemed to make such efforts unnecessary.

Nonetheless, tax reforms continued after the armistice in 1953. Reforms during the early post–Korean War period were focused on meeting the needs of economic reconstruction. In 1956, in order to reduce the disincentive effects of high direct taxes on capital accumulation, rates were reduced on direct taxes and increased on indirect taxes.

When the government prepared its very first economic development plan in 1959, the Three-Year Economic Reconstruction Plan, it decided to reduce most tax rates and simplify tax administration. In 1960, when the first Republic of the Liberal Party was overthrown by the students' April Revolution and the second Republic of the Democratic Party was established, tax reform was undertaken to meet budgetary demands and promote industrial development. The 1960 tax reform also reduced rates for direct taxes while increasing rates for indirect taxes. In order to promote exports and capital accumulation, the new government increased various tax exemptions and deductions geared to those activities.

In summary, the tax reforms during the 1950s involved a conversion of the tax system from wartime to peacetime, more reliance for revenue on indirect taxes than on direct taxes, and some consideration of incentive measures to promote saving and investment. Despite all these reform efforts, the tax system during the 1950s can be criticized for inequities in the tax burden, adverse effects on business activities due to excessive taxes on business income, and arbitrary assessment and collection methods.

The Development-Supporting Period

The early 1960s were disruptive years of social upheaval and drastic changes in economic policy. The Chang Myon regime that came into power immediately following the Student Revolution of April 1960 managed the national economy until it was overthrown by the military coup in May 1961. After about three years of rule by the military junta, through to the end of 1963, the nominal civilian government, with the same military personnel at the center of political power, emerged from the general election in early 1964. Apart from an important constitutional change in October 1972, no major changes in political power occurred until 1979.

The military government that came to power in 1961 made a drastic change in the direction of economic policy. In response to changing conditions confronting the Korean economy in the late 1950s and early 1960s economic policy shifted away from reconstruction and stabilization. The economic growth strategy epitomized in successive Five-Year Economic Development Plans starting in 1962 was to achieve the highest growth possible with two principal objectives: (1) to shift from an inward-looking strategy to an export-led industrialization strategy and (2) to increase public and private saving. Some of the policy measures were unsuccessful and even deleterious, and specific details of several reform programs required major adjustments as circumstances changed.

Given Korea's long history of weak compliance, corruption, and inept administration, the military government immediately forgave all existing penalty claims for past tax delinquencies by enacting the Temporary Measure Law for Tax Collection and the Special Measure for Tax Delinquency, but at the same time pledged to deal more harshly with future cases. Furthermore, a tax reduction was rendered for voluntary filing of tax returns on personal income, corporate income, and business activity taxes. To assist taxpayers in filing tax returns voluntarily, a tax accounting system was instituted through enactment of the Tax Account Law.

Even though revisions in both the content and administration of the tax laws had taken place in almost every year since 1961, major tax reforms were undertaken around the beginning of each Five-Year Economic Development Plan simply because the dynamics and inner workings of each tax reform had to be closely integrated with development planning.

Therefore, it is no surprise that major tax reforms occurred during the years 1961, 1967, and 1971.

Tax reform of 1961. To correct the inefficiency resulting from the unilateral assessment by the tax authorities on the basis of presumption, the unilateral assessment system was abolished. In its place a bookkeeping system was introduced which required taxpayers to keep correct and honest records. In the effort to raise tax revenue to meet burgeoning development needs, heavy reliance was placed on indirect taxes. Tax rates on alcoholic beverages, entertainment activities, and luxury consumer goods were increased.

Two of the most important reforms in fiscal management since independence were the establishment of the Economic Planning Board (EPB) in 1961 and the Office of National Tax Administration (ONTA) in 1966. The EPB was organized as a powerful national planning agency by absorbing the Bureau of Budget from the Ministry of Finance and the Bureau of Statistics from the Ministry of Home Affairs. The chief of EPB was elevated to the position of deputy prime minister, and had the extraordinary power of preparing the annual budget in formulating development plans.

The creation of ONTA marked the beginning of a new era in tax administration. Before the establishment of ONTA, the Bureau of Taxation in the Ministry of Finance was responsible for both tax policy and tax administration. After reorganization, it became responsible for tax policy, whereas ONTA, as an agency independent of the Ministry of Finance, assumed the responsibility for administration of domestic national tax laws. In addition to this separation of major functions, the reorganization upgraded the status of the tax administration agency and expanded investigation and inspection activities. ONTA had a clear mandate to increase tax collections and it received full support from the political leadership. With a strengthened mechanism for tax administration, the government now had a powerful instrument for monitoring business performance and applying pressure on businessmen to comply with the policy objectives.

Tax reform of 1967. Major changes in personal income tax reform included the partial introduction of the global income tax system, an increase in the exemption level for wage and salary earners, and an increase in the number of tax brackets from five to seven. Adoption of a partial global tax system with progressive tax rates was designed to increase the tax burden of high-income groups.

Tax reform of 1971. The stated specific objectives of reform included reduction in the tax burden on low-income earners, provision of tax incentives to encourage savings, and improvement in horizontal equity. In line with these objectives, tax rates on wages and salaries and business income were lowered and the basic exemption level increased. Interest income from bank deposits, previously exempted, became subject to withholding taxation at a rate of 5 percent. Corporate income tax rates were also revised downward and the investment tax credit was expanded to cover a wider range of businesses and industries.

The Heavy Industry Promotion Period

Having achieved rapid growth during the decade covered by the first and second economic development plans, the economy underwent dramatic structural changes during the 1970s. Two oil shocks and the resulting changes in the level and structure of world demand critically affected the course of structural adjustments in the open Korean economy. The development strategy as envisioned in the third and fourth Five-Year Economic Development Plans included the modernization of rural areas and, more importantly, the promotion of heavy and chemical industries. A concrete goal policymakers pursued with remarkable consistency in the 1970s was often expressed by the phrase "deepening of the industrial structure," which simply meant increasing the relative importance in the economy of heavy industries such as iron and steel, machinery, petrochemicals, and transportation equipment.

Although the establishment of the equitable tax system had been one of the major aims of tax reforms up to the end of the 1960s, each tax reform emphasized specific measures to enhance economic growth rather than further social equity. From the early 1970s onwards, an increasing concern was placed on the redistributive function of the tax system.

In the history of Korean tax reforms, 1975 and 1977 are important years: an almost global personal income tax was introduced in 1975 and the VAT was implemented in 1977. Until 1974, the personal income tax system had been largely schedular. Tax was levied separately on five categories of income: wages and salaries, business income, rents, interest and dividends, and miscellaneous income. The rate structure was progressive for wage, business, and rental schedules and proportional for the other two schedules. In addition, a global tax was levied at progressive rates on high-income earners. The tax reform in December 1974 replaced the schedular income tax system of six bases and four rate structures. Under the new system almost all income was subject only to global tax; however, special rates for certain types of income not aggregated in total income were devised to foster savings and the development of the capital market.

As part of the government's efforts to develop a more efficient and modernized tax system, streamlining a complicated system of excise and turnover taxes was a natural and even inevitable step following the adoption of the global income tax in 1975. The consensus among Korean tax experts since the early 1970s had been to make the global income tax the central direct tax with the VAT as its counterpart in the indirect tax system. After a long preparation starting in 1971, the government undertook a drastic overhaul of the existing indirect tax structure in 1976.

The introduction of the VAT as a substitute for a plethora of then existing indirect taxes was part of a large scale reform in 1976, including eighteen newly enacted or amended taxes. The introduction of the VAT was in response to a complex set of political, economic, administrative, budgetary, and tax goals. The reasons for the introduction of the VAT, as announced by the government, included the simplification of the indirect tax system and its administration, the promotion of exports and capital formation, and the preservation of the neutrality of indirect taxes.

The Basic Law of National Taxes was enacted to clarify the legal basis of taxation, to promote fairness in tax administration, and to protect the rights and interests of taxpayers. Under this new law the National Tax Tribunal was established as an independent agency to handle tax appeals. Revisions in the Tax Exemption and Reduction Control Law (TERCL) restricted tax exemptions and reductions to major and strategic industries and allowed firms to select only one of the three tax incentives available: direct exemption, investment tax credit, and special depreciation. The excess profits tax measure was extended beyond the original one-year period.

Soon after the tax changes in 1974 the defense tax was enacted to support national defense. In the face of the tense international political and security situations involving the Korean peninsula just after the Nixon doctrine was announced, the tax did not arouse much opposition. Most taxpayers of internal direct and indirect taxes, customs duties, and local taxes, and advertising sponsors were subject to defense tax rates of 0.2 percent to 30 percent on the basis of, respectively, taxable income, import prices, telephone charges, and advertisement rates. The defense surtax was legislated as a temporary earmarked levy with an original expiration date of 1980, but was postponed twice until 1990.

The 1976 tax reform also included adjustment of the national and local tax systems to expand the tax base of local governments. The local entertainment and restaurant tax was transferred to the national government to be consolidated with the newly introduced VAT, whereas the national registration tax was transferred to the local governments.

The Structural Adjustment and Liberalization Period

The worldwide recession following the second oil crisis, combined with several structural problems in the domestic economy, caused negative growth in 1980, the first time since 1956. This decline in output was accompanied by high inflation and significant deterioration in the national balance of payments. Recognizing that the domestic setback was due to a chronic inflationary spiral and inefficiencies induced by extensive government intervention in the economy during the 1970s, the government, starting in 1980, undertook a series of wide-ranging structural reforms aimed at reinvigorating the market mechanism, promoting competition in all sectors of the economy, and liberalizing the nation's external trade policies.

Unlike the 1960s and 1970s, there were no major tax changes between 1980 and 1987. This is not to say that there were no minor revisions of tax laws or that there was no need or demand for tax changes. On the contrary, throughout the 1980s the demand for tax reforms from all sectors of society was greater than ever before and the debate on tax reforms grew. As a result, the government established the Commission on Tax Reform. In early 1980 tax changes followed the major tenet of the government's efforts toward structural adjustment and liberalization, along with some measures to deal with short-term economic problems.

To combat economic recession, the 1980 tax changes sought to stimulate investment and boost business confidence. A lower flexible rate was introduced with respect to the capital gains tax, which was believed to be very effective in stimulating the housing construction industry. The range of capital gains tax rates on real estate transactions was reduced from a range of 50 to 80 percent to a range of 35 to 75 percent to stimulate residential construction. In addition, a 30 percent cut in special consumption tax rates was provided to encourage purchases of selected consumer durables. The withholding tax rates on interest and dividend income were increased from 5 to 10 percent.

The 1982 revisions of tax laws and policy issues involved three changes: (1) lowering of personal and corporate income tax rates, (2) streamlining of the tax incentive system, and (3) partial movement toward the "real-name" system for financial transactions.

In 1982 the government took its very first step to revamp the tax incentive system by announcing, as a guideline, the principle of low tax, low exemption. Furthermore, the government adopted the "functional" or "indirect" approach in contrast to the previous industry-specific or "direct" approach. Under the functional approach preferential tax treatments were based on the development of technology and manpower training regardless of the nature, size, or sector of business. A distinctive feature of tax incentive schemes during the 1980s is the growing dominance of the functional approach over industry-specific incentives.

One of the heated political issues in 1982 with important implications for the tax system was whether to continue the no-real-name financial transactions system, which allowed depositors or security holders to open accounts under a fictitious name. It was introduced by the Secrecy Law for Deposits in 1961 to encourage savings, but had long been criticized for generating inequities in tax burdens and stimulating curb market financial activities.

In early 1982, a national financial scandal involving anonymous accounts caused heavy pressure by some key economic policymakers to pass the bill banning the use of fictitious names. The debate lead to a political compromise under which the date of implementation for a real-name system would be left to the president after the bill was passed. In the meantime, income from financial assets was subject to differential taxation, depending on whether financial transactions were conducted under the taxpayer's real or fictitious name. Income from financial assets carrying no real name was to be subject to taxation at the rate of 22.63 percent until the end of 1985 and at the rate of 28.5 percent thereafter, while a tax rate of 16.75 percent was levied on income from financial transactions using a real name.

Tax laws revised from 1983 through 1990 emphasized continuous structural adjustment and liberalization. Minor revisions over several years included further streamlining of the tax incentive system, reducing tariffs, abolishing tax privileges for foreign investors, strengthening the tax incentives for research and development, establishing small and medium businesses, and installing antipollution equipment.

It is worth nothing that from the early 1970s onwards increasing concern was placed on the redistributive function of the tax system. Although the personal income tax system was not fully developed enough to carry out the redistributive function, various tax measures were introduced in almost every tax reform in order to alleviate the tax burden of low- and middle-income brackets. However, during the period from 1982 to 1987 no adjustment was made for personal exemption in the personal income tax.

The income tax reform of 1988, which increased the personal exemption level and reduced the number of brackets and the marginal tax rates, undermined the already weak base of the income tax system and may contribute negatively to distributive equity in the long run. Emphasis on social development early in 1980 resulted in the provision of various tax incentives for housing construction, manpower development, and employee's stock holding.

The centerpiece of tax reform in the late 1980s was aimed at controlling land speculation through tax measures. Speculation on land has been a serious economic and social problem since the beginning of industrialization. The explosive increase in the price of land since the early 1980s has caused many social economic problems, interfering with efficient use of resources and severely distorting the distribution of income and wealth. To prevent speculation and to improve income distribution a new tax on excessive landholding was established in 1986, and made effective in January 1988.

In September 1988 an ad hoc committee was established to study the ownership of land and draft a proposal on various measures to deal with pending land problems. The proposals by the ad hoc committee included tax and nontax measures, including a new system of assessing land, upward adjustment of the assessment ratio, establishment of the aggregate land tax (the global landholding tax) and the excessive land value tax, introduction of a ceiling on the holding of residential land, and adoption of the system to retake development profits.

One of the major reasons for the poor performance of property-related taxes was the unrealistically low and extremely uneven assessment of real assets for tax purposes. The government announced a schedule to raise the landholding tax assessment to 60 percent of the actual market price by

1992. Under the aggregate land tax, which replaced the property tax on land and the excessive landholding tax, all land owned by individuals and corporations is classified into three groups: (1) properties to be taxed under the main global scheme, (2) properties to be taxed under the secondary global scheme, and (3) properties to be taxed separately at flat rates. Nationwide landholding of an individual or a corporation is added together and taxed at progressive rates. The aggregate land tax is a personal tax but administered as a local tax. With the introduction of the aggregate land tax, the traditional property tax now only applies to aircraft, building and structures, mining lots, and vessels. The excessive land value tax is a tax levied on accrued (not realized) net capital gains in excess of normal gain (national average rate of land price increase) at a rate of 50 percent every three years. But its tax base is not comprehensive as only a limited portion of land is integrated into the taxable base. The effects of these radical measures to stabilize the real estate market and eradicate speculation remain to be seen.

Another pillar of the tax reform debates during the 1980s has been whether to promote the so-called real-name system of financial transactions. The legality of fictitious names has long been criticized for its failure to develop a more unified taxation of different sources of income and for providing safe harbor for curb market activities, the most typical underground economic activity.

It had not been an urgent issue, however, until a financial scandal that shook the nation in 1982. The economy was hit so hard by this incident that a law disallowing financial transactions with fictitious names was passed in 1982 by the National Assembly with a proviso that the actual implementation of the system be entrusted to the president.

In the presidential election of 1987, all candidates promised to activate the real-name system when elected. In 1988 the government announced a schedule to implement the system and established a working group in the Ministry of Finance. However, in early 1990 the economy performed poorly and political opposition by the rich led the government to suspend implementation of the real-name system. At present nobody knows what the future holds for the real-name system.

Tax Administration and Tax Policy Formulation

Tax administration and enforcement have been a constant source of public discontent. It is believed by many that the problems plaguing the tax system are administrative. The gap between tax law and practice, indicative of the shaky state of tax administration, must be closed in order to achieve a reasonable degree of efficiency.

There are three government organizations responsible for tax administration: the Office of National Tax Administration assesses and collects domestic national taxes; the Office of Customs Administration assesses and collects customs duties; and local governments—provinces, special cities, cities, and counties—administer local taxes, but must seek approval for their decisions from the Ministry of Home Affairs.

The establishment of ONTA in 1966 marked the turning point in tax administration. Ever since the establishment of ONTA, the political leadership has expressed a strong personal interest in its effectiveness by appointing the director. One key element in the improvement of tax administration has been the insulation of the tax collectors from political interference.

Whereas the influence of political favoritism has been largely eliminated from the tax administration, economic penalties have remained a part of tax enforcement, providing a powerful instrument for monitoring business performance and applying pressure to businessmen to comply with national policy objectives. It is estimated that administrative improvements since the establishment of ONTA have accounted for a more than 50 percent increase in the central government tax revenue in the late 1960s.

Public discontent with the tax administration stems mainly from the public perception that ONTA has sometimes infringed on taxpayers' basic rights in collecting more taxes. However, it must be pointed out that the lack of systematic recordkeeping on the part of taxpayers has made tax administration difficult and evasion easy, even under the best of circumstances.

The tax office has often used a system of tax targets or quotas broken down by geographical areas and tax type. Furthermore, the government applied a system of awards and penalties to tax officials for exceeding or falling short of these targets. Tax collectors working under quotas imposed by their superiors frequently imposed high assessments initially to negotiate a lower rate in return for a personal gift from the taxpayers.

In its effort to improve tax compliance the government adopted several measures, including the voluntary disclosure program and the "green-return" system. In order to encourage a new era of voluntary taxpayer compliance tax delinquencies before 1966 were forgiven. Once a corporation attains the status of a "green-return corporation" it enjoys various privileges including assessment of tax based on tax reports alone, generous treatment in counting expenses, installment payment of taxes, and special depreciation allowances. Every year the government awards public recognition to the "greatest" taxpayers and tries to stimulate an awareness among the public of the importance of paying taxes.

Tax enforcement is highly selective on the one hand and very comprehensive on the other. Despite broad-based efforts to administer tax compliance in a nondiscretionary manner, in practice tax enforcement has been a highly selective. Firms can be audited for noncompliance with various government edicts. It has usually been the case that a company is investigated when it does not follow the order by the government or when it does not comply with certain regulations. Such selective enforcement has led the management of Korean companies to believe that they are penalized for ignoring government commands on dubious political grounds even when good and sufficient grounds are found for penalizing the company for noncompliance.

At the present time the government is formally obliged to examine global income tax returns from all sources. Such comprehensive auditing is costly and unnecessary. Since the initial processing of tax returns has been computerized, it should be possible to rely on spot audits based on a statistically based random sample. The remaining audits could be targeted on returns with unusual characteristics. Spot auditing would permit staff to devote more time to particular auditing problems, such as taxation of interest and dividends.

The VAT is ordinarily regarded as self-enforcing because of the way it is usually administered. Korean experience with the VAT, however, suggests that the supposedly built-in self-enforcing character of the VAT, which permits the matching of the tax credits of one taxpayer against the tax payments of another, is illusory or, at best, a much overrated advantage because invoices can be falsified. Although taxpayers do have an incentive to request invoices for their purchases in order to increase their input tax credit, this incentive is in many instances counterbalanced by the desire to suppress both purchases and sales records in order to avoid not only the VAT but also income taxes.

In Korea, as in all other countries, there are problems taxing the self-employed. Faced with the difficulty of verifying information supplied by such taxpayers, ONTA adopted a very arbitrary method called the standard assessment.

The standard assessment method had been widely used in Korea for both the income tax and the VAT. A set of guidelines is established for each major economic activity on the basis of whether the income or sales for any individual taxpayer can be estimated in a relatively objective manner. Basically, all the assessing officer has to do is obtain information on a series of relatively objective indicators, go to the relevant guide, and calculate the tax on the basis of two ratios shown there: (1) the ratio between the indicators on which he has collected information and gross sales, and (2) the ratio between gross sales and net income. Those who wish to rebut the presumed minimum tax have always been allowed to file regular returns.

The ratios in the guides are, in principle, based on careful studies by a small group of experts composed of tax administrators, tax specialists, and representatives from the business community. This group also develops and updates the guidelines. As a rule, different guidelines have been suggested for each trade or profession in each region. Once prepared, the standard assessment guides are published.

Ideally, taxes to be paid calculated according to these guidelines should be on the high side to encourage better recordkeeping. In practice this has not been the case because very few taxpayers have filed a report on the basis of actual income. The standard income ratio system has generally underestimated actual income and tax liability. As a result the use of standard assessment guides has discouraged rather than fostered movement toward the regular system.

Long-term development planning, formulation of short-term economic policies, and tax reform do not occur in a vacuum. There must be organizations and institutions to gather information, formulate, implement and evaluate policies, and bring about the consensus necessary for success. Building a national consensus and receiving proper technical assistance from tax experts are two very important ingredients in successfully carrying out tax reform in any country. The Korean experience is unique and well worth exploring in this regard.

Although the legislative body of government has been solely responsible for enactment of tax laws, it is the administrative body of the government, particularly the tax bureau, that has taken the initiative in every tax reform and designed every detail of each reform. There is a long tradition of authoritarian political systems from previous dynasties to the present political regime. In this case Korean bureaucrats were fully supported by authoritarian political leaders and able to carry out tax reforms successfully. It is also important to note that in economic policy and tax policy in particular decisionmakers and the general public have been very pragmatic and particular. Except for the commitment to anti-communism there have been no ideological biases. This pragmatism allowed the government to use all available instruments to achieve its goals. By being particular, the government was able to apply a certain policy to a limited number of clients in a specific situation.

Even though the government tried to build a consensus not only within the government but also among various sectors of the society with different interests, important tax policy and tax reform decisions were mostly made by a small number of bureaucrats or government officials without much public debate. Until the late 1960s, the plan for tax reform was prepared entirely by government officials, who sometimes sought the opinions and comments of foreign tax experts rather than native experts. Before the 1960s the Korean government relied more on foreign experts.

The post–Korean War tax reform of 1954 was largely based on a report by Haskell P. Wald (1953), a consultant for the United Nations Korean Reconstruction Agency. His recommendations were (1) to increase the total tax revenue to 20 percent of GNP, (2) provide favorable tax treatment for business and corporate income, (3) remove the inequitable and discriminatory aspects of the wartime tax system, and (4) simplify the tax structure and make it more objective and equitable.

When the government prepared the Three-Year Economic Reconstruction Plan in 1959, a tax consultant group headed by Dr. R. Hall provided technical assistance to make reform of the tax system consistent with the economic plan. Richard A. Musgrave provided technical assistance for the 1967 tax reform. In a report written in 1965 he suggested that the

Korean government rely more on indirect taxes than on direct taxes, and suggested the introduction of a differentiated tax rate structure as an inducement to transform closely held family corporations into widely held open corporations. He also suggested distinguishing between "open" and "closed" corporations, and taxing the latter more nearly at personal income tax rates (Musgrave 1965).

The International Monetary Fund (IMF) has been actively involved in the reform of the Korean tax system since the mid-1970s. Initially, this involvement was confined to the design and implementation of a VAT. Subsequently, policy advice on a broad range of tax measures was offered.

Beginning with the tax reform in 1971, the government sought both the advice of tax experts and public consensus on the direction and nature of the tax reform. In 1971 the Ministry of Finance established a Consultative Committee on Tax System to help the minister of finance examine and deliberate on the direction and content for possible tax reform. The committee consisted of thirty members including economics professors, editorial writers of daily newspapers, representatives of the business community, and other tax experts.

The input of this and similar committees thereafter was minimal and unsatisfactory because there was a gap between the perceived and actual role of the committee. For example, members of the committee were not usually consulted beforehand in drafting the tax reform bill. The meetings were usually called in a hurry a few days before the proposal for tax reform was submitted to the cabinet. Drafts of the tax reform bill prepared by the tax bureau were presented to the committee for discussion, but generally passed without any major changes.

It would be less than candid to leave the impression that tax reform and improvement in tax administration faced no problems and opposition. There was resistance from vested interest groups, both public and private. For example, there was passive and active opposition to the introduction of the VAT. Each interest group opposed the new tax for different reasons. Business in general and small businesses in particular opposed the tax because they thought compliance costs would be too high. Even some tax administrators were public about their opposition for similar reasons, pointing out the administrative problems of collecting the tax from retailers and the higher cost of collecting the VAT compared to the taxes it would replace.

With the public's strong demand for tax reform on the one hand and the long-term need for tax reform on the other, the government announced the formation of the Commission on Tax Reform in October 1984. The commission was to carry out a comprehensive study of the Korean tax system and present a tax reform proposal to the government by the end of 1985. The general public favored the formation of the commission.

What distinguished the Commission on Tax Reform of 1984 from the previous Consultative Committees on the Tax System was its organizational structure and the government's willingness to listen. In keeping with the high public demand that prompted the formation of the committee, the government made it clear that it would seek advice from domestic tax experts and nongovernmental experts in future reforms of the tax system.

The organization of the Commission on Tax Reform was two-tiered: The Deliberation Group, an upper echelon in the hierarchy, was to define the broad, general principles of tax reform and solicit public consensus. The Study Group was to study tax reform and complete a comprehensive tax reform plan for final discussion and approval by the Deliberation Group.

The Deliberation Group members represented diverse social backgrounds with influence in the opinionmaking process in Korea. The members were appointed by the Ministry of Finance and included the following professionals: professors, mainly of economics, presidents of major research institutes, representatives of various interest groups, editorial writers and television news editors, ranking government officials at the assistant minister level, and experts with experience in taxation.

The Study Group consisted of four subcommittees: (1) overall tax system and policy, (2) income taxes, (3) consumption taxes, and (4) property taxes. A substantial number of tax experts, mostly professors and some accountants and tax lawyers, were involved. The members of each subcommittee had absolute independence in studying issues assigned to them. Informal meetings were held among the chairmen of the subcommittees to iron out different views and coordinate the work among the subcommittees. The government's views were presented during the informal meetings of subcommittee chairmen. Sometimes these views were ignored by the study groups, but in most cases, they were accepted with some reservation.

The interim report of the Study Group was presented to the Deliberation Group on July 30, 1985 and the final report of more than 800 pages was completed and published in December 1985 under the names of both study groups of the Commission on Tax Reform. A summary was written by members of the Study Group and published by the Ministry of Finance. The unabridged version, containing all working papers, amounting to several thousand pages, was not released.

The original schedule set up by the Ministry of Finance was to solicit the final report by the Commission on Tax Reform and finalize the whole process of tax reform by the end of 1986 so that the new tax reform would be effective as of January 1987. However, drastic changes in the political situation prevented the government from submitting the tax reform bill as scheduled.

Although it is too early to evaluate the organization of the Commission on Tax Reform and its effects on the tax reform efforts, formation of the commission proved effective in establishing a channel of communication between nongovernmental tax experts and those in government. This new willingness on the part of government to make full use of domestic tax experts and nongovernmental experts in formulating tax policy will remain a unique experience in the exercise of tax reform in Korea.

In 1992 the Korean government established the Korea Tax Institute (KTI), a policy-oriented think tank affiliated with the Ministry of Finance, to meet the need for comprehensive research on tax policy. It remains to be seen how the Commission on Tax Reform and KTI will interact and complement each other. It is most likely that the KTI will provide technical analysis and the Commission on Tax Reform will channel communication between taxpayers and government officials.

Taxation and Resource Allocation

Taxation is a process by which resources are transferred from the private sector of the economy to the government. In this process tax policy affects the allocation of resources and the distribution of income. Tax policy, in interaction with other economic policies, can exert a significant influence on economic growth, price stability and unemployment, and the balance of payments.

In the long term, growth requires an increase in investment. Tax policy affects the rate of economic growth by affecting the rate of investment. The tax system is believed to have contributed significantly to high economic growth in Korea although the actual measures employed are not unknown in other industrial or developing countries. Indeed, many similar tax and nontax measures have been used to create investment incentives and promote business activities in other countries.

The most important aspect of tax policy in connection with Korea's rapid industrialization has been the tax incentive system. The government applied various tax incentives, such as a preferential depreciation allowance and tax reductions and exemptions, to induce the private sector to engage in certain economic activities deemed desirable for industrial development and export growth. Enterprises engaging in heavy and chemical industries and export industries have benefited most from these tax incentives.

In the early phases of industrialization, the tax incentive system played only a limited role, largely because of prevailing market imperfections, the relatively small share of the industrial sector, and the unorganized tax administration. In the 1960s, the core of Korea's industrial policy was to promote exports and design tax incentives to promote foreign exchange–earning activities. In the 1970s, during the course of the so-called heavy industrialization period of the Korean economy, incentives for export promotion were actually reduced. Although direct policies such as credit allocation continued to play a major role during the decade, tax incentive policies began to influence the allocation of resources to match the increasing reliance of the government on market forces.

One very important characteristic of Korean investment incentive policies is that they were, in most cases, employed as a means of affecting the sectoral allocation of investment resources. In a few cases they were also employed as countercyclical policy measures.

Evaluating these tax incentives is very important and interesting from a theoretical and policy point of view. Although sophisticated models have been developed, there is limited data on the interaction of tax variables with nontax variables, so it is difficult to evaluate the full effect of tax incentives on the economy. To our dismay, few studies have been conducted in Korea to investigate the impact and effectiveness of tax incentives. The following sections review the evolution of the tax incentive system and summarize statistical data and available empirical studies.

Major Tax Incentives

Among the extensive tax incentives applied the most important are special depreciation, tax holidays, and investment tax credits.

In addition to the general depreciation system, a special depreciation system was introduced in the corporate tax law in 1962. This included a 20 percent rate for machinery and equipment in manufacturing industries that operated for more than sixteen hours daily, which was extended to heavy equipment in construction industries in 1969. During 1975–82 a special depreciation rate of 80 percent was allowed for mining, fishery, manufacturing, and construction industries, while a special depreciation rate of 40 percent was applied to other industries. During 1983–85 the special depreciation rate was reduced to 20 percent for machinery and equipment in the mining, manufacturing, electricity, and water and gas industries, and heavy equipment in construction industries.

Since 1986 the special depreciation system has allowed different rates for machinery and equipment in the mining and manufacturing industries according to the operating hours per day. For example, a special depreciation rate of 20, 30, and 50 percent has been respectively applied to firms operating twelve to fifteen hours, fifteen to eighteen hours, and more than eighteen hours a day. In 1993 this special depreciation system was extended to the computer industry.

Tax holidays introduced in 1949 exempted all corporate tax for the first industrial group (consisting of petroleum refining, shipbuilding, iron and steel refining, copper refining, cement manufacturing, and chemical fertilizer) during the first five years of normal operation of those business, and for the second industrial group (consisting of mining and plate glass industry) during the first three years. In 1954 the tax exemption given to the first industrial group was reduced from five to three years, while the tax exemption to the second industrial group was reduced from three years to one year. In addition, for the next two years the corporate income tax allowed an exemption of two-thirds and one-third of profits, respectively. In 1963 the mining industry moved from the second industrial group to the first industrial group, and the corporate tax was exempted for the first industrial group for four years. Petroleum refining and the chemical industry were excluded from tax holidays in 1966.

The investment tax credit replaced tax holidays in 1967 because the latter had distorted firms' investment patterns by discouraging investment for improving industrial facilities and expanding production facilities. The investment tax credit allowed a 6 percent tax credit for automobiles, chemical fertilizers, electricity generation, food processing, iron and steel, machinery, petroleum chemical products, pulp, shipbuilding, synthetic fibers, and the electronics, mining, and construction industries. In 1974 the investment tax credit was absorbed into the special tax treatment for strategic industries.

After the oil shock in 1974 tax holidays, the investment tax credit and special depreciation were integrated into the special tax treatment for strategic industries. The strategic industries included chemical fertilizers, electricity generation, iron manufacture, machinery, mining, petroleum chemicals, shipbuilding, water and gas, and the electronics, and iron and steel industries. Each firm in a strategic industry was allowed to choose only one type of special tax treatment. Tax holidays exempted 100 percent of corporate tax for the first three years and 50 percent for the next two years. The investment tax credit allowed an 8 percent tax credit for machinery and buildings (10 percent tax credit for Korean-made machinery and buildings). The special depreciation system allowed a 100 percent depreciation rate.

With the revision of tax law in 1982, tax holidays were abolished completely and the investment tax credit was limited to the machinery and electronics industries and reduced from 8 to 6 percent. Petroleum chemicals, iron and steel, chemical fertilizer, electricity, and water and gas were excluded from enjoying the investment tax credit.

In 1982 5 percent of tax-free reserves for investment was given a 100 percent special depreciation allowance. In this context, special depreciation remained the most important preferential tax. In 1986 special depreciation disappeared as an element of special treatment, and all special tax treatments for strategic industries in the TERCL were eliminated. Following the elimination in 1986 of special tax treatments for strategic industries, the special depreciation system was substantially scaled back in the corporation tax law, but there was still a 10 percent investment tax credit for facilities that save energy, reduce pollution, and improve safety and transportation systems, and for machinery in small and medium-size industries.

Korea's complicated tax incentive system includes programs for business reorganization, countercyclical investment incentives, energy conservation and environmental protection, export promotion, foreign investment, key industries, local industrial development, research and development (technical innovations), resources development, social welfare, and small and medium-size firms. The incentives for export promotion, key industries, research and development, and small and medium-size firms seem to have been the most important in terms of size and structural implications, so they will be examined in more detail.

Tax incentives for export promotion. From the earliest stage of economic development, Korean policymakers fully recognized the importance of foreign exchange. Naturally, foreign exchange–earning activities enjoyed substantial tax privileges, especially in the 1960s. From 1953 to 1960, all income from ships and aircrafts involved in international transportation activities had been tax exempt. In 1960 a fully fledged tax incentive system for export promotion was introduced. A 30 percent corporate tax exemption was given on income from export business, and a 20 percent exemption was given to income from the sale of goods and services to foreign military forces based in Korea and from foreign currency income derived from the tourist industry. In the following year, this trend was further reinforced by raising the exemption rate to a uniform 50 percent for all foreign exchange–earning activities.

This incentive system played an important role in the export-oriented industrialization period in the 1960s. A 50 percent exemption on corporate taxes sounded drastic and could well have provoked counteractive protectionist measures from some importing countries, even though the absolute level of the effective corporate tax rate on exporters after allowing the exemption was still high because of the high inflation rate, relatively low tax depreciation rate, high discount rate, and high statutory tax rate (30 to 45 percent plus surcharges at the margin). However, the effective tax rate differential between the exporting and domestic sectors created by the exemption was substantial, and resources shifted to the export sectors.

This system was replaced in 1973 by two tax-free reserve systems: (1) a reserve for losses in the export business, and (2) a reserve for overseas market development. Under the overseas market development reserve system, 1 percent of the value of total exports can be deducted from taxable income for tax-free reserves, and after a two-year grace period the amount is evenly spread over the following three years and added to taxable income. The export loss reserve system worked on a similar basis except for the method of calculating the maximum reserve amount, where the reserve amount could not exceed the lesser of either 1 percent of total export value or 50 percent of the profit from export business.

In 1977 another tax-free reserve system—reserves for price fluctuation—was added to the list of tax incentives for export promotion. The maximum amount that can be reserved for taxable income deduction is 5 percent of the inventory asset value at the end of the accounting period. The reserved amount is added to the taxable income after a one-year grace period.

Finally, export incentives have been provided in the form of a special depreciation system. Although the special depreciation system first became effective in 1962, export industries began to receive special benefits from this system in 1963. Machinery and equipment directly employed by foreign exchange–earning activities could be depreciated at rates 30 percent higher than the corresponding statutory depreciation rates. Effective from 1967, two different special depreciation rates were applied to the export share of total sales. If the share was greater than or equal to 50 percent, 30 percent special depreciation was allowed. If the rate was less than 50 percent, the applicable rate was 15 percent. Effective from 1971, the special depreciation rate for firms whose export shares are less than 50 percent was calculated by applying the multiplicative formula, 30 percent times the export share.

From the early 1970s onwards, however, the government tried to reduce the scope of export incentives. The replacement of the 50 percent corporate tax exemption by two tax-free reserves in 1973 was the most significant change in this direction. In 1975 the system of prior tariff exemptions on imported inputs used in export production was changed to a drawback system.

Tax incentives for key industries. With the creation of the modern corporate tax system in 1949, a generous tax holiday was provided for the industries deemed "important" to national economic development, including heavy industries such as chemical fertilizers, basic metal, machinery, petrochemicals, and shipbuilding. The major tax reforms of 1967 replaced the tax holiday incentive with an investment tax credit system, while generally maintaining the list of privileged sectors. In the early 1970s, incentives for heavy industrialization were reinforced. To cope with the serious economic recession, an extensive set of investment incentives was introduced in 1972. It was not, however, a pure countercyclical measure; rather, it reflected the strong commitment of policymakers to heavy industrialization. In the 1974 tax reform, all major incentives to promote key industries were unified and rearranged under the title of "special tax treatment for key industries" in the TERCL. The special treatment provided three options to qualified firms in the selected heavy industries: tax holidays, the investment tax credit, and special depreciation. Another major tax reform implemented in 1981 abolished the tax holiday option altogether and limited eligibility for the investment tax credit option to the machinery and electronic industries. Hence, the special depreciation system became the most important tax incentive for key industries. Table 6.6 summarizes details of institutional changes in the preferential tax treatment of investment in strategic industries.

Tax incentives for research and development. Details of the evolution of tax incentives for R&D efforts are summarized in table 6.7, which shows that preferential tax treatment for R&D has become a major form of tax incentive and that the functional tax scheme has replaced the industry-specific incentive scheme.

With the passage of time, the spectrum of tax incentives has increased to include the following:

- An allowance for reserves for technological development up to an amount equal to 1.5 percent of the gross receipts

Table 6.6 Preferential Tax Treatment for Strategic Industries

	Before 1974	*1975–81*	*Since 1982*
Strategic industries	Shipbuilding Iron and steel Chemical fertilizer Power generation plant Synthetic fibers Motor vehicles Machinery Electronics Industrial chemical Pulp	Petrochemicals Shipbuilding Machinery Electronics Iron and steel Petroleum refining Power generation plant Chemical fertilizer Aviation	Naphtha cracking Iron and steel Machinery Electronics Shipbuilding Aviation
Preferential treatment	• Investment tax credit • Income tax deduction • Corporation of business tax exemption	Direct forms of preference • Investment tax credit • Income tax deduction • Corporation or business tax exemption	Direct forms of preference • Investment tax credit Indirect forms of preference • Preferential • Depreciation • Tax-free reserves for investment
Changing features	• Types of preferential tax treatment different for each benefiting industry • Direct forms of tax credit or tax exemption preferred	• Tax holidays rather than tax exemption • Direct as well as indirect forms in operation • Industries receiving preferential tax treatment formally defined as strategic industries	• Indirect forms of preferential treatment becoming dominant over direct forms

Source: Ministry of Finance.

Table 6.7 Preferential Tax Treatment for Research and Development Expenditures

	Until 1974	*1975–81*	*Since 1982*
R&D expenditures receiving preferential treatment	• Foreign engineer's wage income • Corporation tax of research institutes • Venture business	• Foreign engineer's wage income • Corporation tax of research institutes • Reserved funds for technology development • Venture business	• Foreign engineer's wage income • Corporation tax of research institutes • Reserved funds for technology development • Research facility investment • Job training expenses • Venture business
Preferential treatment	• Tax credit or exemption	• Tax credit or exemption • Tax-free reserves for expenses in technology development	• Tax credit or exemption • Tax-free reserves for expenses in technology development • Preferential depreciation
Characteristics of tax incentives	• Preferential treatment applying only to a few selected and exceptional cases	• Tax preferences generally applying to R&D activities in indirect form such as tax-free reserves	• Pan-industry type functional approach tax preferences rather than industry-specific one • Indirect forms of tax preferences such as tax-free reserves and preferential depreciation

Source: Ministry of Finance.

- A 10 percent tax credit on the expenses for the development of technology and manpower
- A tax credit of 10 percent on the additional expenses for technology and manpower training
- A tax credit or special depreciation on investment in facilities for research and development
- A tax credit for investment in a venture business
- An exemption for foreign engineers' wage income
- A 50 percent income tax deduction for technology service businesses for the first six years.

Tax incentives for small and medium-size firms. When preferential tax treatment is not constrained by industry specification, it is called a functional tax incentive. The functional incentive scheme is better than the industry-specific incentive scheme at correcting market failures. In recent years the government has favored the functional tax incentive scheme over the industry-specific scheme, which implies fewer tax distortions in the allocation of resources. In the following pages, functional tax incentive schemes for small and medium-size firms (SMFS) and research and development (R&D) are summarized.

During the 1960s and 1970s the tax incentives geared specifically to promote SMFS were nominal. In 1968 a 30 percent special depreciation was granted to SMFS operating in the mining or manufacturing sectors for investment in machinery and equipment. The special depreciation rate was raised to 50 percent by the tax reform implemented in 1977.

In the 1981 revision of the TERCL, which went into effect in the following year, the investment reserve system was made available to SMFS. The annual reserve limit was 15 percent of the book value of fixed business assets evaluated at the end of the previous accounting period. This amount is deducted from taxable income and, if after a grace period of four years the actual investment expenditures exceed the reserve amount, it is evenly spread over the following three years and added to taxable income. If actual investment expenditures are below the reserve amount, the difference is immediately added to taxable income in the fourth year. The SMFS eligible for this privilege are in the construction, fishing, manufacturing, mining, and transportation industries.

For a market economy to function properly, it must be competitive. Competition depends on the presence of many small firms and the absence of overwhelmingly large ones. Although the large firms are still small by international standards, conglomerates have been a dominant force in domestic production and employment. In the 1960s and 1970s most tax incentives were targeted at large firms to capitalize on economies of scale. In early 1980s, the government adopted a new policy of helping people with entrepreneurial inclinations and know-how to establish themselves as independent businessmen with a series of tax incentive measures.

Tax incentives for the establishment of SMFS include those for newly organized firms and those for companies investing in newly organized firms. When SMFS are newly established in agricultural or sea districts to run businesses in the construction, fishing, manufacturing, mining, or transportation industries, or when SMFS are organized in technology-intensive industries, personal income tax is exempted for the first four years and reduced to 50 percent for the subsequent two years. Furthermore, the property tax on the business assets of newly organized SMFS is reduced by 50 percent for five years and the acquisition tax and the registration tax are reduced by 50 percent for two years.

Tax incentives for companies investing in newly organized SMFS include (1) nontaxation of capital gains, (2) tax-free reserves for investment loss, and (3) separate taxation of dividend income at a 10 percent rate. With respect to capital gains accruing from the transfer of real estate for business purposes at the time of merger, the personal income tax and the special surtax on corporation income are waived.

Given that most of the firms eligible for industry privileges have been large, despite the above-mentioned incentives for SMFS, the argument that the SMFS have not been treated on an equal footing with the large firms appears plausible. Whether the recent change in policy direction will lead to an increased role for SMFS in the economy and to greater competition remains to be seen.

Effective Tax Rates and Cost of Capital

In order to examine the role of various tax incentives in allocating resources among industrial sectors it is necessary to calculate the effective tax rate by sector. Using a modified version of the effective tax rate formula developed by D. W. Jorgenson and M. A. Sullivan, Kwack (1984) estimated the effective tax rate by sector (table 6.8). According to his estimates, which incorporated a detailed account of complicated tax incentives over time, the relative size of incentives provided to key industries was substantial, particularly during the latter half of the 1970s.

Table 6.8 also shows that a typical Korean firm paid corporate income taxes at an extremely high effective rate in spite of the numerous and complicated tax incentives during this period. It should be noted that the effective tax rate is affected not only by the statutory tax rate and tax incentives, but also by the discount rate and the inflation rate. The major reason for the high effective tax rate shown in table 6.8 is the increase in the relative price of capital goods and a tax depreciation system based on historical cost. Although it is difficult to evaluate the net contribution of all kinds of tax preferences given to the heavy industrialization effort during the 1970s, it is reasonable to conclude that tax incentives were actively employed in the course of industrial development.

Table 6.8 Effective Marginal Tax Rate by Sector
(percent)

Sector	*1973*	*1975*	*1978*	*1980*	*1981*	*1982*	*1983*
Processed food, beverage, and tobacco	50.6	55.1	42.8	46.7	57.1	58.6	39.5
Textile, leather, paper, and printing	49.8	54.3	42.1	46.1	56.2	57.6	38.7
Construction materials	50.3	54.6	42.8	46.5	56.6	58.1	39.7
Chemical products							
General	48.9	54.2	41.1	45.3	55.8	57.1	37.6
Special	46.3	38.8	29.5	32.0	42.4	50.8	34.8
Basic metal and metal products							
General	49.0	53.2	41.9	45.5	55.0	56.4	38.1
Special	46.9	38.7	31.0	32.9	42.6	50.8	35.8
Machinery, electrical, and electronic equipment							
General	49.3	53.7	42.0	45.8	55.6	57.0	38.4
Special	41.7	39.1	30.9	33.0	43.0	51.2	36.0
Statutory maximum tax rates	40.0	40.0	40.0	40.0	40.0	38.0	33.0
Inflation rates in capital goods market (3-year moving average)	17.0	33.7	8.2	19.4	24.3	21.5	9.7

Note: General rates are for firms that do not qualify for special tax treatment; special rates are for firms that do qualify for special tax treatment.
Source: Kwack (1984).

According to a study by Jun Young Kim (1993) summarized in table 6.9, the effective tax rate in the corporate sector fluctuated over time because of different tax policy measures. The effective tax rate was 32.7 percent in 1966 , 45.4 percent in 1970, dropped during the late 1970s, and was on a upward trend since the mid-1980s. The effective tax rate was lower than the statutory nominal rate. The largest discrepancy between the effective tax rate and the nominal tax rate occurred in 1977, when the nominal rate was 53 percent and the effective rate was 35 percent.

Examination of the effective tax rate by type of asset shows that the effective tax rate of transportation equipment was higher than other assets. The effective tax rate of funds from debt financing was negative while the effective tax rate on new shares was substantially higher than the statutory nominal tax rate. The discrepancies of the effective tax rates among sources of finance have decreased over time because of tax reforms in 1980s, which lowered the negative effective rate associated with debt finance.

One very common and accepted method of examining the effects of tax incentives on corporate investment behavior is to estimate the marginal cost of capital. There are several different approaches to deriving a cost of capital formula for a given economy using a specific type of financial and tax system.

Using the Hall-Jorgenson method, Kwack developed a cost of capital formula applicable to the Korean economy based on the assumption that the financial resources have been allocated efficiently through the perfectly competitive curb market. His model incorporates not only tax incentives such as tax holidays, the tax credit, and tax-free reserves, but other unique features of the economy including the collateral loan system and export loan system.

Table 6.9 Effective Tax Rate by Asset and Source of Finance
(percent)

Asset	Finance source	1966	1970	1975	1980	1985	1989
Buildings and structures	Debt	–26.3	–143.8	—	–181.0	–54.3	–52.9
	New shares	69.9	76.0	77.5	75.6	72.1	73.4
	Retained earnings	37.5	46.4	47.9	47.6	43.4	43.8
	Average	32.1	44.2	33.5	36.5	38.1	39.3
Machinery and equipment	Debt	–154.8	–236.3	—	–141.3	–135.8	–98.8
	New shares	71.8	76.7	79.5	76.5	73.8	75.9
	Retained earnings	37.1	42.9	47.9	45.0	39.9	42.5
	Average	31.3	40.9	37.9	38.1	36.4	40.2
Transportation equipment	Debt	–248.2	–202.8	—	–118.3	–123.2	–77.5
	New shares	73.0	79.4	82.0	79.2	77.5	79.5
	Retained earnings	38.6	49.9	55.9	53.1	47.7	50.3
	Average	30.8	47.3	41.5	45.7	42.7	46.8
Overall average	Debt	–142.9	–173.4	—	–165.9	–104.9	–84.1
	New shares	71.6	76.9	79.4	76.6	73.7	75.8
	Retained earnings	37.3	44.7	48.7	46.4	42.6	43.5
	Average	32.7	45.4	38.0	41.0	39.2	42.8
Statutory nominal corporate tax rate		38.5	45.0	53.0	53.0	43.7	43.7

— Not available.
Note: Statutory tax rate includes corporate tax rate, defense tax rate, and resident tax rate.
Source: J.Y. Kim (1993).

An empirical estimation of the cost of capital by Kwack is presented in table 6.10, which shows the trends of the cost of capital by type of asset, a proxy for the neutral cost of capital, and a measure of distortion due to tax incentives. The estimated results reveal several interesting features. The absolute level of the cost of capital is rather high because it assumes the curb market is efficient and that firms resort to the curb market for the financing of investment at the margin. Among the types of investment assets the estimated cost of capital is highest in transportation equipment and lowest in building and construction.

The degree of distortion due to the tax system can be measured by the difference between the estimated cost of capital and the neutral or distortion-free cost of capital, where the latter is by definition the discount rate plus economic depreciation minus the expected inflation rate. One salient feature of table 6.10 is that the tax incentive system provided substantial support for investment in machinery and equipment in the early 1960s, the 1972–74 period, and the 1978–81 period, as shown by the negative figures of tax distortion. With the introduction of the 1981 tax reform, effective in 1982, the degree of tax distortion rose substantially.

Table 6.11 provides a summary of a most recent study by Jun Young Kim, which examined the cost of capital by type of asset and source of finance (1993). The cost of capital in the corporate sector (bottom line of table 6.11) was high in the late 1960s, declined in the early 1970s, and then increased from the mid-1970s to a peak in 1980. The decline in the early 1970s was due to the 1972 emergency measure and the 1974 tax reform. Since the early 1980s the cost of capital has been on a downward

Table 6.10 Cost of Capital by Type of Asset and a Measure of Distortion
(percent)

	Estimated cost of capital		
Year	*Building and construction*	*Machinery and equipment*	*Transportation equipment*
1963	.365	.426	.430
1964	.392	.423	.448
1965	.484	.517	.564
1966	.529	.601	.717
1967	.497	.581	.694
1968	.524	.589	.669
1969	.570	.649	.742
1970	.491	.603	.665
1971	.438	.541	.612
1972	.297	.401	.488
1973	.223	.300	.415
1974	.265	.324	.414
1975	.314	.371	.489
1976	.309	.385	.495
1977	.263	.371	.496
1978	.263	.416	.542
1979	.239	.429	.551
1980	.234	.414	.524
1981	.168	.318	.419
1982	.239	.375	.462
1983	.176	.297	.384

a. Real discount rate plus economic depreciation rate.
b. Estimated cost of capital minus neutral cost of capital.

Table 6.11 Cost of Capital by Types of Assets and Source of Finances
(percent)

Asset	*Finance source*	*1966*	*1970*	*1975*	*1980*	*1985*	*1989*
Buildings and structures	Debt	11.3	6.6	−1.2	4.0	8.1	6.5
	New shares	69.6	67.1	58.0	78.7	44.8	37.3
	Retained earnings	33.4	30.0	24.8	36.7	22.1	17.7
	Average	27.4	25.0	17.0	26.9	18.9	15.3
Machinery and equipment	Debt	8.2	3.0	−2.4	3.0	5.3	5.0
	New shares	74.1	69.0	63.1	82.0	47.8	41.3
	Retained earnings	33.2	28.2	24.8	35.0	20.8	17.3
	Average	26.5	22.7	17.4	26.0	17.7	15.0
Transportation equipment	Debt	6.0	3.2	−2.3	4.6	5.6	5.6
	New shares	77.4	78.0	71.8	92.5	55.6	48.6
	Retained earnings	34.0	32.1	29.3	41.0	23.9	20.0
	Average	26.3	26.4	19.3	31.4	20.2	17.6
Overall average	Debt	8.6	3.8	−2.1	3.4	6.1	5.4
	New shares	73.5	69.6	62.7	82.2	47.6	41.0
	Retained earnings	33.3	29.1	25.2	35.9	21.4	17.6
	Average	27.8	25.4	18.0	28.7	19.1	16.2

Source: J.Y. Kim (1993).

Cost of neutral capital[a]			Measure of distortion[b]	
All assets	Machinery and equipment	All assets	Machinery and equipment	All assets
.391	.546	.507	–.120	–.116
.407	.585	.556	–.162	–.149
.502	.578	.545	–.061	–.043
.600	.597	.546	.004	.054
.578	.595	.526	–.014	.042
.582	.591	.533	–.200	.049
.657	.552	.491	.097	.166
.586	.539	.466	.064	.120
.529	.500	.440	.033	.089
.402	.422	.362	–.021	.040
.305	.330	.285	–.030	.020
.330	.362	.327	–.038	.003
.394	.349	.316	.022	.078
.392	.371	.328	.014	.064
.376	.369	.315	.002	.061
.408	.420	.351	–.004	.057
.398	.437	.345	–.008	.053
.380	.442	.353	–.028	.027
.296	.333	.261	–.015	.035
.350	.308	.240	.067	.110
.278	.277	.218	.020	.060

Source: Choi and others (1985).

trend. The machinery and equipment enjoyed lower cost of capital than other assets. The most striking feature of table 6.11 is that the Korean tax system discriminates heavily against new shares. For all assets the cost of debt was lowest and the cost of new shares was higher than that of retained earnings. The discrepancies in the cost of capital among sources of finance had been declining throughout the 1966–89 period.

Tax Policy and Economic Growth

Given that Korea relied heavily on tax policy in general and tax incentives in particular for the sectoral resource allocation, one might ask about the relationship between economic growth and tax policy. Unfortunately there are only a few empirical investigations of this relationship for Korea.

One study by Trela and Whalley (1992) shows the effects of the tax structure on growth and another study by Evans (1988) examines the effects of tax rate and tax distortion on economic growth. Trela and Whalley show how the tax structure was modified over time to support Korea's economic growth strategy and they developed a general equilibrium model to estimate the contribution of the tax structure to the growth rate. They concluded that the changing tax system had probably facilitated rather than fueled high growth. The average annual increase in gross domestic product (GDP) attributable to tax policies during 1963–82 is small, only 0.54 percent or less than 10 percent of actual average annual Korean growth in real GDP. When they broke down the relatively small contribution of tax policies to growth into two separate effects—direct tax deductions (mainly corporate tax rebates for exports) and indirect tax exemptions (rebates of sales and excise taxes on exports)—they found that indirect tax exemptions had contributed far more to economic growth than had direct tax measures.

In two papers which investigated whether a simple neoclassical model can explain output and investment in Korea, Evans (1988, 1990) reported regression results showing the relationship between tax policy and output. According to these results, increasing the average tax rate by 1 percentage point lowers output by 3 percentage points relative to the trend, while the resulting increase in marginal tax rates falls primarily on consumption as private investment is affected much less than consumption. Moreover transferring one unit of resources from taxpayers to the recipients of transfer payments reduces output by 1.13 to 1.19 units.

Taxation and Income Redistribution

In view of Korea's rapid economic growth the present distribution of income is judged by many observers to be equitable by international standards. Relatively equal distribution of income has been traced to several factors, including the leveling of Korean society by the Japanese, postliberation land reforms, asset destruction during the Korean War, confiscation of illegally accumulated wealth after the downfall of some previous regimes, a labor-intensive strategy during the early development stage, and the even distribution of higher education.

Available figures on the distribution of income are summarized in table 6.12. The overall distribution of income improved during 1965–70, deteriorated somewhat during 1970–76, and then improved again during the 1980s, as evidenced by the fact the Gini coefficient fell from 0.344 in 1965 to 0.332 in 1970, rose to 0.391 in 1976, and then fell to 0.336 in 1988. Improvements in the distribution of income during 1965–70 were due to the creation of employment through labor-intensive, export-led growth strategies, while the deterioration in income distribution during the late 1970s was caused by large capital-intensive investment and severe inflation. Improvement in the overall size distribution of income can be attributed to price stability, the realignment of the industrial incentive system, and several government measures targeted at raising the welfare of low-income groups.

The rather equitable distribution of income and generally low level of tax burden that have existed so far do not imply that the general public is satisfied with the present distribution of income and tax burdens.

Table 6.12 Income Distribution
(percent, except for the Gini coefficient)

Income decile	1965	1970	1976	1980	1985	1988
1	1.32	2.78	1.84	1.56	2.06	2.80
2	4.43	4.56	3.86	3.52	4.02	4.46
3	6.47	5.81	4.93	4.86	5.32	5.47
4	7.12	6.48	6.22	6.11	6.39	6.39
5	7.21	7.63	7.07	7.33	7.47	7.35
6	8.32	8.71	8.34	8.63	8.76	8.43
7	11.31	10.24	9.91	10.21	10.21	9.79
8	12.00	12.17	12.49	12.38	12.41	11.64
9	16.03	16.21	17.84	15.93	15.42	14.58
10	25.78	25.41	27.50	29.46	28.29	29.09
Bottom 40 percent (A)	19.34	19.85	16.85	16.06	17.71	19.12
Top 20 percent (B)	41.81	41.62	45.34	45.39	46.71	43.67
Ratio of A/B	0.4626	0.4716	0.3716	0.3538	0.4052	0.438
Gini coefficient	0.3439	0.3322	0.3908	0.3891	0.3631	0.35

Source: Korea Development Institute and Economic Planning Board data.

Koreans are very equity oriented and equality minded. The skewed distribution of wealth and resulting unequal distribution of income have remained a thorn in the side of economic policy. People are very critical of the high concentration of industrial fortunes in the hands of the wealthy few and excessive holdings of land and dwellings by the rich. Recently there has been an increasing concern with inequities in the tax burden.

Recent reports on wealth and land distribution by the Korea Development Institute and the Public Concept of Ownership of Land Committee (PCOL) suggest that there might be a discrepancy between statistics and the actual or perceived state of income distribution. According to Kwon (1990), the distribution of wealth was far worse than that of income. The Gini coefficient of wealth distribution is 0.58, while that of income distribution is 0.35. As table 6.13 shows distribution of financial assets is relatively more concentrated among the well-to-do, with the upper 10 percent of sample households owning 61 percent of total financial assets, and the bottom 30 percent with no financial assets. The top 30 percent of assets holders own 72.1 percent of financial and real assets in

Table 6.13 Distribution of Wealth and Landownership, 1988
(percent)

Income decile	Financial assets	Real assets	Total assets	Land
1	0.00	0.00	0.005	0.000
2	0.00	0.01	0.013	0.002
3	0.00	0.02	0.022	0.003
4	0.00	0.03	0.034	0.005
5	0.02	0.05	0.049	0.010
6	0.04	0.07	0.066	0.019
7	0.06	0.09	0.093	0.032
8	0.10	0.12	0.120	0.054
9	0.17	0.17	0.170	0.108
10	0.61	0.44	0.431	0.769

Source: Kwon (1990).

nominal terms. The report by the PCOL shows that the Gini coefficient of landownership distribution is 0.85, and the top 10 percent of holders own 76.9 percent of privately held land. It is quite likely that the unequal distribution of financial and real assets (in particular land) severely enhances income inequality, partly because the assets are the main sources of income, and partly because the rapid price increase of real assets such as land and housing generate a lot of unearned capital gains.

Even though enhanced welfare through more equal distribution of income and wealth has been a centerpiece of every Five-Year Economic Development Plan, this has not been measured. Although the litany of equity or better income distribution was always one of the stated objectives in all tax reforms, the Korean tax system does not have a capacity for income redistribution. It retains a significant regressive element, as all the empirical studies examined below quite strongly suggest. The method most frequently relied upon to improve equity in the distribution of the tax burden is increasing tax exemptions for the personal income tax for low-income groups. However, this approach has weakened the role of personal income tax, thereby reducing the redistributive capacity of the whole tax system.

The public's major discontent lies in their belief that governmental activities themselves have generated rather than reduced income inequalities. Government policies responsible for worsening income inequities include the following: (1) subsidies implicit in huge sums of domestic and foreign loans that have accrued mainly to big business at lower effective interest rates, thus contributing to asset formation among those in the high-income brackets; (2) tax policy that has relied heavily on consumption taxes rather than on income or property taxes and has made excessive tax concessions to capital income; and (3) development first–oriented policy that resulted in large windfall gains or capital gains, which were not captured by the national treasury through the tax system.

From the distributive point of view horizontal equity as well as vertical equity is of considerable importance. The typical Korean taxpayer's dissatisfaction lies not in the fact that tax burdens are unbearably high, but in the fact that he pays more tax than others with the same income. Three major factors generate both horizontal and vertical inequities in the tax burden: (1) relatively heavy reliance of tax revenues on indirect taxes, (2) the provision of various tax incentives along with the incomplete globalization of personal income tax, and (3) the erosion of the tax base due to tax evasion and underground activities.

Since the government depends heavily on indirect taxes for its revenue, the incidence of the indirect tax burden plays a large role in determining the regressiveness or progressiveness of the overall tax burden. As discussed before, the share of indirect taxes in the total tax revenue is more than 65 percent. Since, as discussed below, the burden of indirect taxes is quite regressive, the overall tax burden is regressive. The special consumption tax, introduced to inject some progressiveness into the consumption tax burden, is actually regressive.

One may claim that the various tax incentives to stimulate investment, exports, and economic growth have benefited many low-income earners, not so much by reducing their tax burden, but by creating employment. Many special tax provisions introduced into tax law to accelerate economic growth benefited only selected groups of taxpayers. The special application of reduced tax rates to certain capital incomes favored high-income classes against low-wage and salary earners. Likewise, special exclusions, deductions, and tax exemptions eroded the tax base and provided higher tax benefits for the high-income class.

Within the income tax system, the tax burden is concentrated on the middle-income class and on wage and salary earners. This is because the government has traditionally mobilized domestic resources for development by reducing the level of consumption through the use of indirect taxes. Expanding the scope of exemptions in the personal income tax schedule under various tax incentives, which was a ritual in all major and minor tax reforms, has put a heavier income tax burden on the middle-income class. The separate and inadequate taxation of interest and dividend incomes, in particular, has contributed to the disparity in income distribution and the concentrated tax burden on the middle-income class. The problem exists for two reasons. First, most forms of interest and dividend incomes are subject only to a final, separate withholding tax rate of 5 or 10 percent. Second, a significant amount of such income is unreported, and effectively free of tax. This loss of revenue represents a serious departure from both horizontal and vertical equity in income taxation, and welfare losses due to the inefficiencies generated.

As table 6.14 shows, 99.7 percent of interest incomes reported by ONTA were separately and finally taxed at the lower withholding rates in 1992. Only 0.6 percent of interest income was included in the global income tax base in 1992. Only presumptive dividends, dividends from major stockholders of open corporations, and dividends from closed corporations were subject to global taxation and these together accounted for only 40.9 percent of reported dividends in 1992. The separate rates of 5, 10, and 20 percent (including defense and education surcharges and the inhabitant tax) on interest and dividend incomes is very high for low-

Table 6.14 Separate and Global Taxation of Interest and Dividend Income

	1982		1986		1992	
Taxation	*Amount (billion won)*	*Percentage of income subject to tax*	*Amount (billion won)*	*Percentage of income subject to tax*	*Amount (billion won)*	*Percentage of income subject to tax*
Interest income						
Separate taxation	1,812	97.7	2,819	99.9	136,712	99.4
Global taxation	43	2.3	19	0.1	843	0.6
Dividend income						
Separate taxation	161	58.3	238	66.5	6,747	59.1
Global taxation	115	41.7	119	33.5	4,662	40.9

Source: Ministry of Finance data.

Table 6.15 Comparison of Factor Income Reported on Personal Income Tax Returns versus National Income Statistics
(billion won except where noted)

	1977			1983		
Income	National income (A)	ONTA returns (B)	Ratio of B/A (percent)	National income (A)	ONTA returns (B)	Ratio of B/A (percent)
Compensation of employee	5,845	4,004	69.2	255,875	191,855	75.0
Income from property	1,886	594	31.6	77,316	24,811	32.1
Rent	670	111	16.6	23,772	2,799	11.8
Interest	844	282	33.4	49,247	19,794	40.2
Dividend	372	134–201	35.9–54.0	4,347	2,218	51.0
Incomes from unincorporated enterprises	4,517	2,572	56.9	—	—	—
Nonagricultural	1,956	926	47.2	27,952	21,053	75.3
Agricultural	2,561	1,648	64.4	—	—	—

— Not available.
Source: IMF (1979); Commission on Tax Reform (1985).

income people because their income level is well below the bracket with marginal tax rate at that level, but relatively low for the rich because their marginal tax rate usually exceeds the separate rate.

Erosion of the tax base arises from two sources: nontaxation or exclusion of some incomes from the tax base and tax evasion. Income tax on capital gains from the transfer of real properties is administered as a separate tax and capital gains from stock exchanges are accorded nontaxable status. Such nontaxation of capital gains from selling stocks is a major cause of the substantial erosion of the personal income tax base and provides the greatest advantage to high-income taxpayers. It also results in large disparities in the tax burden among the different income classes.

The issue of tax evasion is particularly important given that the tax system fails to capture various types of income reported in the national income accounts (table 6.15). One very important point that table 6.15 shows is that there are big variations among types of income in the "capture ratio," defined as income reported on tax returns as a percentage of factor income in the national income account. For example, in 1983 only 40.2 percent of interest income and 51.0 percent of dividend income reported in the national income account were subject to taxation. In particular, tax is levied on only 11.8 percent of rental income.

Business income and tax are particularly vulnerable to underreporting. The magnitude of underreported business income and tax estimated by Roh (1992) is shown in table 6.16. The estimated ratio of underreported income and tax to the actual income and tax actually paid, according to the expenditure method, were in the range of 8 to 25 percent and 13 to 39 percent in the late 1980s.

The study of the effect of a particular tax or a tax system on the distribution of income or economic welfare is tax incidence analysis. The key question is who actually bears the burden of the resources transferred to

Table 6.16 Ratio of Underreported Business Income and Tax to Actual Amounts
(percent)

Year	Income ratio: At average income	Income ratio: Upper limit	Tax ratio: At average income	Tax ratio: Upper limit
1986	12.1	24.8	18.2	39.0
1987	11.3	11.3	17.7	17.0
1988	7.9	14.8	12.8	21.4
1989	11.9	18.4	19.8	24.5

Source: Roh (1992).

the government by tax. To estimate tax incidence a detailed microeconomic database relating the level and distribution of consumption of specific commodities is required. Since these data are not available, empirical studies have been made using rough estimates of underlying variables and highly simplified analytical assumptions. Accordingly, the empirical studies available are based on different data and different assumptions. Major results are summarized in tables 6.17, 6.18, 6.19, and 6.20.

The study by Heller (1981) of the redistributive impact of the tax system in 1976 shows that regardless of assumptions chosen for the shifting of taxes, the burden of the tax system relative to income is roughly proportional up to the top 10 percent. More specifically, Heller finds that under a regressive set of assumptions the poorest 10 percent pay about 16.4 percent of income, an almost constant percentage up to the ninth income decile, and then rising to 21.6 percent in the highest income decile. The global income tax, the gift and inheritance tax, the assets revaluation tax, and the farm land tax are the principal progressive elements of the tax system. Indirect taxes such as the VAT, the special consumption tax, the liquor tax, monopoly profits tax, and customs duties are regressive (table 6.17).

Table 6.17 Overall Tax Burden in Korea
(tax burden as a percentage of income)

Taxes	Income decile: 1st	2nd	3rd	4th	5th	6th	7th	8th	9th	10th
Heller										
National taxes 1976										
Progressive assumptions	12.67	13.78	11.69	13.63	13.04	13.81	13.35	12.70	14.02	22.79
Regressive assumptions	14.35	14.97	12.62	14.88	14.22	15.02	14.55	13.53	14.59	20.10
Local taxes 1976										
Progressive assumptions	0.99	0.86	0.76	0.82	0.93	1.05	1.06	1.28	1.60	2.16
Regressive assumptions	2.06	1.60	1.18	1.44	1.52	1.72	1.55	1.68	1.78	1.51
Han										
National taxes 1970	13.10	12.10	12.20	11.70	11.60	14.90	15.60	15.20	17.10	25.70
National taxes 1976	15.70	13.40	12.90	13.00	13.00	13.40	13.40	14.50	16.40	22.80
National taxes 1978	20.40	16.30	14.70	14.30	13.70	13.60	13.20	13.30	13.70	20.20
National taxes 1980	28.00	19.90	17.60	16.70	15.70	15.30	14.90	14.70	14.80	20.60
Shim and Park										
National taxes 1984	27.33	21.76	20.48	19.76	19.19	19.80	19.97	21.61	23.57	29.00
National taxes 1986	35.17	20.08	21.25	20.45	20.38	20.45	20.93	21.90	24.10	29.83

Source: Heller (1981), Han (1982), and Shim and Park (1989).

Heller's study shows that under more progressive assumptions the direct tax burden (not shown in table 6.17) is only 0.9 percent of the income in the lowest decile, 3.65 percent in the ninth decile, and 11.7 percent in the highest decile. Under more regressive assumptions, the direct tax burden rises from 2.17 percent of income in the lowest decile to 8.84 percent in the highest decile. The movement from the schedular income tax to the global income tax increased the income tax burden of the first, sixth, and tenth deciles of the population by a small margin while for all other deciles the burden decreased rather drastically.

According to a study by Han (1982), the overall tax burden in Korea varies irregularly (see table 6.17). Han's study, for all the years examined, found a U-shaped pattern of tax incidence, with effective tax rates for both the poor and the rich being higher than for the middle-income groups. In the U-shaped incidence pattern, the richest group was shouldering a substantially heavier effective burden than the middle group until 1976, but thereafter the poorest 40 percent income decile was shouldering a higher tax burden than the remaining income decile groups. The U-shaped pattern of tax burden is also confirmed by a study by Shim and Park (1989). In the study by Han direct taxes and indirect taxes exhibit a very uniform pattern of incidence. The former are progressive along the entire range and the latter are regressive in all income brackets.

Empirical analyses show that the distributional impact of the property tax is rather sensitive to the assumptions made about the shifting of taxes in question (table 6.18). According to Heller's study the property tax burden is more or less proportional except for the top two deciles under the assumption of full shifting of the tax burden to consumers and somewhat regressive under the assumption that the property tax is borne by recipients of property income. In contrast, Kim's study (1987) shows that regardless of the "old" or "new" views, the burden exhibits a skewed U-shaped pattern of tax incidence, implying that both the rich and the poor bear a higher burden than the middle-income groups.

Like other taxes, the VAT has distributive properties whereby its burden may fall more heavily on some sections of society than on others. Perhaps the most controversial aspect of the VAT, when it was under consideration, was its possible effect on the distribution of tax burdens. The regressive nature of the VAT continues to be hotly debated.

A comprehensive VAT is regressive since lower-income taxpayers consume a higher proportion of their income. A number of studies have been carried out to estimate the distribution of VAT burdens. The results of these estimates are summarized in table 6.19. Using household income and expenditure surveys, all these studies base the distribution of VAT burdens on consumption patterns and the estimated rate of taxation on each category of consumer goods.

As table 6.20 shows, these studies hold that the VAT is regressive and differing only in the degree of regressiveness. According to Heller (1981), the VAT is regressive, with the burden declining from 5.55 percent of in-

Table 6.18 Burden of Direct Taxes and the Property Tax
(tax burden as a percentage of income)

	Income decile									
Taxes	*1st*	*2nd*	*3rd*	*4th*	*5th*	*6th*	*7th*	*8th*	*9th*	*10th*
Heller										
National taxes 1976										
Corporate income tax[a]	0.04	0.23	0.16	0.20	0.20	0.33	0.28	0.31	0.42	5.03
Corporate income tax[b]	2.12	1.89	1.41	1.93	1.82	2.02	1.99	1.58	1.54	2.17
Schedular income tax	0.04	0.17	0.10	0.17	0.52	0.83	1.81	1.58	4.08	6.07
Global income tax	0.05	0.07	0.04	0.06	0.26	0.88	1.42	1.54	3.23	6.67
Han										
Direct taxes										
1970	2.3	2.4	2.7	2.4	2.5	5.9	6.7	6.4	8.4	16.8
1976	0	0.3	0.5	1.1	1.5	2.2	2.6	4.1	6.7	13.7
1978	0	0.5	0.9	1.2	1.6	2.2	2.6	3.4	4.7	13.1
1980	0	0.5	1.0	1.6	1.8	2.3	2.8	3.5	4.7	11.6
Shim and Park 1985										
Direct taxes	—	2.05	2.11	2.60	2.73	3.75	4.04	5.66	7.61	13.16
Corporate income tax	1.85	1.48	1.51	1.48	1.70	1.65	1.85	1.88	2.33	2.88
Na and Hyun 1991										
Personal income taxes										
All households	0.91	1.58	2.14	2.44	2.86	3.85	4.87	5.77	7.14	10.26
Nonfarm households	1.10	1.85	2.49	2.75	3.24	4.36	5.60	6.67	8.32	11.75
Heller 1976										
Property tax[c]	0.15	0.16	0.12	0.15	0.14	0.16	0.15	0.16	0.20	0.56
Property tax[d]	0.54	0.43	0.34	0.37	0.37	0.39	0.33	0.35	0.29	0.14
Kim 1985										
Property tax[e]	0.54	0.42	0.39	0.35	0.38	0.42	0.42	0.45	0.54	0.64
Property tax[f]	0.29	0.25	0.22	0.20	0.27	0.34	0.37	0.41	0.60	0.84

— Not available.

Note: The following tax burden assumptions are made in each case:

a. Recipients of corporate property income bear the tax burden.

b. Burden is shared at 37.5 percent on wage earners, 37.5 percent on consumers, and 25.0 percent on recipients of corporate property income.

c. Consumers of housing services.

d. Recipients of property income.

e. Old view of property tax incidence.

f. New view of property tax incidence.

Source: Heller (1981); Han (1982); Shim and Park (1989); Na and Hyun (1993); and Kim (1987).

come at the lowest decile to 3.91 percent at the highest. The burden is lower in the farm sector than in the nonfarm sector, with the relative burden becoming lighter in the upper farm-income deciles. Whereas Oh's (1982) analysis claims that the distribution of the VAT burdens is only slightly regressive, Han's (1982) study concludes that it is strongly so. According to Han's estimate, the effective burden of the VAT on income for the highest decile is about 40 percent of that for the lowest decile. The corresponding figure based on Oh's study is around 70 percent. Those in the lowest decile pay 9.38 percent of their income as tax according to Han and 3.56 percent according to Oh. People in the top income decile are estimated to have tax burdens on income of 3.82 percent and 2.60 percent, respectively (Han 1982; Oh 1982). The regressive nature of the VAT burden is the result of using current income as the measure of ability

Table 6.19 Burden of Domestic Indirect Taxes
(tax burden as a percentage of income)

	Income decile									
Taxes	*1st*	*2nd*	*3rd*	*4th*	*5th*	*6th*	*7th*	*8th*	*9th*	*10th*
Heller										
All households										
Pre-VAT regime	7.42	7.29	6.30	6.73	6.76	7.14	6.73	6.43	6.22	7.00
Post-VAT regime	7.81	7.57	6.29	7.15	6.85	7.22	6.80	6.20	6.20	6.85
Nonfarm households										
Pre-VAT regime	7.64	7.83	8.14	7.57	7.73	7.92	7.73	8.31	7.49	7.00
Post-VAT regime	8.39	8.40	8.70	8.29	8.07	8.15	8.03	8.38	7.71	6.85
Oh										
Based on 1976 data										
Pre-VAT regime	5.93	4.75	4.91	4.79	4.61	4.67	4.51	4.47	4.53	4.12
Post-VAT regime	4.93	4.07	4.22	4.13	3.98	4.01	3.87	3.82	3.92	3.54
Based on 1978 data										
Pre-VAT regime	5.82	5.15	5.14	5.09	5.07	4.99	4.90	4.81	4.78	4.54
Post-VAT regime	5.71	4.98	4.97	4.96	4.93	4.85	4.75	4.68	4.62	4.34
Han										
1976	15.70	13.10	12.40	11.90	11.50	11.20	10.80	10.40	9.70	9.10
1980	28.00	19.40	16.60	15.10	13.90	13.00	12.10	11.20	10.10	9.00
Lee	7.34	4.27	3.95	3.89	3.59	3.47	3.37	3.25	3.10	2.85
Shim and Park	17.66	13.57	12.55	11.67	10.99	10.72	10.48	10.47	10.14	9.65
Na and Hyun 1991										
All households	3.86	3.58	3.52	3.42	3.68	3.88	3.69	3.91	4.05	3.12
Nonfarm households	4.39	3.71	3.64	3.68	3.91	4.19	4.02	4.17	4.52	3.57

Source: Heller (1981); Oh (1982), Han (1982); Lee (1987); Shim and Park (1989); and Na and Hyun (1993).

to pay. When the VAT burden is expressed as a fraction of current consumption, a proxy for permanent income, the VAT is slightly progressive, with the burden from 2.47 percent at the lowest decile to 2.76 percent at the highest as shown by Lee (1987) in table 6.20.

A variety of indirect taxes were replaced by the VAT and the special consumption tax. Therefore, it is worth ascertaining whether the VAT was more or less regressive than the taxes it replaced. Table 6.19 summarizes the burdens of domestic indirect taxes before and after the reform. Empirical studies done to date yield mixed results.

According to Oh and the data from 1978, the distribution of the tax burden by income decile became slightly less regressive after the tax reform. However, studies by Heller and Han indicate that the VAT was more regressive than the taxes it replaced. Two explanations can be offered to account for the relatively small change or the worsening of the distribution of the indirect tax burden as a result of the shift from the pre-VAT to the post-VAT regimes. First, the VAT system was designed to be quite regressive. Second, the tax rates and tax base of the special consumption tax, which was concurrently introduced to supplement the VAT, were insufficient to tax higher incomes more heavily. All the studies reviewed find the burden of the special consumption tax proportional to income or even somewhat inversely related to income.

Table 6.20 Burden of the Value Added Tax
(percent)

	Income decile									
Taxes	*1st*	*2nd*	*3rd*	*4th*	*5th*	*6th*	*7th*	*8th*	*9th*	*10th*
Heller 1976										
All households	5.55	5.19	4.19	5.00	4.67	4.84	4.79	4.04	4.11	3.91
Nonfarm households	5.94	5.63	5.82	5.75	5.46	5.38	5.51	5.31	5.02	3.91
Farm households	4.80	4.02	3.42	3.27	3.13	2.89	2.73	2.46	2.22	—
Oh										
Based on 1976 data	3.62	2.90	2.98	2.94	2.86	2.85	2.76	2.73	2.79	2.42
Based on 1978 data	3.56	3.10	3.07	3.05	3.05	2.99	2.91	2.86	2.77	2.60
Han 1978										
Farm households	8.44	5.96	5.14	5.07	4.24	4.18	3.73	3.53	3.17	2.90
Nonfarm households	9.38	7.50	6.70	6.40	5.99	5.69	5.38	5.06	4.67	3.82
Lee 1984										
Income base	4.31	2.46	2.30	2.27	2.18	2.10	2.05	2.20	1.97	1.88
Consumption base	2.47	2.51	2.55	2.63	2.64	2.63	2.60	2.64	2.67	2.76
Na and Hyun 1991										
Nonfarm households	3.77	3.20	3.05	3.07	3.21	3.31	3.23	3.39	3.45	2.72
Farm households	2.45	1.87	2.24	1.77	1.73	1.71	1.63	1.72	1.43	1.06

— Not available.

Source: Heller (1981); Oh (1982); Han (1982); Lee (1987); and Na and Hyun (1993).

The incidence studies reviewed above vary in their estimates of the distributive effect of the VAT itself and the comparative effects of the VAT and previous indirect taxes. All these studies indicate, however, that the VAT is regressive and that the replacement of the previous indirect taxes with the VAT and the special consumption tax has not lessened the regressive nature of indirect tax burdens.

The empirical studies of the distribution of the tax burden lead one to conclude that the overall burden of tax among income classes is more or less regressive and that the tax system is not working as a principal policy instrument to narrow income differentials. When account is taken of income received by the rich from their underground activities as well as benefits of preferential tax treatments, the distributional effects of the government tax policies are probably more regressive than the tax policy might suggest.

To center all efforts on equity at the cost of efficiency or growth would be as grave a mistake as to center all efforts on growth or efficiency at the cost of equity. However, there is room for significant improvements in both equity and efficiency by reorganizing the tax structure. A movement toward greater reliance on direct taxation of both income and wealth should be effective in establishing a more progressive tax burden with minimum distortion of relative prices. If equitable distribution of income and wealth is a desirable objective, strong political determination will be needed to make income and property taxes, not consumption taxes, play the major role in the Korean tax system.

Some Major Issues

The need to review the long-run overall tax structure and reform of the tax system is stronger now than ever before for several reasons. First, the rapid growth of the economy involves massive structural changes that would produce many unintended distortions in the allocation of resources if the tax system remains unchanged. The tax system needs to be adapted to current and expected developments in the Korean economy. Second, there are structural problems—namely, the regressiveness of the overall tax burden—that should have been corrected some time ago, but have remained untouched for various reasons. Third, in order to meet the increasing demand for social welfare programs, more revenue is needed. This requires fundamental changes in the tax system and a more streamlined tax administration.

Major policy issues within the Korean tax system have been more or less clearly identified and agreements reached for the direction of future tax reform. Increases in government expenditure should be met by expanding the tax base within the present tax system. Various special tax provisions that erode the base of the tax system should be phased out. The underground economy, estimated to be somewhere between 20 to 30 percent of GNP, should be tapped for tax revenue. The tax system should rely more on direct taxes for revenue. Also, renewed efforts to increase the nontax component of the government's revenue structure are needed as fees and charges represent only 1 to 2 percent of central government revenue.

Discussions of tax reform have centered around several important issues, which are summarized next.

Insignificance of Personal Income Tax

The personal income tax plays an insignificant role in raising tax revenue in Korea, as in most other countries. As table 6.21 shows, revenue from the personal income tax in 1991 was 3.1 percent of GNP and accounted for 16.8 percent of total tax revenue. During the 1970s and 1980s the share of personal income tax in GNP or total tax revenue has been declining despite the 1974 reform and the introduction of the global income tax system. Personal income tax as a percentage of GDP and total taxation is 11.7 percent in Korea compared with the 30.1 percent average of OECD countries. Although insignificant as a contributor to revenue, the personal income tax has received the most attention from tax officials and the general public because continuing inflation requires frequent adjustments to exemptions and bracket rates, and because structural problems with the personal income tax have begged correction.

Several major defects have weakened the personal income tax system as a revenue-raising instrument and as an income redistribution sys-

Table 6.21 Personal Income Tax Revenue

Year	Personal income tax as a percentage of			
	GNP	National income	Total tax revenue	National domestic tax
1955	1.4	1.5	12.0	18.3
1960	0.9	1.0	7.3	10.8
1965	1.4	1.6	16.8	27.9
1970	3.1	3.6	21.2	29.8
1975	1.9	2.4	12.8	19.6
1980	1.7	2.3	10.1	18.0
1985	2.0	2.6	10.9	19.8
1986	2.0	2.5	11.4	15.3
1987	2.0	2.6	11.6	21.6
1988	2.3	3.0	13.1	23.6
1989	2.5	3.2	13.6	23.4
1990	2.8	3.5	14.2	24.7
1991	3.1	3.9	16.8	27.5

Source: ONTA (various years); Bank of Korea (various years b).

tem. The major problem with the global income tax is that the tax base is not fully comprehensive and different types of income are subject to different rates. In reality, the global income tax is far from global; it retains many schedular rates and is highly complicated. Capital gains realized from selling stocks and some generous fringe benefits to employees are not taxable under the personal income tax code. There is a separate low tax on real estate–related capital gains, retirement income, timber income, and dividend and interest incomes, available only to more affluent members of society. Interest income from life insurance savings, imputed income from owner-occupied housing, capital gains on securities, interest income from government bonds, and many kinds of fringe benefits such as cars, drivers, lunches, and housing allowances are fully tax exempt. A significant portion of interest and dividend incomes is taxed at a separate lower withholding rate. Farm income is subject to the farmland tax, not the personal income tax, which has a different structure of exemptions and tax rates at the local level.

If four basic aims are to be achieved—revenue generation, equity, efficiency, and simplicity—it is essential to broaden the personal income tax base by bringing all sources of income under the tax net and consolidating all sources of income with comprehensive global income.

Explanations for the failure to develop a more unified taxation on different sources of income is both economic and institutional. The economic argument relates to the fear that full globalization would either discourage savings and investment or channel funds into unproductive, speculative, or untaxed forms of investment. Some maintain that nontaxation of capital gains on securities should continue until the capital market is fully developed and, given Korea's relative shortage of capital, more preferential treatment of savings should be provided. The institutional argument is that the current practice of allowing bearers of bonds and securities to have fictitious names is bound to result in a huge loss of interest and dividend income in the global income tax base. It will be virtually impossible to fully globalize interest and dividend incomes unless there is a law to limit the use of fictitious names on bearer bonds and securities. In the absence of a record of ownership, ONTA has no way to monitor securities transactions. The holders of no-name bearer shares, secure in the knowledge that their stock ownership cannot be discovered, confidently evade not only tax on income, but also tax on gifts and inheritances. This also precludes the possibility of taxing capital gains on securities.

The level of personal deduction and exemption has always been a subject of heated political debate largely because the benefit to the taxpayers is direct and tangible. As a result of maneuvering by politicians to secure political gains and a mistaken belief by policymakers that increases in personal exemptions and deductions are the best ways of relieving the tax burden of wage and salary earners, the level of personal exemption and deduction has been raised every year since the tax reform in 1974.

Issues in the Value Added Tax

In hindsight, the introduction of the VAT in 1977 has been a success. It has worked relatively well, and in some cases much better than anticipated. Complaints has been few although some have been loud. The Korean experience during the past ten years shows that practice is not as simple as theory. Criticisms or complaints should be regarded as a measure of its success rather than failure. From the policy point of view, recent discussions concerning the VAT have revolved around three issues: (1) coverage of the VAT, (2) the level and structure of the tax rate, and (3) the treatment of small taxpayers.

One recurrent question is whether to extend the VAT to sales that are currently exempt. The widespread use of exemptions is based on the desire to reduce the regressiveness of the VAT burden. The goods and services currently exempt leads one to question the appropriateness of some items on the list, such as financial services, independent professional services, and government-provided goods and services that compete with commercial operations.

Based on tax neutrality grounds, it has been suggested that some commercial activities by semigovernment bodies should not be exempt. The major problem with imposing the VAT on financial services is the difficulty of calculating the correct tax base. Since 1982 Korea has imposed a tax of 0.5 percent on gross receipts of banking and insurance companies, not under the VAT, but under the education tax. Most tax experts agree that all independent professional services, such as those provided by doctors, lawyers, accountants, and architects, should be taxed.

A single rate of 10 percent has been used since the introduction of the VAT. Although suggestions to use differentiated multiple rates to reduce the tax burden on low-income groups has been made, rate differentiation has been resisted on the grounds that multiple rates do not achieve redistributive objectives and actually complicates administration and compliance. Even though Korea does not have a multiple rate VAT system, the addition of excise taxes such as the special consumption tax and the alcoholic beverages tax on top of the VAT produces the same effect.

The 10 percent VAT rate is lower than the reduced rate of some countries. Given the fact that the share of VAT in GNP and the proportion of revenue from the VAT relative to that of domestic taxes on goods and services are fairly low compared with other countries, as shown in table 6.5, there is room to raise more revenue by raising the tax rate. However, Korean taxpayers have been very hostile to this proposal.

A major criticism of the VAT has been the burden on business, particularly on small business, to keep adequate records and file tax returns in the prescribed format. Under any form of general sales taxation, small businesses are granted special treatment in order to reduce their administrative burden. Small businesses, called special taxpayers under the Korean VAT system, are those with total sales of less than 24 million won a year. Unlike general taxpayers, whose tax base is value added, a 2 percent tax rate applies to annual sales of small businesses. In 1986 special taxpayers represented about 70 percent of total VAT payers, whereas revenues collected from them accounted for about 4 percent of total VAT revenues.

There has been debate about whether the current threshold between special taxpayers and general taxpayers should or should not be increased. The temptation to increase the limit due to political pressure by special taxpayers, numbering about a million, should be resisted because the aim is not to give them more favorable tax treatment, but rather a simplified system that approximates true tax liability without imposing an intolerable burden on either the taxpayer or the tax administration.

Real-Name System of Financial Transactions and Tax Policy

In 1961 the capital-hungry military government legalized the use of pseudonyms in financial transactions in an effort to encourage saving. This system of anonymous ownership helped increase the supply of loanable funds to financial institutions and made sense when the economy was starving for investment funds. In the long run, it was like opening Pandora's box. Many of these private transactions may be improper or illegal (bribery, illegal political contributions, real estate speculation, and tax evasion). In fact, a national financial scandal broke out in 1980 when a number of leading corporations were found to be involved in fraud and curb market financial transactions.

With pseudonyms, it is easy to hide financial transactions from the tax authorities. As a result, taxation of certain categories of individual income (business income, rental income, capital gains on real properties), and inheritance and gifts becomes particularly difficult to administer. This system of anonymity was repeatedly criticized for generating inequities in the tax burden and providing a safe harbor for curb market activities, the most typical underground economic activity in Korea.

At last, on August 12, 1992, the new government launched, without any advance notice, a special presidential decree requiring real names on all financial accounts. In addition, interest and dividend income would be included in the global income tax base starting in 1996 and an examination of taxing capital gains from securities would follow after 1997.

Real Estate Speculation and Tax Policy

Real estate speculation has been a difficult problem for a long time and the government has placed a high priority on containing real estate inflation. Furthermore, the government was seriously concerned about the concentrated distribution of real assets and the severe shortage of urban land for housing and business. Since inflation in the real estate market has far exceeded that in the commodity and labor markets the skewed distribution of land and buildings has been the single most important source of the ever-widening gap between the haves and the have-nots. In addition, the rapidly rising price of housing has contributed to serious social instability.

The policy instruments mobilized to head off further speculation included the denial of interest deduction, a land speculation–control tax, heavy taxes on capital gains arising from real estate, excess land capital gains tax, global land tax, tax on excessive holding of residential land, social and political pressure on the corporations and individuals with excessive landholdings, and so forth.

Since taxes relating to the transfer and ownership of properties have hardly been an important source of the tax revenue, they have made little or no contribution to lessening prevailing inequities in the distribution of assets. Taxation of wealth or properties should form an important element of the overall tax structure, not for revenue purposes, but to complement income taxation, ameliorate the significant unequal distribution of wealth, and ensure efficient resource allocation.

As a means of curbing speculation on real properties, the government has always relied on capital gains tax on real properties rather than the property tax on real properties. The capital gains tax on real properties replaced the land speculation control tax in 1975. At first the tax rates were flat at 50 percent on land and 30 percent on buildings. The present tax rates vary widely, depending on the type of assets, the length of ownership, and whether assets are publicly registered or not.

Not surprisingly, the capital gains tax on real properties proved to be totally ineffective in controlling speculative real estate purchases. The high tax rate on capital gains on real properties may have a psychological impact on the real estate market, dampening speculative initiatives.

However, a more important effect of imposing tax at high rates has been to defer real estate sales in the hope that taxes would be ultimately reduced. Artificial blocking of real property transactions by taxation results in a market shortage in buildings and land, thereby raising rather than reducing prices of real properties. The general public has sometimes been misled when the government increased or decreased tax rates of capital gains on real properties in order to cool the heated economy or to recover from a business recession.

The use of capital gains tax on real properties to reduce short-run speculative purchases of real estate was inappropriate. The roots of high inflation during the rapid industrialization period should have been dealt with at their source. Speculation on real estate was just a symptom. The best tax instrument to curb speculative buying of real properties would have been a property tax imposed directly on ownership rather than the capital gains tax on real properties, which is imposed at the time of ownership transfer. Both administrative difficulties on the tax collector's side and widespread tax evading efforts by the taxpayers nullified the government's efforts to cool speculation on real estate via heavy taxation on capital gains from real properties.

A better way to curb the increase in prices of real properties due to the perennial supply shortage and lessen prevailing inequities in the ownership of real estate would be to introduce an annual, comprehensive, personal tax on the combined property holdings of the individual. Instead the government in 1986 adopted the excessive landholding tax in 1986 (effective January 1988). The government's interim proposal was an annual personal land tax as a supplement to the local property tax. The rate structure of the excessive landholding tax is progressive, ranging from 0.5 to 5.0 percent. However, since its tax base is not comprehensive, only a limited portion of land is integrated into the tax base. Few experts believe that the excessive landholding tax will discourage real estate speculation on land or reduce the high concentration of landholding in the hands of the wealthy few.

Three real estate–related laws to promote the "Public Concept of Ownership of Land" became effective in 1990. These laws impose a ceiling on ownership of residential land, a tax on profits from regional development projects, and a tax on profits from excessive landholdings. There is a ceiling on ownership of residential land for each household and a tax on the amount of land held in excess of the legal limit. The limit is 200 pyong (660 m^2) in Seoul and its five direct jurisdictional cities, and 300 pyong (990 m^2) in the next six largest cities. Business corporations are not allowed to own residential land unless they are in the rental housing business.

The law is designed to encourage the owners of large residential landholdings to sell and hopefully expand the ownership of land and lower the price. However, only 7,300 households are subject to this limit in the six largest cities, and there is no guarantee that the released land will be used for residential purposes. Landowners can also choose to retain holdings and incur the tax burden.

Tax on profits from regional development projects is levied on the increase in land value associated with such projects as residential land development, land readjustment, zoning changes, and industrial site development. The tax base is the difference between the "posted" land value at the completion and initiation of such projects, less costs of development projects, less the "normal" increase in land value based on the interest rate on a one-year savings deposit and the national average rate of increase in land value. The tax rate is 50 percent.

Tax on excessive profits from landholding is a tax on capital gains realized from land for any reason and regardless of the type of land being sold. The tax applies to vacant lots, vacation homes, factory sites or land attached to employee training and education facilities held by business corporations exceeding the legal limit, corporate-owned land not in proper business use, arable land owned by absentee landlords, and golf courses.

The tax is levied wherever the increase of land value during a one-year period exceeds 150 percent of the "normal" rate or the rate of increase during a three-year period exceeds 100 percent of the "normal" rate. The tax is 50 percent of the base, which is defined as the increase in the value of land assessed by ONTA, less capital expenditure and normal capital gains. This tax is controversial because it is levied on accrued gains rather than realized gains. There is also a technical issue of determining whether a plot of land is used for proper business purposes or not.

The introduction of the excess land capital gains tax highlights the urgency of controlling real estate inflation and the accompanying real estate speculation and distortions in wealth distribution. Given that the excess land capital gains tax is extremely cumbersome to administer and fraught with shortcomings in compliance and administration, its introduction also demonstrates the lack of effective and politically acceptable tax instruments for achieving policy objectives. The intensity of the government antispeculation effort is also evident in the current state of the taxation of capital gains on real property, where the lowest marginal tax is 40 percent.

It is not clear whether the excess land capital gains tax is an effective instrument for achieving the policy objectives of the law. For instance, the excess land capital gains tax may not be effective or easy to administer in the case of a nationwide inflation in land prices, and the tax may induce inefficient use of land by providing an incentive for socially premature development of idle land.

Lessons and Conclusion

The dynamics and inner workings of industrialization in Korea can be described as follows: the policy direction and incentives provided have generally proved correct. International market conditions have been favorable. The political leadership has been committed to economic development. That commitment has been translated into action by the bureaucracy and private firms. Development plans and strategies exist in almost every developing country, but Korea has been able to put these plans and strategies into practice.

Korean policymakers and the general public have been very pragmatic and displayed no ideological biases. This allowed the government to use all available instruments in making policies to achieve its goals. In the policy planning process, particularism also prevailed, permitting the government to apply a certain policy to a limited number of specific cases. Of course, trial and error adjustments in the absence of an ideological basis had the drawback of causing frequent changes in policy.

The Korean experience affirms the productive role of the market and the importance of correct price signals. Scarce resources were allocated by correct market price signals to dynamic entrepreneurs operating in a competitive environment. The government let the businessmen exploit favorable opportunities by providing appropriate incentives or eliminating disincentives for their activities. Although the Korean government influenced the activities of private firms and the use of economic resources, competition was allowed to flourish whenever possible.

With regard to the role of the market, Korea has also learned from experience. The structural imbalances that developed in the late 1970s were caused in part by the arbitrary selection of strategic industries and the distortions caused by these policies. As the economy grows in size and complexity, excessive intervention by the government will not be a good substitute for the market mechanism, which is better at achieving efficient allocation of resources. The increasing complexity of the Korean economy will demand even greater decentralization of economic decisionmaking.

Characteristics of Korean fiscal policy during the industrialization period include a relatively small public sector, adherence to the balanced budget, comparatively low taxes, relatively low taxes on capital income, liberal use of tax incentives for investments, heavy reliance on indirect taxes, insignificant property taxes, increased public savings, relatively little spending for redistributive social services, and budgeting for significant resources for industrial development.

Tax policy in Korea reveals a pattern of development that differs from most countries during the modernization process. Korea chose a low-tax effort, minimal taxation of the foreign trade sector, reduced reliance on income and profits taxes, and it did not make income distribution a major goal of tax policy.

Korea had already practiced supply-side economics long before it gained prominence in theoretical and policy discussions. The fact that fiscal policy was geared to the policy line of supply-side economics is borne out by the low share of tax and expenditure in GNP, heavy reliance on indirect consumption taxes, extensive tax incentives for saving and investment, and less emphasis on welfare spending.

The tax system has evolved historically through a continuing series of compromises, adjustments, and reform, which have attempted to achieve some level of efficiency and equity while raising a given revenue yield. Until the mid-1960s the tax system did not adequately serve economic policy objectives. The intention was there and institutional reforms did take place, but enforcement was lax and structural changes did not go far enough. It is difficult to identify precisely the role that taxation has played in the development of the Korean economy. Although there have been a large number of changes in the tax code and in tax administration during the period of rapid economic growth, more fundamental reform of the tax system is needed to make it less distortionary.

The most important tax policy in connection with rapid industrialization has been the use of tax incentives to induce the private sector to engage in economic activities deemed desirable for industrial development and export growth. Enterprises engaging in heavy and chemical industries and export industries have benefited most from these tax incentives. Recent tax changes focused on containing real estate speculation—using increasing taxes on land transfers rather than on holdings of land and real estate—have not had their intended effect. Another major problem is tax administration and compliance, exacerbated by a "no real-name system" that prevents the identification of assets and income from financial holdings.

The current tax system is not without structural problems. The overall tax burden is regressive. The challenge ahead is to restore the progressiveness of the tax burden while generating more revenue to meet increasing demands for social programs. All the discussions above clearly indicate that another major tax is needed to resolve clear-cut problems in the present tax system.

References

Bahl, Roy B., C. K. Kim, and C. K. Park. 1986. *Public Finances During the Korean Modernization Process*. Cambridge, Mass.: Harvard University Press.

Bank of Korea. Various years (a). *Economic Statistics Yearbook*. Seoul.

Bank of Korea. Various years (b). *National Accounts*. Seoul.

Choi, Kwang, Kazuhisa Ito, Taewon Kwack, Eigi Tajika, and Yuji Yui. 1985. *Public Policy, Corporate Finance and Investment: The Experiences of Japan, Korea and Taiwan*. Tokyo: Institute of Developing Economies.

Commission on Tax Reform. 1985. *Final Report of the Commission on Tax Reform in Korea*. Seoul: Ministry of Finance (in Korean).

Economic Planning Board. 1980. *Handbook of the Korean Economy*. Seoul: Economic Planning Board.

———. Various years. *Major Statistics of the Korean Economy*. Seoul: Economic Planning Board.

Evans, Paul. 1988. "The Effects of Fiscal Policy in Korea." *International Economic Journal* 2(Summer): 1–14.

———. 1990. "The Output Effects of Fiscal Policy in Korea." In Jene K. Kwon, ed., *Korean Economic Development*. New York: Greenwood Press.

Han, Seung-Soo. 1982. *Empirical Analysis of the Tax Burden in Korea and Theoretical Analysis of Optimal Tax Burden*. Seoul: Korean Economic Research Institute (in Korean).

Heller, Peter S. 1981. "The Incidence of Taxation in Korea." International Monetary Fund DM Series No. 14. Washington, D.C.

IMF (International Monetary Fund). 1979. "Korea: Taxes in the 1980s." Washington, D.C.

Kim, Myung-Sook. 1987. "Distributive Effects of the Property Tax in Korea." *Korea Development Review* 9(Winter): 119–36 (in Korean).

Kim, Jun Young. 1993. "Tax Policy, Cost of Capital, Investment and Saving in Korea." Paper presented at the Conference on Taxation and Economic Growth (Asian Miracle), August 2–3, Monterey, California.

Kwack, Taewon. 1984. "Budget Policies and Investment Allocation." In K. Choi and J. W. Kim, eds., *National Budget: Goals and Priorities*. Seoul: Korea Development Institute (in Korean).

Kwon, Soonwon. 1990. "Korea: Income and Wealth Distribution and Government Initiative to Reduce Disparities." Working Paper No. 9008: 1–32. Korea Development Institute, Seoul.

Lee, Kye-sik. 1987. "Burden of Indirect Taxes in Korea." *Korean Journal of Public Finance* I(March): 127–57 (in Korean).

Ministry of Finance. Various years. *Government Finance Statistics in Korea*. Seoul: Ministry of Finance.

———. 1967. *The White Paper on Tax Reform*. Seoul (in Korean).

———. 1972. *An Outline of the 1971 Tax System Reform*. Seoul (in Korean).

———. 1975a. *Defense Tax Law*. Seoul (in Korean).

———. 1975b. *Summary of Revised Tax Laws*. Seoul (in Korean).

———. 1977. *Value-Added Tax*. Seoul.

———. 1979. *History of Korean Tax System*. Seoul (in Korean).

———. 1992. *Korean Taxation*. Seoul.

Musgrave, Richard A. 1965. "Revenue Policy for Korea's Economic Development." USOM/KOREA, September 1965.

———. 1967. "Suggestions for the 1967 Tax Reform." Unpublished report submitted to the Ministry of Finance, March 14, Seoul.

Na, Sung Lin, and Jin Kwon Hyun. 1993. *The Effects of Taxes and Benefits in Korea*. Korea: Tax Institute.

National Statistical Office. Various years. *Major Statistics of the Korean Economy*. Seoul: National Statistical Office.

Oh, Yeong-Cheon. 1982. "An Evaluation of the Tax Reform for a Value-Added Tax in Korea, with Special Reference to the Distribution of the Tax Burden, Administrative Efficiency and Export." Ph.D. dissertation, New York University.

ONTA (Office of National Tax Administration). Various years. *Statistical Yearbook of National Tax*. Seoul: Office of National Tax Administration.

Roh, Keesung. 1992. "The Estimation of Under-Reported Business Income Tax." *Korea Development Review* 14(Fall) (in Korean).

Shim, Sang-Dal, and Inn-Won Park. 1989. "Public Finance and Income Redistribution." *Korea Journal of Public Finance* 3(March) (in Korean).

Trela, Irene, and John Whalley. 1992. "The Role of Tax Policy in Korea's Economic Growth." In Takatoshi Ito and Anne O. Krueger, eds., *The Political Economy of Tax Reform*. Chicago: University of Chicago Press.

Wald, Haskell P. 1953. "Report and Recommendations for the Korean Tax System." United Nations, Korean Reconstruction Agency.

7

Mexico's Protracted Tax Reform

Francisco Gil-Díaz
Wayne Thirsk

Mexico has witnessed a flurry of tax reforms in the past two decades. The tax system today bears scant resemblance to the system in place twenty years ago. Below we describe and evaluate this tax reform activity. We begin by outlining the main characteristics of Mexico's tax system prior to its extensive reform. We next look at the driving forces behind these reforms and delineate both the objectives and the constraints that have shaped the direction of tax reform. Subsequent sections evaluate the sequence of reforms introduced by different federal administrations. The final section summarizes lessons learned.

The Prereform Period

From 1945 to 1970, Mexican tax reform was guided mainly by administrative practicality. By the 1970s the system was a patchwork mess, particularly so in the case of indirect taxes. The income tax was schedular, but its design (if not its scope) was probably as equitable and as progressive as it could be. A thorough reform began in the 1970s with sharp increases in excise taxes from initially low levels and continued with a wide-ranging structural reform that centered on the value added tax (VAT) and on attempts to nudge the system toward an ideal global income tax.

The personal income tax, until 1973, and the corporate income tax, until 1965, had progressive schedules that differed according to the source of income. In some cases, progressivity was simple, as in the case of financial interest payments to individuals. From 1965 to 1971 these were taxed according to a schedule that started at a rate of 2 percent to a maximum of 10 percent on annual interest rates above 15 percent. Later, in 1973, interest rates below a certain threshold were exempted on the probably correct presumption that mostly lower income people held such deposits. At the same time, a two-tier interest withholding tax was introduced with a lower rate for those who chose to combine their interest income with other taxable income.

Overall the business income tax was progressive—since the income of small firms and partnerships was taxed progressively as profits at the entity level and as wages at the personal level—if paid out as such to family members although presumably at lower average rates. However, some of that income would go untaxed if extracted as vicarious consumption through the accounts of the firm. The last possibility was more likely for small family firms. Corporations, on the other hand, also faced an element of double taxation on distributions as they were taxed on paid dividends via a noncreditable dividend tax.

The small number of public corporations could, in principle, avoid the 10 percent tax on dividends through retaining earnings and realizing untaxed capital gains. Even though corporations are the main shareholders in the stock market, only individual shareholders can claim the existing exemption for stock market–related capital gains. Furthermore, corporations and individuals have generally preferred cash payments in a thin market where capital gains may not be easily realized by selling shares.

The federal sales tax was introduced rather late in 1948 to partially replace the stamp tax, a relic from colonial days that finally became history with the reforms of 1979. A turnover tax was adopted with all the associated distortions and random incidence, but it was simple to administer in an environment where illiteracy was still rampant. A few excise taxes were also gradually adopted on goods or services with low price elasticities of demand and that were also easy to administer, such as gasoline, alcohol, tobacco, and telephone services. Excises were imposed even on nonconsumer goods, such as automobile tires and glass, which were produced by only one or two manufacturers nationwide. The subsequent imposition of the turnover sales tax (TST) on top of these excises made the cascaded tax effect on top of some final consumption goods even more acute.

Up until 1962 Mexico had an inheritance tax. This tax was repealed by congress because it was felt that, as a practical matter, the tax had an impact mostly on the middle class where there was still wealth to bequeath, mostly real estate, but little legal sophistication to avoid the tax through the use of trusts availed by wealthier taxpayers.

In the 1940–70 period, there was an important relative and absolute increase in tax effort as total taxes went from 6.5 to 10 percent of gross domestic product (GDP). This 3.5 percentage point increase was largely the result of better administration, the introduction of a streamlined turnover tax in the late 1940s, and rising income tax burdens. As table 7.1 shows, the structure of the economy was moving toward manufacturing and services and away from the primary nonminerals sector and this structural shift increased the tax to GDP ratio. Foreign trade taxes gradually lost importance, not because protection as a whole was relaxed, but because quantitative controls were preferred to tariffs.

Table 7.1 Federal Government Tax Revenues and Social Security Contributions
(percentage of GDP)

			Tax revenue					
					Indirect taxes			
					Trade taxes			
Year	Total fiscal burden	Social security contributions	Total tax revenues	Direct taxes	Total	Import	Export	Other indirect taxes[a]
1938	5.0	—	5.0	0.6	4.4	1.3	0.4	2.7
1939	6.1	—	6.1	0.6	5.5	1.5	1.0	3.0
1940	6.4	—	6.4	0.8	5.6	1.3	0.9	3.4
1941	7.0	—	7.0	0.6	6.4	1.6	1.1	3.7
1942	6.6	—	6.6	0.8	5.8	1.0	1.2	3.6
1943	7.9	—	7.9	1.8	6.1	0.8	1.7	3.6
1944	6.2	0.2	6.0	1.7	4.3	0.7	1.0	2.6
1945	6.3	0.3	6.0	1.5	4.5	0.8	1.0	2.7
1946	6.1	0.3	5.8	1.5	4.3	0.9	0.9	2.5
1947	6.0	0.3	5.7	1.6	4.1	0.8	0.6	2.7
1948	6.3	0.4	5.9	1.5	4.6	1.0	0.6	3.0
1949	8.1	0.5	7.6	1.9	5.7	1.0	1.4	3.3
1950	7.9	0.5	7.4	1.9	5.5	1.1	1.3	3.1
1951	8.6	0.5	8.1	2.3	5.8	1.3	1.6	2.9
1952	8.0	0.5	7.5	2.4	5.1	1.1	1.3	2.7
1953	7.4	0.5	6.9	1.9	4.9	1.1	1.1	2.7
1954	7.2	0.5	6.7	1.8	4.9	1.0	1.4	2.5
1955	8.3	0.6	7.7	2.3	5.4	1.1	1.8	2.5
1956	8.2	0.6	7.6	2.5	5.1	1.2	1.4	2.5
1957	7.3	0.7	6.6	2.4	4.2	0.9	1.1	2.2
1958	7.4	0.8	6.6	2.2	4.4	1.2	0.9	2.3
1959	7.4	0.9	6.5	2.2	4.3	1.2	0.9	2.2
1960	7.5	1.1	6.4	2.3	4.3	1.2	0.7	2.2
1961	7.5	1.2	6.3	2.4	4.1	1.0	0.6	2.5
1962	7.8	1.3	6.5	2.5	4.0	0.9	0.6	2.5
1963	8.2	1.5	6.7	2.6	4.1	1.0	0.5	2.6
1964	8.4	1.5	6.9	3.0	3.9	1.0	0.5	2.4
1965	8.3	1.8	6.5	2.3	4.2	1.3	0.4	2.5
1966	8.6	1.5	7.1	3.0	4.1	1.2	0.4	2.5
1967	9.2	1.7	7.5	3.2	4.3	1.5	0.3	2.5
1968	9.7	1.8	7.9	3.4	4.5	1.3	0.3	2.9
1969	9.8	1.8	8.0	3.5	4.5	1.3	0.3	2.9
1970	9.9	1.8	8.1	3.5	4.6	1.4	0.2	3.0
1971	10.0	2.0	8.0	3.4	4.6	1.2	0.2	3.2
1972	10.7	2.4	8.3	3.7	4.6	1.2	0.2	3.2
1973	11.3	2.5	8.8	3.8	5.0	0.9	0.1	4.0
1974	12.0	2.7	9.3	4.0	5.3	1.0	0.2	4.1
1975	13.7	2.9	10.8	4.4	6.4	0.9	0.1	5.4
1976	13.0	3.0	10.0	4.8	5.2	0.9	0.2	4.1
1977	15.0	3.4	11.6	5.1	6.5	0.6	0.3	5.6
1978	15.3	3.3	12.0	5.7	6.3	0.6	0.1	5.6
1979	16.2	3.5	12.7	5.6	7.1	0.9	0.1	6.1
1980	18.3	3.3	15.0	5.8	9.2	1.0	(0)	8.2
1981	18.1	3.2	14.9	5.8	9.1	1.1	(0)	8.0
1982	18.4	3.4	15.0	4.9	10.1	0.9	(0)	9.2
1983	20.1	2.8	17.3	4.2	13.1	0.5	(0)	12.6
1984	19.1	2.6	16.5	4.2	12.3	0.5	(0)	11.7
1985	19.1	2.8	16.3	4.1	12.2	0.6	(0)	11.6
1986	17.5	2.7	14.8	4.3	10.4	0.7	(0)	9.7
1987	17.8	2.5	15.3	3.9	11.3	0.7	(0)	10.6
1988	21.6	2.6	18.5	4.8	13.6	0.4	(0)	13.2

a. It includes revenues from taxes on fuel, tobacco, alcohol, telephone, production, and services. Up to 1979 it includes the turnover sales tax and since 1980 the VAT, natural resources, and others. From 1977 on, revenues from taxes on domestic consumption and on oil exports have not been included in the natural resources tax.

Source: Bank of Mexico and Treasury Department data.

Thus, the tax system prior to the active tax reform years included a schedular income tax on individuals, a haphazard mosaic of indirect taxes, and a declining role for foreign trade taxes. Table 7.1 traces the evolution of federal tax revenues over the half-century from 1938 to 1988. From 1938 to 1968 total tax revenues as a fraction of GDP displayed no perceptible trend. Since then the tax effort has more than doubled, reflecting an expanded public sector. Almost all of the increased tax effort is attributable to heavier indirect taxation. Although it is not shown in table 7.1, within the group of direct taxes the relative contributions made by the personal and corporate income taxes have normally been about the same.[1]

Tax Reforms: Impetus, Goals, and Constraints

Tax policy from the 1940s to the mid-1970s was dominated by practical and administrative considerations. As the tax system was modernized, the pace and direction of change managed to keep the system simple and avoid excessive tax burdens. The latter was feasible because the size of government stayed small.

During the period 1945–70 the government pursued a highly protective, inward-looking development strategy. Real economic growth rates averaged over 6 percent annually between 1950 and 1970, amidst a climate of stable price levels and exchange rates. The public sector's appetite for resources began to grow, as did its capacity to satisfy that appetite. Government spending as a proportion of GDP showed little tendency to rise during this period (Urzua 1993). However, despite this long period of enviable economic growth, the government faced growing criticism in the late 1960s because of a perceived deterioration in income distribution and relatively slow employment growth in the modern sectors of the economy. To promote greater income equality and faster employment growth, the government increased the size and activity of the public sector.

From 1971 to 1975 total government spending rose rapidly from 20.5 to 30 percent of GDP with the bulk of the growth occurring in the parastatal sector. State-owned enterprises grew tenfold over this period, from 84 to 845 in number (Urzua 1993). As described by Gil-Díaz (1987, 1990), the government tried to expand the size of the personal income tax base and enhance its progressivity. Although these measures were partially successful, revenue-enhancing tactics such as high excises and improved enforcement could not keep up with rapidly accelerating revenue needs. Blessed with new oil discoveries and the pleasant prospect of imminent oil wealth, the government showed no reluctance in financing high levels of public spending through foreign borrowing and the inflation tax. By 1975 the fiscal deficit had swelled to 10 percent of GDP and annual inflation had reached 27 percent.

In the wake of the second oil price shock and rocketing oil revenues from PEMEX, the state oil monopoly, the López-Portillo administration purchased more private enterprises and embarked on a large-scale public investment program. Total public sector spending grew from 31 percent of GDP in 1978 to 41.3 percent in 1981. Then calamity struck. Oil prices collapsed in 1982, world interest rates escalated dramatically, massive capital flight ensued, and the foreign lending sources dried up. Mexico could no longer service its foreign debt obligations and, after suspending payments on this debt, entered into a series of negotiations with foreign creditors to restructure its debt.

The de la Madrid administration (1982–86) responded to this macroeconomic crisis by sharply curtailing public expenditures, increasing excise taxes and the VAT, and privatizing a number of public enterprises. Despite these efforts, falling terms of trade and negative net transfers abroad contributed to a sharp recession. Real GDP fell for the first time since 1932 and inflation soared to triple-digit levels. Aggregate investment plummeted 28 percent between 1982 and 1986 and real minimum wages declined by 40 percent.

When the Salinas government came to power in 1987, it clung to the path of restoring macroeconomic stability by slashing government spending and continuing the sale of most state-owned enterprises. At the same time, the economy was pointed in a new direction of freer trading relationships, privatization of most government enterprises, and deregulation of domestic prices for gasoline, electricity, and food. New reforms in combination with a series of tax reform initiatives begun by previous administrations lowered public spending and raised revenues, which pared the public deficit to almost 3 percent by 1990 and trimmed the inflation rate to 30 percent.

Tax reform was front and center on the political agenda throughout this period of macroeconomic disarray. Initially tax reform was viewed as a requirement for sustaining the planned growth of the public sector. However, the tax authorities were aware of the large distortions and inequities in the tax system and of the fact that they could not simply raise rates without magnifying these faults. More revenue would have to come from an improved structure of both direct and indirect tax bases.

The scope of tax reform was further enlarged to include indexation of the tax system in an attempt to avoid the inflationary erosion of real tax revenues and the distortions caused by the taxation of nominal rather than real incomes.

The traditional tax reform goals of achieving greater vertical and horizontal equity, more neutral tax treatment, and simpler administration and compliance were not ignored by the tax officials. Accepting the conventional wisdom of tax reform, tax reformers emphasized how the development of broader direct and indirect tax bases and their taxation at lower marginal rates would serve equity and neutrality goals. We will examine in later sections how successive tax reforms have propelled the tax system along this path.

Acutely conscious of the potential for domestic taxes to distort trade and capital flows, the tax authorities did not want to hamper the economy's international competitiveness so they unilaterally aligned several features of the tax system with its most significant competitor, the United States.

International interdependence and the mutual interaction of tax systems greatly influenced tax design in Mexico. Proximity and economic integration present both constraints and opportunities. Limits on choice may generally apply either to the base or to tax rates. If the government imposes excessive taxes on consumption or imports, revenues may fall because of substantial commercial diversion to smuggling. In the first half of the 1970s, mink coats were taxed at 30 percent rather than the previous 10 percent, causing collection on this item to fall one-third from its previous level. However, Mexican consumers continued to buy mink coats in U.S. cities close to the northern border.[2] Similar tax evasion occurred when a jewelry luxury tax was increased from 10 to 30 percent.

The international movement of goods, services, and factors present economic restrictions within which national tax reform has had to operate. It would be ideal to design optimal taxes in a world context. However, this would only work if all national governments agreed to terminate tax havens and coordinate their legislation and tax management.

In order to understand the implications of tax reform on the openness of the economy it will be useful to examine two broad categories of the balance of payments: the trade balance and the capital account. Starting with capital flows, the economy has always been open (except for three months in 1982), making it vulnerable to monetary and tax policies in other countries. Also Mexico shares a long border with the largest capital market in the world.[3]

Taxing income on financial interest is difficult because in many countries—such as the United States—this item is exempt for nonresidents. Taxing interest income would cause the gross interest to adjust upwards so that the interest rate, net of taxes, would remain competitive with respect to other countries. In this case, a tax on interest income is paid by debtors, who are mostly small enterprises and not-so-rich individuals who do not have access to international credit markets. To the extent that the tax is not fully shifted forward, it penalizes domestic financial intermediaries, thus reducing their size below the level they would otherwise have.

Corporate tax rates must not exceed the rates of foreign direct investors. If the corporate income tax exceeds international standards, investment (domestic and foreign) will be reduced unless the excess portion of the tax is shifted. In this case, however, the excess portion of the tax is likely to be paid by consumers and/or labor rather than by capital (Fernández 1980). Nonalignment of nominal tax rates may induce a shifting of existing corporate tax bases among different tax jurisdictions, while international variation in effective tax rates could have consequences for the location of real investment expenditures. Also, the fact that corporate capital abroad is taxed means that corporate tax rates should not be below international rates or foreign treasuries will tax the difference. If the tax is lower, collections are transferred abroad without attracting foreign investment. Recent reforms in 1987 and 1989 have brought corporate rates much closer to the international level. One must also take into account that state and local income taxes exist in other countries and that, at least in the United States, tax credit provisions are important. Recent changes in U.S. tax legislation have modified the creditability of foreign taxes, limiting it by type of income and on a country-by-country basis and, in some cases, even on a firm-by-firm basis.

Net wealth taxes are easily evaded. Because capital has international mobility, truly wealthy individuals who are familiar with the various options for investment and shelters abroad are not seriously affected by wealth taxes. Also, these taxes affect owners of estates and medium-income individuals only on a once-and-for-all basis.

The economy is also open with respect to foreign trade. Even when imports were subject to strict quantitative controls from 1960 to 1985, illegal trade made the economy more open and competitive. Cavazos Lerma (1976) provides a measure of this influence in his study of implicit nominal protection in Mexico (price differences between U.S. and Mexican goods). The study shows that most Mexican products were, during the first half of that year, below the cost, insurance, and freight price (CIF) plus the import duties because of smuggling activity.

Early 1970s to the Present

In the 1970s, the aim of policymaking was to obtain a comprehensive global income tax. Before 1970, a practical compromise between this half-hearted ideal and the limitations of reality had been intuitively reached but without any underlying theory to support it.

The Echeverria Administration, 1972–75

During President Echeverria's electoral campaign there were passionate discussions about the need for a radical tax reform in general and income redistribution in particular. Out of all the political jostling and shoving of his first two years in office emerged two basic reforms, one aimed at more comprehensive income taxation and the other at harmonizing excise and sales taxes.

Reforms on interest income. The first reform was largely cosmetic and affected financial interest income. It established a higher withholding for those who opted not to aggregate their interest income with other income sources. Practically everybody opted for the lower withholding formula since the threshold income level at which it paid to globalize interest income voluntarily was relatively low. The effect of this higher tax on interest income initially reduced the real tax yield from deposits, but ultimately created higher nominal interest rates and induced a leveling off of the growth rate of financial intermediation.

In fact, the net revenues from this tax were likely to be nil or possibly even negative. The government was and is a prime borrower of internal funds, so the higher tax was reflected in higher costs on its internal debt. If government debt represented half of financial intermediation, on average one-half of the higher tax boomerangs as additional government expenditures if the bank's profit margins remain constant. If the other half of financial intermediation consists of loans to corporations, half of the increase in the grossed-up interest rate ended up as a lower corporate income tax. In such circumstances, the government netted only one-quarter of the tax increase. If financial markets had grossed-up the higher tax in full, financial intermediation may not have suffered as much.[4]

If interest had been grossed up, savers would have ended up with a higher gross income with net income staying the same. Thus the tax would have been borne by the users of credit or, most probably, by the consumers of nontraded goods and services. If an incidence measure did not take this effect into account, the apparent burden of the tax on savers would be merely an optical illusion. The best option may have been not to tax interest income at all.

On the other hand, not taxing interest earned by individuals who lend directly to a firm would lead to companies and their owners circumventing the corporate income tax through "back-to-back" operations via interest deductibility on corporate loans. Therefore, the existence of a corporate income tax ends up justifying the tax on interest income with all its inconveniences. The tax increases on financial interest during President Echeverria's administration produced little revenue and discouraged deposits channeled to domestic financial intermediaries—the main internal source of funding for the government and the private sector. Greater recourse to foreign borrowing by both sectors was the ultimate outcome.

The globalization of income from rental housing. The second income tax reform concerned with the globalization of rental income from housing is more difficult to assess. Rents had been subject to a separate tax according to a schedule with a maximum marginal tax rate of 10 percent. In 1973 rents were made cumulative to income, making it difficult to document whether rental housing paid higher or lower taxes after the reforms. Since the schedular tax could not be evaded the way the global tax is today, by fragmenting income among several minors (grandchildren, children, nieces), it is likely that it produced a higher yield before the reform.

However, private sector analysts have presumed that the decline in the stock of rental housing and the reduced after-tax yield from rental housing was the result of this tax measure. This allegation seems plausible since the tax reform coincided with a drop-off in investment in rental housing despite the fact that the pace of private investment was brisk. However, the tax base was enhanced in subsequent years because rental housing is generally an unleveraged investment and it was adversely affected by the appearance of high rates of inflation that considerably lowered the real value of depreciation allowances. But the low investment situation continued into the late 1970s and 1980s. This remained the case even though reforms were introduced in 1979 to fully index depreciation allowances for past and future inflation or, alternatively, to permit a blind deduction of 50 percent of gross income if the investor preferred not to itemize deductions.

The lack of investment in rental housing is thus probably more a result of the highly inflationary environment from 1972 onwards, coinciding with a regulatory structure that made the maintenance of real rents practically impossible, rather than the result of high taxes (Creel 1988). The rental situation in the Federal District was also aggravated by legislation that permitted long contract disputes (up to four years) with the nominal rent fixed in the interim, as well as by a 1984 reform that set a ceiling on rent increases equal to 65 percent of the increase in minimum wages.

But here, as with other income items, the globalization ambitions of the law contrast sharply with administrative capacity (although it can be argued that better administration is, at least in principle, within reach). It is also a fact that the ever-smaller supply of rental housing in the three largest metropolitan areas (Mexico City, Guadalajara, and Monterrey) is being preferably diverted to foreigners, which considerably diminishes the probability of a contract "lock-in" for the owner. Moreover, these transactions are easier to hide from the tax collector since foreigners, who eventually depart, can make a noncontract rental, or a contract with a rental below the actual paid value, reasonably safe to operate from a legal standpoint. Aside from administrative considerations, the situation that prevailed in rental housing, if compared to the ideal comprehensive income definition, presents the biggest inequity and distortion because of the exclusion of rental income from owner-occupied housing from the base of the tax.

Adding income from rented property in the tax base and reforming taxes on financial interest income were the most important reforms of the Echeverria administration (1970–76). Other notable minor reforms included an increase in the individual income tax rate from a 35 percent maximum marginal tax rate to 50 percent. Despite a progressive schedule, corporations actually paid a flat 42 percent. By the end of the Echeverria administration, higher inflation rates were pushing up effective income tax rates considerably.

Regional indirect tax harmonization. These were delicate times from a confidence standpoint. It would have been foolhardy to push ahead with income globalization, so the government prudently went no further.[5] However, important reforms in indirect taxation coordinated sales taxes between the states and the federal government to harmonize the motley array of rates and tax treatments across the thirty-two federal entities (the Federal District and the states). The new coordination law introduced the TST along with state-federal revenue sharing. This paved the way for the introduction, some years afterward, of a federal VAT requiring just such a tax-sharing system to be politically acceptable.

The TST let each state keep a share of the tax it collected (administration was shared but tax collection remained with the states). Under the TST revenue-sharing agreement, regional sales by a firm located in a particular state were allocated to other states according to the distribution of sales. When the VAT was introduced some years later a formula already existed for revenue sharing among the states, making it easier to redesign the distribution system to avoid huge revenue imbalances between states.

The Echeverria government wanted to ensure an equal base between states for the TST. To do so, it was necessary to legally restrict the ability of states to raise taxes on other goods and services. Of course, it was not easy for governors to surrender their power to raise a tax to pay for a particular spurt of spending. However, the government held out an enticing carrot.

Most big states had a local TST rate of 1.2 percent compared to the federal rate of 1.8 percent. So the federal government offered each state a federal TST rate of 4 percent with revenues to be shared almost equally. A state could choose not to adjust its local taxes, but then it could not share in federal tax and would have to superimpose its own local taxes on top of the new tax. Given these circumstances, it is not hard to understand why every state entered the new system.

The federal tax burden increased considerably during this administration to 3.1 percent of GDP, which was equal to the gain realized in the last three decades. This sharp increase was due mostly to the increase in the TST and excise taxes (alcoholic beverages, cigarettes, electricity, soft drinks, and telephones). Both the TST and excises had been extremely low by international standards and probably remained so even after these substantial adjustments. The only exceptions were some Lafferian luxury rates of 30 percent on jewelry and furs, which provided less, rather than more, revenue after being imposed.

The López-Portillo Administration, 1976–82

Despite the tax improvements during President Echeverria's term, falling real taxes and rapidly rising government expenditures kept the public deficit rising. The government had embarked on a expansionary fiscal program, giving subsidies to the private sector and creating expensive marketing agencies (for coffee and tobacco) and large new parastatals (for fertilizer and steel). This led to a balance of payments crisis and devaluation of the peso in 1976.

The incoming government of President López-Portillo (December 1976) quickly stabilized the economy thanks to a combination of luck, higher oil revenues, and greater business confidence. The president made conciliatory gestures to the whole nation, including the private sector, and, as an initial macroeconomic measure, tightened the budget.

Even though the economy grew rapidly and inflation was brought under control by 1978, rapid growth in public expenditures (mainly on oil, seaports, and large petrochemical plants) led to a rise in the real exchange rate. By 1982, the appreciation in the real exchange rate was somewhat greater than had existed before the September 1976 foreign exchange crisis. But on the whole the economy had been adequately managed up to the end of 1980. The sharp increase in public expenditures had been matched by an even sharper increase in revenues, thus (leaving aside the ever-increasing foreign debt financing of the fiscal gap) improving the public debt ratios to GDP, both external and internal.

Given such a healthy macroeconomic setting—a strong economy and fast growth, virtually pursued by foreign bankers—why bother with a tax reform? In fact, the favorable revenue environment made it politically difficult to even suggest an increase in the tax load or a "structural" reform. Nevertheless, a political commitment to reform had been made in the days of President Echeverria's administration and the Treasury Secretary, David Ibarra, who combined a grasp of economics with great ability as a negotiator, understood that there was a unique political opportunity to bring about the globalization of income tax and a reduction in resource allocation distortions through structural reform of the tax system.

Another important player was the Undersecretary of the Treasury, Guillermo Prieto. His job combined the administrative and interpretative functions of the U.S. Internal Revenue Service with the tax policy role of the U.S. Treasury. He was also an economist who was able to translate the array of policy options into administratively viable proposals, an unusual circumstance given that policy formulation is often at odds with the administrative acceptance of new policies, particularly if the design intends to change the way administrators operate. A similar human equation was to be decisive in pushing through the reforms of 1987, as we will see.

The reforms carried out in various stages from 1978 to 1982 included some important changes to the income tax of individuals and corporations. For example, in January 1979 a VAT was introduced and sweeping changes were made to simplify and incorporate excises. Congress approved these changes to go into effect one year later, together with the new act on tax coordination with the states.

The new VAT and the reform on excises. The legislation was delayed one year to let taxpayers become acquainted with the new law and the accounting, computer, and administrative requirements of the new VAT. Lawyers, accountants, and tax administrators took full advantage of this respite to learn the new system. But the private sector (erroneously but perhaps understandably) believed the government lacked resolve to put the tax in place (probably because the administration itself was not united behind the proposals although treasury officials presented position papers and international evidence and comparisons to persuade its opponents of the worthiness of the proposals up to the last moment) and continued to lobby against its implementation.

The arguments presented by the private sector against the VAT had been raised on previous occasions. However, based on the experience of other countries the treasury was convinced that the effect of the tax would be at most a slight once-and-for-all increase in the price level. Moreover, the treasury regarded the likely economic and evasion control effects of the new set of indirect taxes of overriding importance.

In the end the treasury secretary managed to persuade the president that the inconveniences were outweighed by the advantages and the VAT proposal went ahead. But doubts lingered and rising inflation (26.3 percent in 1980 versus 18.2 percent in 1979) was popularly attributed to the new tax. Another vulnerable point of the new VAT was that, unlike modern European versions, the tax was to be itemized on a separate line on all sales slips. This visibility made the public more irritable, particularly when the rates had to be increased later.

Diverse groups were interested for their own particular reasons in the effects of the new tax. Separate negotiations were conducted both before the law was enacted and in the year before it went into effect to take their points of view into consideration and make amendments.

Perhaps the most important and protracted discussion involved the general rate of the VAT. The new rate had to match revenue previously raised by the earlier 4 percent TST, plus the TST's special 30 percent luxury rates, and 25 federal and 300 state excises. The best estimate by the treasury was a general VAT rate of 12.7 percent versus the 10 percent rate proposed by the private sector. When it was realized that the gains from reduced tax evasion would very likely compensate for the lower rate, the treasury agreed with the private sector to set the rate at 10 percent, thus eliminating the biggest negotiating obstacle.

Other issues involved the amount of credit for taxes embodied in previous investments, the reduction in excises required for a pyramided VAT with the same incidence as the previous rate, and negotiation with the states regarding a new revenue-sharing formula. Of these issues, the most interesting and important one is related to revenue sharing. The old system that gave them a share of the amount they collected was considered impossible to calculate under the VAT, so a procedure was negotiated whereby a general-sharing fund was created as a percentage of total federal tax collections. From that fund, in the first year, each state would receive a percentage calculated as the amount of federal transfers received over the past three years divided by total federal transfers made to the states in the same period. The federal government had been previously sharing almost 50 percent of the TST plus varying percentages of various excises.

After a transition year with shares allocated through constant coefficients, a new formula was supposed to go into effect in which those states producing a greater percentage of increasing revenues than the average percentage increase would be awarded a larger coefficient in the following period. However, the constant coefficients applied during the transitional period were almost impossible to eliminate later. Since the amount of funds to be distributed was a fixed percentage of federal revenues, those states that expected their coefficients to fall withheld information, for example on the regional distribution of sales of interstate firms, and short-circuited the process.

Such behavior put the revenue-sharing system into a yearly crisis when states that had been raising their tax efforts understandably expected to be rewarded. The solution was to raise the overall percentage, and thus the size of the fund, to keep everybody happy under a system of constant coefficients. Naturally this led to an indifferent administrative attitude on the part of most states and to an increasing hemorrhage of federal funds (federal transfers to the states went from 11.2 percent of federal tax revenues in 1979 to almost 17.3 percent in 1986).

The lack of tax collection effort by some states contributed to the laggard behavior of the VAT in the years after the transition (1982–86), but it is difficult to sort this out from high inflation, lower imports, and higher exports, which also contributed to a relative loss of VAT revenues. The administrative complications of the VAT arose, of course, because the states had relinquished their rights to impose local sales taxes in order to participate in a federal tax but maintained administrative responsibilities which gave them authority and presence with their constituents. Thus arose the incentive problem. The solution had to combine some administrative role for the states because the national VAT had to coexist with local governments keen on keeping some of their tax prerogatives. This was partially corrected by the 1987 reforms with a zero-sum formula that gives states 30 percent of the VAT they collect with no ceiling on their share. They responded immediately with zealous collection efforts.

The rate structure of the VAT went through three stages. It was simple and elegant when first introduced; along with the basic rate of 10 percent, there was a zero rate for some agricultural goods and for some basic foodstuffs, and a 6 percent rate for the northern frontier strip to take into account the level of sales taxation of neighboring U.S. cities. Unfortunately, the legitimacy of the new tax was eroded by an exemption (though not a zero rate) for sales made through union stores.

The second stage, only one year after its introduction (1980), extended the zero rate to more foodstuffs. Given the degree of political opposition the VAT had aroused before its introduction, the government was quick to compromise on its implementation to please those who felt that inflation had been fueled by the VAT and that the real income of poorer people could be improved by tinkering with the tax rate.

The third stage came about as a response to the fiscal deterioration during 1982–83, though this was also countered in part through strong expenditure reduction measures for the 1983 fiscal year of the entering de la Madrid government. The general rate was raised from 10 to 15 percent, along with a symbolic luxury rate of 20 percent and a lower (6 percent) rate on medicine and on some food items to make the higher rate politically more palatable.

The reforms on the income tax. The income tax also underwent a gradual reform process. From the beginning the three reform objectives for the income tax were to broaden the tax base, reduce distortion of resource allocation, and adjust for inflation.

To broaden its base, dividends were made cumulative to personal income with the introduction of an integration scheme that allowed full credit for the corporate income tax. Capital gains were also included in the individual income tax base, with the exception of gains on stock-listed shares.

Tax policymakers waged a long struggle to bring "special treatments" accorded to agriculture, newspapers, trucking, and construction into normal taxation. They succeeded only in the case of the construction industry, which was taken into the corporate income tax base, albeit with a transitory "blind" deduction of 4 percent of gross income to offset difficulties encountered by construction firms in the countryside in obtaining invoices that fulfilled legal and accounting requirements. Although only a first step, it was an important accomplishment because it reduced overinvoicing through construction industry receipts. Until then, construction firms had been taxed a fixed rate of 3.2 of gross income, which enabled them to inflate construction receipts by overinvoicing and provide other firms an income tax saving at the corporate rate of 42 percent. If the 8 percent profit sharing with workers was included, the net evasion through artificial mark ups in construction expenditures was 46.8 percent (50.0 – 3.2).

Doing away with special treatments would probably have been the most effective possible way to broaden the base. Several objectives would have been achieved simultaneously: (1) improved equity by incorporating a lower-paying sector into the tax base, (2) decreased tax fraud and tax evasion, and (3) better allocation of resources as they flow out of the previously lower-taxed sector into the on-average higher-taxed sectors that have a higher social (tax included) rate of return.

The inflation adjustments made at this time were far-reaching and, while they were incomplete and partial, especially in the case of the corporate income tax, they prepared the way for full indexation. However, in light of the rapid inflation toward the end of the de la Madrid administration, too much revenue was lost before full inflationary corrections could be made to the tax base. Corporate income tax revenues fell from 2.7 percent of GDP in 1980 to only 1.5 percent in 1985.

The only inflation adjustments introduced early enough to have a positive impact were to excises, where the elimination of specific taxes coincided with the 1980 introduction of the VAT. The new ad valorem rates also became effective in the nick of time to prevent a substantial loss in revenues from inflation. Had such a loss occurred, any effort to compensate through a continuous uplifting of the specific rates would have been imperfect. Adjustments would have had to allow for imperfect forecasts of inflation while monthly or quarterly would have been administratively cumbersome. Besides, specific taxes tend to be "quality regressive," within a category. For example, a per liter tax on beer fell proportionately less on the highest-priced containers purchased by well-to-do consumers. The same can be said for alcoholic beverages and tobacco.

The inflationary adjustments made to the income tax were related to the "costs" of assets sold (to measure real capital gains) and to the "unleveraged" portion of depreciable assets. A table constructed with the price indices of many years was published in the law, a sort of carry-forward or inverse price index. Full indexation of the corporate income tax was introduced in 1987. In the individual income tax, itemized personal deductions were eliminated in favor of a single personal deduction equal to an annual minimum wage which became subject to indexation. Finally, income tax brackets were adjusted every year to account for inflation.

The de la Madrid Administration, 1982–88

The new de la Madrid government would have faced a fiscal crisis of major proportions even if foreign credit had been readily available. In fact, there was a suspension of foreign credit throughout this presidential term despite the country's compliance with adjustment goals and structural changes that went beyond the goals agreed to with international multilateral lending institutions. Some fresh money was supposedly provided with each negotiation, but the resources foreign banks provided with one hand, they were to take away with the other.

In the first five years of this administration, the total foreign debt rose from $95 billion at the end of 1982 to $98 billion at the end of 1987. The $98 billion is equal to $93 billion in 1982 dollars, a fall of 14 percent in real terms. If the foreign debt is measured net of international reserves, the nominal amounts are $93 billion at the end of 1982 and $85 billion at the end of 1987. The real net foreign debt in this case was $71 billion, a fall of 24 percent in real dollars over the same period.

Faced with such a shortage of funds, the de la Madrid administration slashed government expenditures and increased taxation. The original VAT structure of 0, 6 percent on the northern border, and a 10 percent basic rate had been complicated enough, but the new structure, begun in 1983, included rates of 0, 6, 10, 15, and 20 percent. This made the VAT even more difficult to administer; it caused some of the less advanced states to be overtaxed and increased the ease with which fraudulent claims of excess credits could be made.

However, since the general rate needed to go from 10 to 15 percent and since so many other taxes also had to be raised, it was felt that the VAT needed a harder political sell. Even so, the so-called gains in progressivity were minuscule compared to the administrative inconveniences that ensued. The new general rate may also have been counterproductive because it was high enough to encourage tax evasion, especially given the administrative chaos. The rates of excise taxes were also substantially raised, while a temporary surtax of 10 percent was imposed on personal income.

Even after these reforms, indirect taxation continued to fall short of the rates and coverage prevalent in most other countries. Judging from an international comparison of ratios of excise tax revenue to GDP that shows an extremely low reliance on indirect taxes (Gil-Díaz 1987), the reforms were clearly a move in the right direction.

Despite the substantial increase in excise rates, their contribution to revenues did not increase commensurably. The ratio of excises to GDP was 1.81 percent in 1982 and increased to only 2.39 percent in 1983. The figures from 1984 to 1987 were 2.24 percent, 2.16 percent, 2.76 percent, and 2.48 percent, respectively. The explanation lies partly in the zero rating of exports. Excises generally are not paid on exports, and there is a full rebate on excises and on the VAT paid in previous stages. As imports fell, nonoil exports increased from 2.4 percent of GDP in 1982 to 2.49 percent in 1987, and the taxable base of excises and the VAT contracted.

The Salinas Administration, 1988–89

The outgoing de la Madrid government left a strong stabilization program in place. There was a hefty primary budget surplus, 5 percent of GDP. The nominal exchange rate had been fixed and there was an agreement between labor, business, and government to stabilize prices and wages after the December 1987 alignment. As a result, inflation was brought down from an annual rate of 180 percent in February 1988 to 51 percent in December 1988. Measured in monthly rates, it was brought down from 15.5 percent in January 1988 (an annual rate of 463 percent) to a monthly average rate of 1.4 percent (an annual rate of 18 percent) in the last quarter of that same year.

President Salinas, who took office in December 1988, was resolved to consolidate these gains through the necessary budgetary adjustments. One of his goals was to increase the tax load by 1 percent of GDP to enhance the primary budget surplus. But the attainment of this goal was complicated considerably by declining corporate tax revenues.

The new corporate income tax base was to comprise 60 percent of the previous corporate base in 1989. Its entry into the tax system would have, therefore, reduced the tax on corporations, since in addition to the lower base, the tax rate adopted was 37 percent on the new base (35 percent in 1991) compared with 42 percent on the previous base. But the actual reduction in revenues was even greater, since the private sector had been lobbying for an abandonment of the transition phase. They preferred the new base to start fully in 1989 because the monthly provisional payments using the two overlapping bases had become too complicated.

The government wanted to increase revenue, but at the same time it had agreed to simplify the tax code. These objectives seemed to be mutually exclusive. The government's solution was to introduce a minimum corporate income tax. The minimum tax was supposed to recover the income lost from accelerating the transition and catch the many firms that had so far managed to evade or avoid all or most of the income tax through transfer prices and other tax dodging manipulations.

It was decided that the new minimum tax would consist of a 2 percent tax on the inflation adjusted gross asset value of firms. All assets except shares in other firms were included. The only deductions were loans from other firms and equity investments in other firms. It is easy to see that the two deductions prevented the double taxation of assets and that the base of the tax was actually the sum of all assets if assets are consolidated among firms.

The new system was designed to extract a minimum contribution from those that had remained outside the tax net. It was made creditable against the income tax so it would not double-tax firms. Any excess of the assets tax over the income tax can be indexed, carried forward, and credited against future profits. Because it is a hard tax to evade, costs of compliance should be lower than those of other taxes, though this has yet to be seen. It will also encourage firms to shed unproductive assets, defined as any that do not yield a before-tax real return of at least 5.71 percent per year.[6]

The new tax was the outstanding feature of this reform, but there were other significant changes. The previous integration scheme was eliminated and replaced with a simple tax of 10 percent on dividends paid to individuals. The new tax left the recipients of dividends in the highest income bracket with about the same tax that would have obtained under perfect integration (35 percent + .10 (65 percent) = 41.5 percent), since the maximum personal income tax rate had been lowered to 40 percent. In 1990 the maximum personal income tax was reduced to 35 percent, and the equalizing 10 percent tax on dividends was eliminated.

Thus, corporations that distributed their retained earnings on which they had paid corporate income tax in previous years would have to pay only the corporate income tax on that income. Exempting dividends and unifying corporate and top personal rates were intended to achieve approximate integration of personal and corporate taxes. The tax on profits remitted abroad was also lowered to 35 percent (to 37 percent in 1989 and to 36 percent in 1990) in order to make it congruent with the general lowering of rates and to attract direct foreign investment.

The first two reforms of the Salinas administration—the new tax on firms' assets and the reforms on dividend taxation—were significant. The new asset tax produced an outcry from some private sector representatives, especially in the first part of 1989. But its introduction, besides its economic effects, was quite timely, since without it the revenue loss would have been equal to 1.3 percent of GDP, probably enough to make the stabilization plan fail or at least falter.

By the end of the first year of the Salinas administration, additional reforms had been made that should significantly raise revenues, equity, and efficiency in the future. These have included bringing various groups that are hard to tax more firmly into the fiscal orbit. Taxation of agriculture, cattle raising, cooperatives, and forestry has been modernized through the introduction of a cash flow system that also applies to taxpayers with an annual income of less than 500 million pesos a year. In the new system (optional for the first four years), taxes are incurred when cash is taken out of a productive unit. This is a step toward expenditure, rather than income, tax treatment.

In addition, all special exemptions, including those for trucking, fishing, agriculture, shipping, newspapers, and lumber, were abolished, as was the previous exemption granted to authorship rights, although successful political lobbying from this group reestablished a limited form of exemption.

Finally, the VAT was made collectible on the same monthly return schedule as the corporate income tax. This apparently minor change should greatly improve administrative efficiency, since it will allow for the creation of a national monthly data bank on the VAT, instead of thirty-two unconnected data banks across the country. This will allow the timely cross-checking of information on different individual VAT returns, of VAT returns against foreign trade data, and of VAT returns against income tax returns. Thus, the Salinas administration made a large effort to bring about significant increases in tax compliance and tax revenues. At the same time, it adopted tax rates that were internationally competitive.

Inflation, Revenue Yields, and Indexation

Inflation has had a corrosive effect on all taxes. When the general price level increases, the time lag between when a tax liability originates and when it is collected makes its real value decline. When weighed against zero inflation this effect reached 0.9 percent of GDP in 1987, when the inflation rate was 159.1 percent (table 7.2). Higher inflation causes the ratio of tax collections to GDP to fall because taxes are not collected as soon as they accrue. A numerical example may help to illustrate this phenomenon.

Suppose that, without inflation, a tax base of $100 generates a $15 tax collection on the twentieth of the following month. The tax value at constant prices is, by definition, $15 and the tax base ratio is also 15 per-

Table 7.2 Effect of Inflation on Tax Revenues

Year	Annual rate of inflation (percent)	Tax revenue loss from inflation (percentage of GDP)
1980	29.7	0.41
1981	28.6	0.40
1982	98.8	1.04
1983	80.7	0.88
1984	59.1	0.68
1985	53.3	0.73
1986	105.7	0.80
1987	159.1	0.90
1988	51.7	0.40

Source: Bank of Mexico (1988, p. 268).

cent. Let us then suppose that a 5 percent increase in the price level occurs in the month the tax accrues and in the month the tax is to be paid. The effective tax rate at constant prices will then be 30.75 ÷ 215.25 = 14.28, and the tax burden will have fallen from 15 to 14.28 percent. This drop represents a reduction of 0.5 of 1 percentage point in the tax rate, a significant amount.

In table 7.3, a more realistic example of the so-called Oliveira-Tanzi effect is set out. Real GDP is $1,200, evenly distributed throughout the year. Two taxes—tax "x" and tax "y"—exist. Tax "x" produces $10 per month, which corresponds to the tax base of the previous month. Tax "y" generates $20 on a quarterly basis, in March, June, September, and December. Its base equals that of the three months previous to the month it is paid. Two situations are considered. In the first, there are no changes in the price level. In the second, prices increase at 5 percent per month.

As can be observed, without inflation the tax to GDP ratio is $200 ÷ $1,200 = 16.66 percent. With a 5 percent monthly inflation, the tax ratio becomes 15.9 percent with the assumed collection lags, which results in a reduction of the tax burden of 0.8 of 1 percent of GDP. If inflation is kept

Table 7.3 A Hypothetical Oliveira-Tanzi Calculation
(U.S. dollars)

				Monthly inflation rate of 5 percent		
Month	GDP	Tax (x)	Tax (y)	GDP	Tax (x)	Tax (y)
January	100	10		105.00	10.00	
February	100	10		110.25	10.50	
March	100	10	20	115.76	11.02	21.01
April	100	10		121.55	11.57	
May	100	10		127.63	12.15	
June	100	10	20	134.00	12.76	24.31
July	100	10		140.70	13.40	
August	100	10		147.70	14.07	
September	100	10	20	155.10	14.80	28.14
October	100	10		163.00	15.50	
November	100	10		171.00	16.30	
December	100	10	20	179.60	17.10	32.58
Total	1,200	120	80	1,671.30	159.20	106.04

Source: Authors.

Table 7.4 Numerical Example of the Corporate Income Tax Loss from Interest Deductibility under Inflation
(U.S. dollars)

	Without inflation	*With inflation at 100%*
Asset yield	100.00	200
	15.00	30
Liability interest	50.00	50
(5 percent in real terms)	2.50	55
Taxable profit	12.50	–25

Source: Authors.

constant, the tax ratio will also be kept constant at its new lower level. That is, inflation by itself begins the phenomenon, but once inflation stabilizes at a new (higher) level, the (lower) tax collection ratio also stabilizes. This suggests that one can expect substantial increases in taxes when inflation drops. In fact, such a phenomenon occurred in Argentina, Bolivia, and Israel when instantaneous price stabilization programs were implemented.

Another reason why real tax revenues have been so low has been the fact that the deductibility of nominal interest paid by firms substantially reduces their taxable income in an inflationary environment. A numerical example in table 7.4 illustrates how powerful this effect is. In spite of the fact that real variables are held constant in this example, taxable profits go from $12.50 without inflation to a $25 loss with an inflation rate of 100 percent.

This shrinkage in the tax base is explained by the fact that, with inflation, interest becomes a hybrid compound of interest and debt amortization. If the firm could deduct only that share of its interest that does not include the amortization of principal (which in the example equals $5 when inflation reaches 100 percent), its taxable profit would be $30 – $5 = $25, which, in real terms, equals taxable profits without inflation.

Collection lags became an important issue. Despite a rather short period of only twenty days between the accrual of an excise tax obligation and the monthly payment of it, the average lag became thirty-five days when the fifteen-day lag of the month itself was included. With inflation often averaging 5 percent per month, the Oliveira-Tanzi effect was a prime concern. At such a rate of inflation, it amounted to approximately a 6 percent loss in revenues. The partial indexing that had been introduced back in 1979 aggravated the revenue situation. This had allowed inflation-adjusted depreciation expenditures to the extent that firms were unleveraged, and this was also gnawing at the tax base. The overall effects of inflation on revenue yields are indicated in table 7.2.

Corporate income tax collection thus lost considerable ground. As shown in table 7.5, corporate collection fell from 2.6 percent in 1981 to 1.7 percent in 1982 and averaged 1.6 percent during 1983–87. This outcome was partly due to the Oliveira-Tanzi effect, because the tax was collected in three yearly payments, and partly to the deductibility of nomi-

Table 7.5 Domestic Debt and Taxes of Firms

	Net domestic debt of private corporation			
Year	Current pesos (billions)	Constant (1980) pesos (billions)	Percentage of GDP	Corporate income tax as percentage of GNP
1980	549	549	12.80	2.7
1981	650	505	11.10	2.6
1982	869	340	9.20	1.7
1983	890	192	5.20	1.5
1984	1,400	190	4.90	1.6
1985	2,925	243	6.40	1.6
1986	4,086	165	5.10	1.7
1987	2,085	32	1.10	1.6

GDP is gross domestic product.
GNP is gross national product.
Source: For columns 2, 3, and 4, Bank of Mexico (a); for column 5, Bank of Mexico (b).

nal interest expenditures. As already discussed, interest expenditures balloon when inflation goes up, mostly because they contain an amortization component to compensate for the inflationary loss in the real value of unindexed loans, causing the base of the income tax to erode considerably. It seems that a substantial portion of tax savings was used to reduce corporate debt.

While the income tax fell on average by 1 percent of GDP during the de la Madrid administration, the peso debt of firms, net of their peso financial assets, went from 12.89 percent of GDP in December 1980 to barely 1.1 percent in December 1987 (see table 7.3). Such a clearing of debt was welcome, especially as it allowed corporations to take advantage of the trade opening of mid-1985 to 1987 and of the high real interest rates resulting from the December 1987 stabilization program. Nevertheless, the fall in corporate taxes further weakened the public purse, which had already had to compensate for the 1982 suspension of foreign credit plus the precipitous fall in the country's terms of trade.

The base of the corporate income tax would have fallen even further had it not been for the drastic reduction in the net debt of firms. It may seem puzzling that the peso debt of firms fell at this time, given that they were being allowed to deduct high nominal interest rates. There were probably several reasons for this. First of all, profits increased as a result of a continuous and drastic fall in real wages. There was a 57 percent real fall in manufacturing wages (including noncash payments) from 1981 to 1987. Second, the rate of private investment stagnated in an economy that was static from 1982 to 1987, so the higher profitability was not attracted to new projects. Third, private firms agreed not to distribute profits as one of the covenants of the 1983 debt-restructuring agreements with foreign banks. Fourth, the biggest one-year reduction in corporate debt occurred in 1987, the year in which the base of the corporate income tax was indexed. Finally, also in 1987, the foreign debt of Mexican firms was discounted by 40 to 50 percent. Cash-rich firms obviously took advantage of these discounts and repurchased large portions of their foreign debt throughout 1987 and 1988. The discounts originated in the long-term restructuring of private foreign debt, which took place in the second half of 1986.

To counteract the ravages of inflation on tax revenues, two important modifications were introduced. The first was a drastic reduction in the payment period. In the middle of 1986, firms were required to make twelve payments of the corporate income tax a year, instead of three. The new schedule was tight, as the payments had to be made only ten days after the closing of the month. Payment of the VAT and of excise taxes was also moved forward, from the twentieth to the eighth working day after the end of the month in which were taxes were collected.

Under the new system, firms make their provisional income tax payments by applying to taxable income in the current year the ratio of their profits to taxable income in the previous year. They can ask the treasury to approve a lower ratio when their relative profits have fallen but they most certainly do not ask the treasury to adjust the coefficient up when their relative profits increase. To correct this problem, an important reform required firms to calculate their profit at the middle of their twelve-month taxable period with actual figures on income and expenditures, and to pay any difference between their accumulated provisional payments and their actual liability. They have to repeat the procedure in the twelfth month. The result is that in the fifteenth month, when the final return is due, only minor differences should remain. This procedure substantially corrects for the lag problem.

These reforms reduced the Oliveira-Tanzi effect, but they also considerably increased administrative and accounting costs. A less costly alternative formula was also studied, but it could not be implemented because of legal difficulties. It prescribed the payment of the tax at the old date, multiplied by a representative monthly nominal interest rate times a number related to the average days of delay. In the case of the VAT, for instance, the tax would have been paid on the twentieth day of the following month, plus the tax multiplied by the adjustment factor $i \times (1 + 1/6)$, where i is the previous month's nominal monthly rate of interest.

The other inflationary correction would have required either fully indexing the corporate income tax or introducing an entirely new tax concept based on consumption-oriented cash flow accounting (McLure 1988). This new approach would have completely eliminated the eroding effect of inflation on the base of the tax and would have been simple to administer by taxpayers and authorities. It also had the advantage that taxpayer representatives favored it strongly over the fully indexed alternative. The only defect of the new scheme was that is was not a standard income tax and as such foreign investors might not have been able to obtain a credit for it against their tax liability abroad.

This unfortunate fact left indexation as the only alternative. The inflationary compensation on debt would no longer be deductible (nor would the inflationary compensation on financial assets be taxable) from taxable income, and the purchase cost of fixed depreciable assets would be indexed in calculating depreciation.

To alleviate the complications introduced by indexation, the new system was combined with some of the elements of a cash flow scheme. The greatest simplification was to permit the deduction of all current purchases. This eliminated the complications of inventory costing, which is quite a chore even without inflation. Another simplification was to allow the immediate deduction of the present value of investments that were otherwise depreciable. An annual real interest rate of 7.5 percent was used to bring future depreciation expenditures to the present. An otherwise twenty-year linearly depreciable structure in housing, for instance, was converted into an instant deduction of 51 percent of the investment cost.

However, these simplifications did not bring about immediate administrative relief. It was necessary to phase in the new system to minimize the shock of large gains and losses among taxpayers. After arduous negotiations, it was finally agreed that 20 percent of the new tax would be put into effect each year, beginning in 1987. In 1987 a firm would pay 20 percent of its corporate income tax as generated by the new base and 80 percent according to the old unindexed base. In 1988 40 percent of the new base was considered and 60 percent of the old, and so on. The new tax rate was a flat 35 percent to bring it into line with recent international levels, a reduction from the previous 42 percent flat rate.

If the new indexed base with its cash accounting of purchases and instant depreciation had gone into effect immediately, it would have meant a considerable simplification, even given the need to correct for inflation every month. Unfortunately, in reality the new system's introduction was not that simple.

As we have seen, the phasing in of the new base entailed the coexistence of both accounting systems. But there was one more complication: when partial indexation was introduced in previous years, the unindexed base was maintained in order to calculate the percentage of a firm's profits that, by law, were shared with the workers (8 percent up to 1987, 10 percent afterwards). Profit sharing is based on the old nominal notion of taxable income, but it is not a tax as such. It is derived from a proposal to congress by a special commission set up every ten years. Therefore, even with the new system in place, corporations were still bound by the old system with all it complications (mostly concerning inventory costing).

Inflationary adjustments were also made to the individual income tax. Despite budgetary vicissitudes, there has been a full inflationary adjustment in the schedule of the personal income tax from 1978 to the present. In 1987 the schedule became automatically indexed by making brackets a multiple of the minimum wage. However, even though an important downward adjustment had been performed on average tax rates in the 1979 reforms, by international standards the personal income

Table 7.6 Share of Wages and Salaries and Fringe Benefits on Total Labor Income, Manufacturing Sector
(percent)

Year	Wages and salaries	Fringe benefits
1978–80	73.8	26.2
1981–83	70.1	29.9
1984–86	69.5	30.5
1987–88	66.7	33.3

Source: Treasury Department (various years c).

tax imposed relatively high tax burdens. For example, an income equivalent of twenty times the national average minimum wage, or $26,000 in 1988, was subject to an average tax of 37 percent in the same year. The marginal tax for this bracket was 50 percent. Therefore, the present average and marginal tax levels calculated over money income are still high by international standards and, therefore, likely to lead to high levels of tax avoidance and evasion.

On the other hand, the average burden of the tax would be significantly lessened if income were to include untaxed fringe benefits. In the past few years, these benefits have been continuously broadened in an effort to lower the burden on wage-earning taxpayers. Table 7.6 shows how the nontaxable compensation of individuals has increased from one-quarter of income in the late 1970s to one-third in the late 1980s. This growth in untaxed fringe benefits is the labor market's reaction to inflationary bracket creep.

If the average tax were calculated to include fringe benefits in income, the result would, of course, be a lower average tax. But the absence of such a substantial part of labor income from the tax base tends to force sharp increases in marginal rates and to be a highly distortionary of labor/leisure choices in occupations where the tax is unavoidable. It also makes for considerable horizontal inequities.

The top marginal rate of the personal income tax schedule went up from 55 percent in 1979 to 60 percent during 1980 to compensate politically for the introduction of the VAT. It went back to 55 percent in 1983, down to 50 percent in 1988, to 40 percent in 1989, and to 34 percent in 1990.

Specific Tax Reform Accomplishments

In this section we focus on the substance of tax reforms over the period 1971–90 and document the major accomplishments that occurred in addition to the prodigious efforts undertaken to index the tax system.

A Broader Individual Income Tax Base

The widening of the base of the individual income tax included all kinds of income obtained either in cash or in kind. Thus, capital gains from real estate transactions were included as well as leases, dividends, income from nonfinancial interest, and more forms of labor income. However, income from financial interest, capital gains in stock market transactions, imputed rents from owner-occupied housing, fringe benefits in excess of a multiple of the minimum wage, and authorship rights (until recently) remained exempt from the global individual income tax. There were economic and/or administrative justifications for most of these exceptions.

Lower Marginal Tax Rates

Fiscal authorities, concerned about labor disincentives derived from high marginal tax rates, reviewed the income tax schedule. The schedule was modified, lowering marginal tax rates while barely reducing average rates and, in some cases, even increasing them. Taxpayers who formerly paid a marginal rate of 57.5 percent began to pay a 40 percent rate on the same increase in income. Those who paid a marginal 39.9 percent rate began to pay a 26.2 percent rate.

Subsequently, the highest rate was also increased from 50 to 55 percent, ostensibly to improve income distribution. These rates, the so-called envy rates, which applied to few people and which hardly provided any revenue, were gradually superimposed over the last years, and raised—almost imperceptibly—to extremely high levels. Recent reforms (1987 and 1989) reversed this trend. The maximum rate is now 35 percent and, furthermore, in 1987 the schedule became indexed to the minimum wage.

A Common Deduction for Personal Income

Lower-income individuals who were slightly above the (legally untaxable) minimum wage threshold used to enter the schedule with a high marginal tax. In order that their wage increases should not be taxed at 100 percent when these were only slightly above the minimum wage, these taxpayers were introduced into the schedule by way of a special formula, which resulted in a marginal rate of 55 percent. To solve this problem, personal deductions were eliminated, as were those for spouse and dependents, and replaced by a unique deduction equal to a one-year minimum wage. The new procedure may seem strange, since it neither takes family size nor other family idiosyncrasies into account. Most of the taxpayers who earned between one and four times the minimum wage did not take advantage of deductions they were entitled to, even though they had overpaid their taxes and had balances in their favor.

The new common deduction had several advantages:

- It simplified tax management considerably, which is not an insignificant factor in a country lacking administrative resources.
- It increased automatically when both prices and productivity grew. This is an important characteristic because it eases the automatic tax increases caused by inflation.

- It reduced the tax burden on lower-income taxpayers. The value of the deductions of medium- and high-income taxpayers was reduced by 30 percent. At the same time, the value of deductions for those with low incomes was raised by 200 percent.
- It has probably contributed to the improvement in labor effort and encouraged seasonal double shifts and overtime for workers in the low-income bracket, where the applicable marginal tax rate was reduced from 55 to 3.1 percent.

The sole specific deductions that remain are for medical expenses incurred within the country, funeral expenses, and donations to educational and charitable institutions authorized to issue tax-deductible receipts. To lower marginal tax rates even more and to eliminate the inherent regressivity of the basic deduction, in 1989 the minimum wage deduction was changed into a tax credit equal to 10 percent of the minimum wage. At this rate, minimum wage income continues to be free of tax.

Taxes on Capital Gains

Capital gains have been added to the income tax base. Given, however, that during the 1970s substantial price increases were recorded and that inflation was expected to continue after the reforms, it was necessary to correct asset purchase costs. The solution was to include in the legislation a table constructed with the price levels of the past fifty years. This table is actually a price index—1932 being the base year—which is "updated" on a year-to-year basis.

Capital gains are not recurrent for most individuals and may, therefore, be taxed at an extremely high marginal rate simply by adding overall gains to the rest of taxable income. Therefore, a simple averaging procedure was introduced that eased the impact of accumulation.

In the case of shares, reinvested profits—corrected for inflation—are now added to the adjusted purchase cost, after having deducted any losses and distributed profits. These procedures apply to assets held by both individuals and firms, except that capital gains in stock-listed shares owned by individuals continue to be exempt in order to avoid administrative complications. This exemption has been in force for several years with no fiscal consequences. This is because, if corrections are made for inflation and reinvested profits, the resulting tax base is either minimum or negative although there are times when sharp increases in the stock market index suggest otherwise. Historically, this exemption has allowed the tax consequences of the spread between the personal income tax rate and the corporate tax to be either avoided or deferred because of a strategy that limits dividend distribution and allows shareholders to obtain their income through exempt capital gains. However, as the top personal marginal income tax rate has come to equal the corporate income rate, the incentive to switch the gain into the shares of unlisted companies has disappeared.

The changes introduced in the capital gains tax were intended to improve fairness and to "lubricate" capital market functioning. An efficient capital market is of utmost importance for sound resource allocation. The former tax on nominal gains led to the strangling of asset transactions because individuals and firms preferred to hold on to their assets rather than incur tax on gains that were merely nominal. By showing their portfolio valuations at current market prices and excluding potential liabilities stemming from the tax due on the sale of the assets, firms could also obtain better leverage from bank loans.

Dividend Integration

Dividend taxation, another reform, aimed at improving both fairness and efficiency. Distributed profits were taxed by integrating the individual income tax with the corporate income tax. With a maximum marginal tax rate of 55 percent (35 percent since 1990) on individual income and with corporate taxes and compulsory profit sharing (with workers) taking more than 40 percent of firms' profit, it would have been excessive to tax distributed profits by adding them to personal income.

Initially, the procedure allowed firms to deduct paid dividends just like any other deductible item or cost. It also required a withholding tax at the maximum marginal rates on individual income, creditable by the recipients. The same procedure was applied to interest paid to individuals.

Transfers of dividends among firms were also deductible and cumulative, which allowed interrelated firms to offset losses instantaneously from one side of their operation against gains on the other. This possibility was deemed particularly important within an inflationary environment.

Another reason to adopt an integration formula using a dividend deduction is that in the first stage of the reform effort an integration scheme based on an imputation method was introduced. This proved very complicated for taxpayers and was later abandoned. Subsequently, when corporate and top personal income tax rates were aligned, the dividend deduction was replaced by a dividend exemption.

Given the enactment of these reforms and price stability, the distortion that encouraged debt financing over equity financing was virtually corrected. With inflation, however, the deductibility of real debt amortizations through interest payments remained an enormous attraction to issue corporate debt until the tax treatment of both interest income and expense was indexed in 1988.

Consolidation of Corporate Results

Before the tax reform, major industrial groups could consolidate their profits and losses if they fulfilled certain criteria concerning growth, employment, and exports. The total loss of a subsidiary could be consolidated, regardless of the fact that it was not fully owned. This procedure was abolished and a new law on consolidation of results was introduced. The new rules provide for formal accounting conditions and legal aspects but do not require firms to achieve economic goals in order to be able to consolidate. In other words, any group of businesses either small or large may resort to consolidation. Furthermore, a subsidiary's loss may only be consolidated on a prorated basis. Thus the unjustified subsidy implied in the previous procedure has been eliminated.

Corporate Subsidies

Subsidies to investment, exports, and employment have constituted only a small portion of overall tax revenues. In certain activities, however, they represent a major share of investment. For example, some large cement firms acquired their capital goods practically free as a result of the cumulative effects of several subsidies. Beyond the tax system, however, large subsidies have been granted, like lower energy prices for businesses located in selected zones. In 1982, as a result of arbitrary interest rate reductions decreed by the government and exchange rate subsidies on debt conversions from September to November, one-third of the domestic private debt of the private sector was wiped out. The energy subsidies were eliminated in 1988 as part of the stabilization measures put into effect that year.

Corporate Tax Incentives

During the 1970s, tax incentive policy was aimed at decentralizing industry. Larger incentives were granted to firms located outside the major metropolitan areas. This scheme never worked because of design failures, such as prioritizing areas adjacent to Mexico City by giving incentives to companies located there. This contributed to some extent to the growth of Mexico City. To correct the failures of the previous scheme, a new decree granted tax incentives with the following objectives:

- Increasing employment
- Promoting investment in priority sectors
- Inducing the development of small industries
- Promoting the domestic production of capital equipment
- Inducing greater use of installed capacity
- Promoting balanced regional growth.

The decree granted fiscal incentives to investment that varied according to the region where the investment was to be located and the type of sector. The incentive was a tax credit granted through a special certificate (CEPROFI), which could be used to pay tax liabilities.

Although this scheme was the most important incentive reform during the 1979–82 period, other tax incentive programs were introduced, such as those promoting exports and the development of duty-free zones. Tax incentives were also given to the automobile, cement, publishing, and mining industries. During the 1979–82 period, the revenue allocated to tax incentives grew from 0.69 percent of GDP to 0.83 percent of GDP.

When the de la Madrid government took office in December 1982, it reduced the public sector's deficit by pruning tax incentives. As a consequence, the ratio of tax incentives to GDP declined to 0.2 percent in 1983. As concluded by Sanchez-Ugarte (1988), tax incentives have been ineffective and inefficient for the following reasons:

- Too many objectives have been pursued through tax incentives minimizing the effect of any single one. The several objectives of the CEPROFI scheme weakened the signals that the government intended to send through tax incentives.
- Tax incentives have often been motivated by political rather than economic factors, generating rent seeking actions by economic agents.
- Tax incentives were often given to counteract the effects of macropolicies. For example, tax incentives to promote exports were given when the exchange rate underwent a considerable real appreciation.
- The granting of tax incentives has been uneven and unpredictable, so it is unlikely that they have affected the behavior or planning of those to whom they are aimed.

The 1989–90 reforms eliminated all CEPROFI-related subsidies. The only incentive remaining in the corporate income tax law is the instantaneous deduction of depreciation allowances instead of yearly depreciation. The instant deduction is calculated using an annual 7.5 percent interest rate, an attractive proposition when real interest rates in the economy hover above 20 percent per year. The deduction can be taken only outside of the three large metropolitan areas—Mexico City, Guadalajara, and Monterrey. In this instance, the metropolitan areas have been widely defined to include outlying areas and suburbs.

Introduction of Consumption-Based Value Added Tax

The introduction of a VAT was beneficial in that it replaced the multilayered or cascading turnover tax, which had tended to promote vertical integration and favor large corporations. Another problem with the turnover tax was that the exact refund amount to exporters could not be calculated. Consequently, agricultural and other primary exports were at a disadvantage. Finally, it taxed investment, which the new VAT does not.

Besides economic considerations—which are convincing to economists but perhaps not strong enough to persuade policymakers—the VAT was implemented to modernize the indirect tax structure and to reinforce taxpayers' compliance. The introduction of the VAT removed more than 30 federal indirect taxes and about 300 state excise taxes. At the same time, fiscal administration was remarkably simplified, freeing administrators to concentrate on a unique broad-based tax with a smaller number of rates and exemptions.

The chain effect of the value added data was an additional practical advantage of the new system. Under the former multiple-level tax system, data supplied by medium-size businesses could not be used to control tax observance by their trading firms. In contrast, the VAT is refundable to the business buyer, thus turning immediate transactions into a series of links. This allows for a self control system when final sales are duly reported (or controlled). Under the former tax system, every sale—whether intermediate or final—had to be verified. Under the VAT, a tax omission at an intermediate stage is recouped by the government if a good or service is controlled and taxed at a later stage, for the simple reason that, if the VAT were evaded, there would be no VAT to credit. The fact that the VAT applies to imports further reduces tax evasion. If there are any tax evaders in the economic chain, the imposition of a tax at the border means that the importer has more reasons to be afraid of being caught.

A peculiarity of the VAT is that, although primary activities (agriculture, cattle raising, fishing, and forestry) are exempt, the exemption does attain a nearly zero rate. This makes it unnecessary to refund the tax, which would otherwise be included in the inputs of firms that process primary products. The system is simple: it levies a zero rate tax on producers of major inputs specific to these activities. For instance, producers of seed, agricultural machinery, boats, and fishing nets are treated as exporters; they do not pay the tax and are refunded any tax they paid on their purchases. The VAT on nonspecific primary sector inputs, such as a truck, are refunded upon request by the farmer.

The revenue-raising effects of the VAT and reduced tax evasion were such that, in spite of its having been introduced with an overall rate lower (10 percent) than the equivalent of the taxes it replaced (13 percent), indirect taxes as a ratio of GDP were constant in the year the VAT was first levied. Corporate income tax revenues, however, increased, reflecting perhaps a better compliance by taxpayers as a result of the VAT mechanics.

Nevertheless, the VAT was subsequently revised, with the adoption of a zero rate for unprocessed food. Consequently, indirect taxes became a slightly lower ratio of GDP. Later, the rate structure and its level were again significantly altered to reduce the huge budget deficit incurred in 1982. The new rates were 6 percent for drugs and processed foods, a basic rate of 15 percent, and a luxury rate of 20 percent. These tax reforms, together with the exemptions granted to certain stores, caused the system to lose its original elegance and administrative simplicity plus much of its neutrality.

The 6 percent VAT rate that used to apply in border zones was not fully justified in terms of trade competitiveness. This is because, if the general rate were 15 percent as in the rest of the country, the price of local factors would have to fall in order to avoid the adverse effects a higher tax might have on economic activity on the Mexican side of the border. Furthermore, property taxes in the United States imposed on land, construction, and capital are high. On the other hand, the lower border rate brought about serious control problems, a lower revenue-raising capacity, and inequities. The argument used to justify it was that sales taxes in the neighboring U.S. states (Texas, New Mexico, Arizona, and California) range from 4 to 6 percent. However, the tax differential has been partially and gradually reduced over the years as services have been brought into the general rate under the reasonable presumption that they are dealing mostly in nontradeables.

In 1990 a new revenue-sharing agreement was reached between the federal government and the states in which half of the amount to be shared (18.1 percent of all federal revenue) would be allocated to the states on a per capita basis and the other half would be distributed on the basis of historic shares of federal excises. VAT collections no longer determined a state's revenue entitlement and were taken over entirely by the federal government. This step was long overdue as the states had little incentive to collect a federal tax in which they were only a minority stakeholder. In a federal system it is better that each level of government collect its own taxes. In 1991 the simplicity of a single VAT rate was regained as the basic rate was restored to 10 percent and multiple rates disappeared.

Elimination of Special Tax Bases

Geographical dispersion, social and economic segregation, and assorted administrative problems have favored the development of special tax bases. Before 1989, some taxpayers were required to pay small amounts per fixed asset or per unit produced or sold in lieu of the regular income tax and the VAT. Some of these treatments were only partially justified but remained in place because of political pressures from interested parties such as the construction industry, whose special treatment was eliminated in 1981. Since then, additional restrictions have been imposed to avoid, or at least limit, abuses such as domestic transfer pricing. The relocation of profits through transfer prices occurs because taxpayers with special bases pay an amount independent of their income or their profits. Thus overinvoicing, for instance by agricultural producers to agroindustries or food processors, reduces the tax liability of the latter without increasing the taxes of the former. Until 1990, special tax bases existed for agriculture, fishing, trucking, newspapers, and urban fresh food wholesalers, among others. As explained earlier, taxpayers in these groups are now taxed on a simplified cash flow basis according to the income extracted from the business.

The Asset Tax: A Minimum Tax on Corporate Income

The complications introduced by having two corporate bases generated considerable political pressure to halt the transition, so the new administration proposed to congress that the new base start in 1989. However, the new tax base entailed a reduction in tax collections that, coupled with a lower tax rate, would have considerably reduced 1989 revenues. For this reason, a new corporate tax was proposed.

The base of the new tax was all assets of firms, adjusted for inflation, with no deductions allowed for debt. This gross assets base is taxed at a rate of 2 percent, and offsets the age-old ruse of leveraging a firm to reduce the size of its tax base. The new tax would be complementary to the corporate income tax and would be creditable against it. Furthermore, it could be carried forward three years, inflation adjusted, and used as an offset against future income taxes in case current profits were insufficient.

A minimum real yield on new assets was implicitly assumed under the new law. If the corporate income tax were to equal 35 percent of profits, then a real annual yield of 5.71 percent would allow a full creditability of the income tax. Firms with lower permanent real yields would have to either close or shed their unproductive assets. Furthermore, by being applicable to most government corporations (except those firms in the public sector under the Constitutional Act), the new tax would compensate for the lack of yield on those assets because of ineffective administration and/or insufficient prices. One of the explanations for the low ratio of taxes would thus disappear, and state-owned firms would be forced to reexamine the profitability of their assets.

At the firm level, debts contracted with other firms and equity investment in a firm's portfolio were made deductible to prevent double taxation. The tax created a considerable adverse reaction on the part of the private sector. Most of it was motivated by the desire to convince the population that the government would in due course introduce a wealth tax. It was also motivated, of course, by the fact that the tax would allow for lower tax rates simultaneously with a broader base. After a few months, however, public resistance died out, although the fight against the tax continued in the courts.

The legal battle was inspired by the alleged unconstitutionality of exempting some sectors from the tax, namely financial institutions, taxpayers operating under special bases, and small taxpayers. The reason for exempting financial institutions was, like the VAT exemption, to prevent double taxation, since the assets of financial institutions were included in the assets of other firms. Taxpayers receiving special treatment under the income tax received equivalent treatment under the assets tax for the same reason—inadequate or nonexistent accounting records. In this way the assets tax maintained its complementarity to the income tax.

Taxes on Foreign Trade

Mexico raises relatively small amounts of revenue from foreign trade compared to the average protection levels and given the extent to which the economy is open. This is mainly because of the fact that high levels of effective tariff protection are granted to some sectors of the economy. This is unfortunate, since an additional 1 percent of GDP in taxes could be raised from this source while effective protection would be decreased. The tariff reforms of 1989 were an improvement in this direction. Many products previously taxed at a 0 or 5 percent rate had their tariffs raised, while the maximum tariff remained at 20 percent. This reform raised the ratio of import taxes to 0.7 percent of GDP from 0.5 percent in 1988.

While effective protection probably decreased, this is an issue that merits further research and policy effort, since this is an area where greater revenue and efficiency in resource allocation go together. The reason the full potential was not exploited in 1989 is that some tariff concessions had to be made in order to obtain private sector collaboration in the government's stabilization program.

Stricter Enforcement

Beginning with Finance Minister Aspe's tenure, harsher penalties were imposed on income tax evaders. For the first time the government began to jail tax evaders and issue heavy fines. Evasion of excise taxes and the VAT was also made more difficult in 1992 with the required use by vendors of government-approved cash registers. Table 7.7 presents a snapshot of the major tax reform measures that have been introduced since 1978. Hardly any facet of the tax system remains untouched by reform efforts to create broader personal, corporate, and indirect tax bases and to tax these larger bases at lower tax rates without sacrificing total revenue.

A Partial Assessment of the Tax Reforms

Perhaps because of the rapidity of change in Mexico's tax environment, there have been virtually no recent attempts to appraise the economic effects of tax reform. However, earlier efforts at evaluation suggest that the initial set of reforms resulted in a tax system that is less distortive in its impact on resource allocation and more equitable in its effect on income distribution. Below, we survey the evidence supporting this view.

There are too many insurmountable data problems to allow a study of changes in tax incidence beyond 1981. When recent income-expenditure surveys are published (a new one was started in 1983), a new study may be attempted. Inflation and the more than 40 percent fall in real wage from 1982 to 1988 may have considerably changed consumption and saving patterns.

Table 7.7 Tax Reforms in Mexico, 1978–90

Tax reform measure	*Personal income tax*	*Corporate income tax*	*Indirect tax*
Indexation	Interest income, tax brackets	Cost basis of shares, depreciation, interest expense, inventories	Specific to ad valorem excise taxes
Integration		Dividend deduction, imputation (1983), dividend exclusion (1990)	
Equitable burden sharing	Much higher personal deduction (indexed) geared to minimum wage	Minimum tax	Zero rated agricultural inputs: exempt foodstuffs
Base broadening	Agriculture, authorship rights, construction, newspapers, rental income, trucking	Minimum tax, fewer incentives	Value added tax
Flatter rate structure	Top rate changed from 56 to 35% (1986–91)	50 to 35% (1986–91)	10% single rate VAT to 6–10–15–20% VAT back to single rate 10% in 1991
Simplification	Cash flow tax for certain hard-to-tax groups (1990)	Construction industry blind deduction	Unified turnover tax, replacement of numerous excise taxes with VAT
Administration	Fewer personal	Shorter collection lags	State VAT collection

Source: Authors.

Changes in the distribution of personal income tax burdens (in the years where data allow a comparison) are presented in table 7.8. Before 1979, the first year that the effects of the tax reform were experienced, personal income tax collections concentrated on the lower tax brackets, those receiving incomes of one to five times the minimum wage. As soon as the reform took hold, collections in this range went from 58 percent of the total personal income tax to only 28 percent, while the highest level went up from 8 to 25 percent of the total. This outcome was produced by the reforms discussed above, such as the more generous deduction for lower-income taxpayers plus the new additions to the income tax base. But the new structure of collections may also reflect the effect on real wages of a quickly growing demand for labor. There is no stratified information on wages to evaluate the importance of this effect.

Tax Incidence, 1977 and 1980

In order to infer income tax payments in 1977, data was handled on an individual basis using an iterative tax model. Owing to the lack of an income/expenditure survey for 1980, the relevant analysis was carried out for that year by simulating the new tax regime on the gross income data obtained from the 1977 report. Taxes on expenditures were estimated

Table 7.8 Tax Collection by Income Level as a Share of the Total Personal Income Tax
(percent)

Level[a]	1977	1978	1979	1980	1981
1	0	0	0	0	0
1 to 5	58	57	40	37	28
5 to 10	20	19	32	29	26
10 to 15	14	15	19	18	21
More than 15	8	9	9	16	25

a. Stated as a multiple of the average national minimum wage.
Source: Gil-Díaz (1990).

by applying the relevant tax rates to reported household expenditures. Other taxes, such as social security contributions, import taxes, and the inflation tax, were also assigned to households.[7]

The main features of the estimates are the following: incidence calculations based on a general equilibrium model of tax incidence (Fernández 1980); the inclusion of inflation tax estimates by income level; the use of permanent income as the variable to measure the incidence of taxation; and finally, a comparison of two years, one before and one after the tax reform, in order to evaluate the incidence of taxation by household income level.

It would have been desirable to also analyze the incidence of public expenditures by income level. When Reyes-Heroles (1976) examined the incidence of public expenditures by household income level in 1968, he found a progressive pattern: a higher number of lower-income people benefited from public expenditures and transfers, with benefits declining as income rose. This result is attributable mainly to government expenditures for elementary and high school and for medical attention. Since the structure of government expenditures has shifted in the six years before 1988 toward a greater emphasis on social services, the progressive pattern of public expenditures has probably become more accentuated in recent years.

The results can be analyzed with or without the inflationary tax. In table 7.9 there is a column with subtotals showing the percentage of permanent income taxes without the inflationary tax. In 1977 this tax caused the whole tax system to be approximately proportional up to the ninth half-decile. Other indirect taxes showed a somewhat erratic "J" pattern, but their importance was small enough not to alter the general profile of tax incidence except in providing somewhat greater progressivity at the upper end. Ignoring the inflation tax, overall tax burdens rise from 9.4 percent of income in the first half-decile to 35 percent in the last decile even though the pattern of progression is sometimes uneven. Overall, there is a progressive pattern of tax burdens.

The conclusion is not altered if the inflation tax is considered. Because of the initial regressiveness of the inflation tax (it is strangely U-shaped), the system becomes proportional up to the fourth half-decile,

Table 7.9 Tax Incidence by Half-Deciles

	As percentage of taxes paid from permanent income					
Half-deciles	*Consumption taxes*	*Other indirect taxes*	*Direct taxes*	*Subtotal*	*Inflation taxes (percent)*	*Total (percent)*
1977						
1	6.18	0.82	2.37	9.37	4.21	13.58
2	6.66	0.63	3.29	10.58	4.21	14.79
3	7.13	0.61	3.54	11.28	2.00	13.28
4	7.06	0.51	4.77	12.34	2.00	14.34
5	7.47	0.62	5.26	13.35	2.25	15.60
6	6.81	0.55	6.48	13.84	2.25	16.09
7	7.13	0.64	7.22	14.99	1.39	16.38
8	7.59	0.94	6.71	15.24	1.39	16.63
9	7.39	0.71	7.65	15.75	2.05	17.80
10	7.25	0.84	6.32	14.41	2.05	16.46
11	7.41	0.75	11.77	19.93	1.51	21.44
12	7.52	0.71	10.74	18.97	1.51	20.48
13	6.99	0.78	15.01	22.78	1.73	24.51
14	7.34	0.85	12.55	20.74	1.73	22.47
15	7.07	0.84	15.59	23.50	1.71	25.21
16	7.37	0.86	14.69	22.92	1.71	24.63
17	7.10	0.96	17.31	25.37	2.64	28.01
18	7.88	1.07	13.74	22.69	2.64	25.33
19	7.79	0.95	14.91	23.65	4.29	27.94
20	6.18	2.68	26.22	35.09	4.29	39.38

but then resumes its progressivity. The interpretation of the incidence of the inflationary tax should be taken with a grain of salt, however, because the income/expenditure survey was conducted in 1972, when inflationary expectations were not as high as they became later in the 1970s and in the first part of the 1980s. As a consequence, the share of monetary assets across income levels must have experienced sharp realignments, especially toward lower values in the highest income levels, where the knowledge and availability of money substitutes to avoid the inflation tax are the greatest.

In 1980, the year following the initiation of the tax reform process, substantial changes were made to the personal income tax. As table 7.9 shows, the personal income tax became more progressive and represented a higher share of income in the highest income levels. (The corporate income tax also showed greater progressivity in 1980.) The incidence of the tax on interest income remained unchanged and direct taxes as a whole became more progressive and more uniform than in 1977.

The incidence pattern of consumption taxes, including the VAT, is decidedly more progressive in 1980 than in 1977. If one disregards the inflation tax, the reforms produced a more progressive profile of tax burdens.

Effective Corporate Tax Rates

Fernández (1985) estimates effective corporate income tax rates by sector and firm size. His results, shown in tables 7.10 and 7.11, indicate that the largest businesses paid the lowest effective tax rate—only 16 percent.

Half-deciles	As percentage of taxes paid from permanent income Consumption taxes	Other indirect taxes	Direct taxes	Subtotal	Inflation taxes (percent)	Total (percent)
1980						
1	4.27	0.93	2.33	7.53	6.58	14.11
2	4.33	0.74	3.07	8.14	6.58	14.72
3	4.76	0.69	3.16	8.61	3.14	11.75
4	5.16	0.58	4.17	9.91	3.14	13.05
5	5.70	0.71	4.43	10.84	3.55	14.39
6	4.78	0.63	5.19	10.60	3.55	14.15
7	5.36	0.72	5.85	11.93	2.19	14.12
8	5.73	1.06	4.82	11.61	2.19	13.80
9	5.47	0.79	5.43	11.69	3.25	14.94
10	5.56	0.96	6.83	13.35	3.25	16.60
11	5.63	0.82	9.33	15.78	2.41	18.19
12	5.92	0.84	12.52	19.28	2.41	21.69
13	5.15	0.93	12.24	18.32	2.77	21.09
14	5.90	0.98	15.24	22.12	2.77	24.89
15	5.51	0.98	14.35	20.84	2.73	23.57
16	5.84	1.03	16.14	23.01	2.73	25.74
17	6.07	1.15	17.83	25.06	4.23	29.29
18	6.61	1.31	17.94	25.86	4.23	30.09
19	6.85	1.19	17.37	25.41	6.84	32.25
20	6.18	1.17	29.80	37.15	6.84	43.99

Source: Gil-Díaz (1984).

Equity was thus not well served, since the biggest firms are usually owned by the wealthy. They pay less because they have better banking connections and greater leverage. As far as efficiency is concerned, table 7.11 shows significant variation in effective tax rates among sectors. As the 1986 tax reform gradually takes effect, effective tax rates among the various businesses should converge, tax equity should be improved, and greater economic efficiency should be achieved as capital flows from the prereform lower-taxed sectors to the higher-taxed ones. These distortions in the determination of the taxable base have had a differential effect on firms depending on their degree of access to domestic and foreign credit and on the distribution of their assets between depreciable and nondepreciable types.

Financial Neutrality

Lower corporate tax rates, the exclusion of dividend income and most capital gains from the personal income tax, and the allowance of only real interest expense deductibility have all contributed to reducing the bias in favor of debt over equity finance. The movement toward financial neutrality has produced greater capitalization in the financial structure of Mexican firms at a surprising pace. The explanation for this rapid capitalization probably has more to do with the resolution of the debt crisis than with tax reform. Steep discounts of 50–60 percent in the market value of foreign held bonds prompted many firms to retire their foreign debts.

Table 7.10 Effective Tax Rates on Profits by Economic Sector

Sector	Effective tax rate[a] (percent)
Farming	24
Mining	36
Food, beverages, and tobacco	46
Metallic and nonmetallic minerals	46
Textiles, clothing, and leather goods	38
Chemical industry	46
Other industries	46
Construction, electricity, and transport	46
Services	28
Retail	46

a. Assumes a 60 percent inflation rate.
Source: Fernández (1985).

Table 7.11 Effective Tax Rates by Firm Size

Gross income (millions of pesos)	Effective tax rate (percent)
30–49	46
50–99	46
100–499	46
500–999	42
1,000–2,499	46
2,500–4,999	46
5,000–9,999	37
More than 10,000	16
Total	31

Note: Assumes a 60 percent annual inflation rate and 10 percent profit sharing with workers.
Source: Fernández (1985).

Conclusion

Viewed as a whole, the series of tax reforms between 1970 and 1990 have been moderately successful in raising overall tax effort, guarding the revenue base against inflation, establishing a progressive pattern of tax burdens under the VAT and personal income tax regimes, and in removing several important tax distortions such as the federal and state turnover taxes. Future reforms will likely involve further broadening of the personal income tax base, more effective enforcement, and a strengthening of state and local tax instruments.

Lessons

Mexico's experience with tax reform illustrates several propositions about what the public sector should and should not try to accomplish.

1. *Basing higher public expenditure levels on expected increases in natural resource wealth is a recipe for macroeconomic instability.* Most natural resource prices are extremely volatile and hitching public spending plans to the fortunes of natural resources transmits this volatility to the entire economy in the form of fiscal deficits and rapidly reversed expenditure programs.

Mexico could have avoided its macroeconomic stresses of the 1980s and many of its subsequent tax reform initiatives if it had "banked" the proceeds from petroleum sales and been content to live off the interest income from these proceeds.

2. *Openness to international trade and capital markets imposes strong constraints on what the tax reform menu can offer.* International mobility of goods and financial resources as well as physical capital requires a country to align its tax rates on capital income and commodities with its international competitors. The government, for example, felt compelled to reduce its corporate rate in response to lower rates in capital-exporting countries and to carefully define its corporate tax base in a way that would maintain eligibility for a foreign tax credit. Fear of tax credit disqualification caused Mexico to reject a cash flow business tax as a reform option.

3. *Both partial indexation and partial globalization of the income tax contribute to significant distortions, inequities, and revenue losses.* An example from Mexican experience is the failure to tax fringe benefits as part of labor income. The government has sought to minimize these problems, at considerable administrative cost, by completely indexing all forms of capital income and persisting in its efforts to make most sectors subject to the income tax.

4. *Creation of separate tax regimes invites the formulation of transfer pricing and invoicing schemes to exploit the weak seams between the separate and regular tax systems.* Mexico's shift to presumptive taxation of the construction industry killed the widespread overinvoicing that occurred when the industry was subject to a gross receipts tax. Presumptive taxation is a form of lump-sum taxation that has appeal on both efficiency and equity grounds (Tanzi and Casanegra 1987).

5. *A value added tax is a vastly superior tax instrument in comparison with a turnover tax or a proliferation of excise taxes.* Mexico's successful application of the VAT to replace its turnover sales tax and several exercises ranks as one of its most important tax reform achievements.

6. *Perceptions of the value added tax* as a regressive tax instrument may make it politically impossible to raise the basic rate without, at the same time, accepting multiple VAT rates that complicate administration of the tax. Mexico was able to eliminate multiple rates only by lowering its basic rate.

7. *In a federal system tax collection is best left to the level of government which imposes the tax.* VAT collections were noticeably impeded by allowing the states to collect these federal tax revenues.

8. *A minimum corporate tax on gross assets* may be an effective instrument for taxing corporate incomes, provided it is designed to avoid jeopardizing foreign tax credit eligibility.

9. *Simpler accounting procedures,* such as Mexico's cash flow system, may be needed to reach sectors that are hard to tax.

Notes

1. The government stopped releasing disaggregated income tax data in 1988.

2. Smart-looking female employees of large U.S. department stores would fly to Mexico wearing fur coats and return to Houston or Dallas the same day without them. The costs involved were a fraction of the tax saved.

3. Exchange controls would encounter, among other problems, 275 million northern border crossings a year.

4. If the next-of-tax yield of financial assets is competitive even through the tax on interest income is higher, the financial system may yet suffer. This is because the lending rate will be forced upward, allowing in-roads from informal lenders and borrowers and limiting the forward shifting of the tax.

5. There were rumors and even an article published in the treasury's tax magazine *Numerica* about a proposal to tax wealth. This publication provoked a press uproar and an angry denial from the treasury.

6. The 35 percent of the corporate income tax on assets that yields a gross return of 5.71 percent equals 2 percent although there is a transition in the corporate income tax rate from 37 to 35 percent over the period 1989–91.

7. This section represents a revision of Gil-Díaz (1984) based on new data from the Income/Expenditure Survey made by the Department of Planning and Budget (1977).

References

Bank of Mexico. 1988. "Informe Anual." Mexico City.

———.Various years (a). "Dirección de Investigación Económica." Mexico City.

———. Various years (b). "Indicadores Económicos." Mexico City.

Cavazos Lerma, Manuel. 1976. "Dirección de Estudios Económicos-Hacendarios." Documento Interno, Secretario de Hacienda y Crédito Público (Department of the Treasury). Mexico City.

Creel, Santiago. 1988. "Una Prueba Empírica al Orden Jurídico: La Celebración y Cumplimiento de los Contratos de Arrendamiento para Casa Habitación en el D.F." Available from authors.

Department of Planning and Budget. 1977. "Encuesta de Ingreso Gasto de las Familias." Instituto Nacional de Estadística, Geografía y Informática, Mexico City.

Fernández, Arturo M. 1980. "Taxation in Small Open Economies." Ph.D. dissertation proposal, University of Chicago.

———. 1985. "Teoría de la Tributación en una Economía Abierta y Pequeña: Análisis y Simulaciónes con Referencia Especial al Caso de Méjico." Ph.D. dissertation, Economics Department, University of Chicago.

Gil-Díaz, Francisco. 1984. "The Incidence of Taxes in Mexico: A Before and After Comparison." In P. Aspe and P. E. Sigmund, eds., *The Political Economy of Income Distribution in Mexico*. New York and London: Homes and Meier.

———. 1987. "Some Lessons from Mexico's Tax Reform." In Nicholas Stern and David Newbery, eds., *The Theory of Taxation for Developing Countries*. New York: Oxford University Press.

———. 1990. "Tax Reform Issues in Mexico." In Michael J. Boskin and Charles E. McLure, Jr., eds., *World Tax Reform*. San Francisco: ICS Press

McLure, Jr., Charles E. 1988. "Savings, Investment and Fiscal Harmonization." Paper prepared for the Conference on Dynamics of North American Trade and Economic Relations, University of Toronto, June 12–14. Processed.

Reyes-Heroles, Jesús. 1976. "Política Fiscal y Redistribución del Ingreso." Licenciatura Thesis, Instituto Tecnología Autónomo de Méjico, Mexico City.

Sanchez-Urgarte, Fernando. 1988. "Taxation of Foreign Investment in Mexico: The North American Perspective." University of Toronto, Institute for Policy Analysis, Department of Economics.

Tanzi, Vito, and Milka Casanegra de Jantscher. 1987. "Presumptive Income Taxation: Administrative, Efficiency and Equity Aspects." IMF Working Paper 54. International Monetary Fund, Washington, D.C.

Treasury Department. Various years (a). "Anuarios Estadísticos de IMSS." Mexico City.

———.Various years (b). "Estadística de Finanzas Publicas." Mexico City.

———.Various years (c). "Evolución de la Carga Fiscal del Impuesto Sobre la Renta a Las Personas Físicas, 1978–88." Internal document. Mexico City.

Urzua, Carlos M. 1993. "An Appraisal of Recent Tax Reform in Mexico." El Colegio de Méjico, Department of Economics, Mexico City.

8

Tax Reform in Morocco: Gradually Getting It Right

David Sewell
Wayne Thirsk

When Morocco achieved independence from France in 1956, it inherited a tax system modeled on the French system, one that remained largely intact for the next twenty years. During those two decades, Morocco pursued an import substitution strategy and achieved steady, if not spectacular, economic growth. In the mid-1970s, this era of economic expansion was interrupted by an external shock when the price of phosphate rock, a major export that accounted for roughly one-third of export revenues, more than tripled (table 8.1). As a result, the economy grew unusually rapidly in 1975 and 1976. Government revenue, especially transfers from the state-owned phosphate monopoly, also grew quickly and allowed the government to embark on an ambitious new expenditure program that involved higher levels of both investment and current spending. Concurrently, the Saharan War brought about a large buildup in military spending.

Unfortunately, the export boom did not last long. Even if it had continued for longer, it could not have sustained the enormous concurrent increase in government expenditure. Between 1970 and 1981, government spending as a proportion of gross domestic product (GDP) nearly doubled, rising from 21 to 37 percent. During the same period, government tax revenues as a proportion of GDP only increased from 18 to 23 percent, mainly because of the imposition of higher tax rates across the board. From the late 1970s through the early 1980s, accumulating external debt financed the shortfall between government revenue and spending. During 1980–85, the resulting inflow of foreign savings amounted to about 6 percent of GDP. In 1983, however, the terms of trade turned sharply against Moroccan exports, real interest rates on the external debt climbed rapidly, and the government could no longer rely upon external funding to finance its fiscal deficit.

Table 8.1 Selected Economic and Financial Indicators

Indicator	1970	1971	1972	1973	1974	1975	1976	1977	1978	1979	1980	1981
Annual percentage change												
GDP at constant prices[a]	5.0	5.9	2.2	3.6	14.2	4.2	7.2	6.5	3.3	4.6	3.4	–2.8
Consumer prices, annual average[a]	1.5	4.0	3.8	4.0	17.7	7.7	8.7	12.6	9.6	8.3	9.5	12.5
As a percentage of GDP												
Gross national savings	12.9	14.2	13.6	16.7	24.4	19.6	14.3	18.3	15.9	15.2	17.0	14.5
Gross investment	15.9	15.6	12.6	14.4	20.6	25.2	28.1	34.3	25.7	24.5	24.2	26.1
Central government budget revenue	17.7	16.7	15.9	17.3	22.0	23.3	21.2	23.5	21.2	23.9	22.1	22.6
Central government expenditure	20.9	19.8	20.0	19.4	25.9	32.8	39.3	39.3	32.3	34.0	32.2	36.6
Overall deficit (payment order basis[b])	–3.2	–3.1	–4.1	–2.1	–3.9	–9.5	–18.1	–15.8	–11.1	–10.1	–10.1	–14.0
Overall deficit (cash basis[b])	–3.2	–3.1	–4.1	–2.1	–3.9	–9.5	–18.1	–15.8	–11.1	–10.1	–9.8	14.2
External current account (excluding official grants)	–4.2	–2.2	0.4	1.0	2.5	–6.1	–14.9	–16.7	–10.4	–9.8	–7.5	–12.2
External debt[c]	19.6	22.9	20.9	18.8	17.7	21.0	29.1	44.9	47.3	53.4	51.6	70.0
As percentage of export of goods, nonfactor services, and private transfers												
External debt[c]	103.9	120.6	99.3	75.9	56.7	76.2	123.1	208.0	220.3	238.6	221.2	257.8

a. From 1990 based on a new consumer price index.

By 1985 foreign debt obligations were nearly four times export revenues, debt service absorbed 70 percent of foreign exchange earnings, and Morocco had become one of the most highly indebted countries in the world. As table 8.2 shows, the government tried to trim the fiscal deficit by severely curtailing the amount of capital expenditure. Moreover, between 1981 and 1986, noninterest-related recurrent expenditure fell from 31 to 27 percent of GDP. As a result of these efforts at retrenchment, the overall fiscal deficit on a cash basis dropped from 14 to 6 percent of GDP during the same period.

In the midst of this painful contraction in public sector activity, the government began to view fiscal deficits as the main obstacle on the road to the higher savings and investment rates needed to improve the country's debt-servicing capacity and growth performance. In the short run, the government had managed to finance the fiscal deficit through a combination of more credit from the central bank and the accumulation of arrears to the private sector and some state enterprises. Over the longer term, however, the government felt that it could only achieve higher rates of investment and economic growth through a significantly larger savings contribution from the public sector. As it had pared public capital expenditures to the bone and restrained public sector hiring and wages, the only avenue of adjustment left was to improve the revenue capacity of the tax system.

1982	1983	1984	1985	1986	1987	1988	1989	1990	1991	1992	1993
9.6	–0.6	4.3	6.3	8.4	–2.7	10.4	2.5	3.5	6.2	–4.1	0.2
10.5	6.3	12.4	7.7	8.7	2.7	2.4	3.1	7.0	8.0	5.7	5.2
16.5	17.9	17.9	20.9	21.9	22.3	23.4	20.5	24.7	21.7	21.8	21.5
28.2	24.0	25.3	27.1	22.8	21.1	21.0	23.7	25.1	22.8	23.0	23.1
22.0	21.3	20.9	20.7	18.8	20.9	22.5	22.6	23.8	23.0	26.3	26.4
34.4	33.4	32.1	30.3	24.2	26.8	27.1	28.6	27.4	26.1	28.5	29.2
–12.4	–12.1	–11.2	–9.6	–5.4	–5.9	–4.6	–6.0	–3.5	–3.1	–2.2	–2.8
–9.2	–11.5	–8.1	–8.6	–5.7	–6.3	–5.8	–4.8	–4.6	–3.1	–2.1	–3.5
–12.9	–8.1	–10.8	–7.7	–2.2	–2.0	0.7	–4.6	–2.4	–2.3	–2.2	–2.5
81.3	96.1	111.5	128.4	105.3	110.8	94.7	94.5	90.5	76.8	75.4	78.2
317.4	337.5	358.4	389.1	346.3	351.6	296.9	327.7	270.0	254.9	242.1	260.4

c. Includes debt that is not publicly guaranteed.
Source: International Monetary Fund and Ministry of Finance data.

Even before responding to the pressures for more revenue, the existing tax system had already begun to show signs of being stretched to capacity. From 1970 to the early 1980s, tax rates had escalated rapidly as the authorities attempted to satisfy demands for more revenue. During this period corporate income tax rates were raised from 43 to 56 percent; the average import duty had nearly doubled, going from 15 to 28 percent as revenue garnering surtaxes were imposed on both incomes and imports; the sales tax on goods was hiked from 15 to 17 percent in 1982 and to the even higher rate of 19 percent in 1983; and the sales tax on services was raised from 7 to 12 percent. At the same time, however, the investment codes and numerous exemptions created both dwindling sales and income tax bases and ample opportunities for tax avoidance. In an entirely unplanned fashion, Morocco had transformed itself into a narrow-based, high-tax rate country and had become ripe for a comprehensive reform of its tax system.

At this juncture, the International Monetary Fund (IMF) began to take a hand in shaping the reform process. Morocco had successfully negotiated a series of stand-by loans with the IMF in the early 1980s, and among other things, these agreements called for Morocco to move quickly to lower its fiscal deficits. Prior to that, the government had requested a technical assistance mission from the IMF to review the tax system's shortcomings and offer suggestions on how it might be improved. The result-

Table 8.2 Tax Effort and Government Expenditures
(percent)

Year	Government revenue/GDP	Current spending/GDP	Current spending/GDP
1973	19.4	16.6	8.1
1974	25.0	23.0	7.0
1975	26.8	29.8	11.8
1976	22.0	20.0	18.0
1977	24.7	25.6	16.6
1978	24.0	20.0	14.0
1979	24.3	26.6	9.0
1980	26.1	28.8	7.4
1981	28.4	31.2	10.0
1982	26.8	29.1	12.9
1983	26.2	30.1	8.2
1984	25.8	30.0	8.3
1985	24.8	28.7	6.2
1986	24.6	26.5	4.1
1987	23.8	25.5	3.5
1988	22.5	19.7	6.4
1989	22.6	20.7	7.4
1990	23.8	19.0	7.2
1991	22.9	19.5	6.3
1992	25.9	19.8	7.0

Source: Mateus (1988a); International Monetary Fund and Ministry of Finance data.

ing mission conducted a thorough and highly detailed analysis of the tax system and made a sweeping set of proposals for its reform. This report and later IMF consultations acted as the primary catalyst for tax reform and laid out the reform agenda for the next decade.

Morocco's efforts to come to terms with its fiscal crisis have been determined, persistent, and eventually successful. The country has made sustained and steady progress in its efforts to achieve macroeconomic stability during the past decade and a half, despite exogenous shocks and adversities, such as further adverse movements in the terms of trade (for example, from increases in oil prices), regional conflicts, and recurrent droughts. Tight fiscal discipline and a series of structural reforms have restored financial equilibrium. In 1993 the authorities reduced the deficit to 3.5 percent of GDP on a cash basis (on a commitments basis to 2.8 percent of GDP) and restored the convertibility of the dirham. Inflation also came down from the double-digit levels of the first half of the 1980s to annual levels of around 5 percent in the 1990s. The current account deficit also declined sharply, and external debt declined from a peak of 128 percent of GDP in 1985 to 78 percent in 1993. Finally, Morocco managed to maintain healthy annual rates of growth throughout much of the adjustment period. The authorities have managed to restore macroeconomic stability without undergoing the social and economic upheaval many other countries that have undertaken structural adjustment have experienced. The authorities are now undertaking market-oriented reforms related to trade liberalization, the financial area, the public enterprise sector, and the fiscal and regulatory environments.

Structural reform was not confined to fiscal measures such as tax and expenditure reform. Although a full account of the nonfiscal adjustment measures is outside the scope of this chapter, a brief review is in order. Monetary and exchange policies obviously played important roles in stabilization policy. More generally, a large element of the longer-run adjustment strategy was a switch from an inward-oriented to an outward-oriented growth policy. This, in turn, involved shifting resources from the government sector to support the growth of private sector investment and reducing government controls to allow market forces to play a greater role in resource allocation. In addition, by eliminating trade monopolies; reducing tariff and nontariff barriers; rationalizing, and more recently privatizing, public enterprises; and deregulating and liberalizing the financial sector, the authorities improved economic incentives. The government even attempted to rethink its governmental process: it promoted decentralized decisionmaking not only in the sectors producing goods and services, but also to a limited extent in government (Sewell 1994).

The subsequent sections outline the role of tax reform as a central element in this broad reform initiative. We begin by documenting in greater detail the main characteristics and defects of the tax system prior to reform. Afterwards, we describe the tax reform measures and evaluate Morocco's progress in reforming its taxes. As part of this assessment, we highlight the revenue and allocative effects of the failure to eliminate an array of tax incentives, paying particular attention to the tax breaks enjoyed by housing. Finally, a concluding section attempts to identify some of the lessons of tax reform that emerge from the Moroccan experience.

Main Features and Flaws of the Tax System before 1986

Income Taxes

Table 8.3 summarizes the recent evolution of the tax structure. Before the reforms, direct taxes on income and profits accounted for about a quarter of total tax revenue. The two most prominent components of direct taxes were the tax on business profits and the tax on wages and salaries. The profits tax used to be about twice as important a revenue generator as the tax on wages and salaries. By 1986, however, these two taxes contributed nearly equal shares of total revenue. In large part, this convergence was due to the failure to index the tax on salaries and wages, which resulted in significant bracket creep and higher revenues brought about by ongoing inflation of approximately 10 percent per year.

Other sources of direct taxation contributed relatively little to total revenue. Agriculture, for instance, used to be taxed on a presumptive basis—the tax was a fixed, nominal amount that depended on location and type of crop—but the amounts raised were low and were progres-

Table 8.3 Tax Structure
(percentage of total tax revenue)

Tax	*1972*	*1980*	*1986*	*1992*
Taxes on income and profits	20.7	24.0	25.8	30.0
Agriculture tax	1.6	0.34	0.0	..
Wages and salaries	4.6	9.25	12.3	15.0
Business profits tax	13.6	12.3	10.5	14.0
Urban property tax	0.7	0.16	0.15	..
Tax on dividends	0.0	0.48	0.56	1.0
National Solidarity Fund contribution	—	1.47	2.5	..
Tax on property transfers	3.0	2.0	1.95	..
Taxes on domestic goods and service	38.2	29.3	28.2	34.0
Sales tax	12.8	13.8	15.2	25.0
Excise tax	22.2	13.1	11.1	..
Business license tax (*patente*)	2.5	1.4	1.3	9.0
Tax on international trade	32.3	40.7	40.6	21.0
Customs duties	13.2	8.9	9.9	..
Special tax on imports	2.7	15.0	8.4	..
Sales tax on imports	16.4	12.5	17.3	..
Petroleum levy	n.a.	—	—	10.0
Other taxes				
Stamp tax	2.8	1.7	1.9	..
Registration fees	1.1	0.76	1.36	0.5

— Not available (tax was not in place).
.. Negligible.
Note: Nontax revenue, consisting mainly of dividends from public enterprises, administration fees, and a special petroleum tax, were 9 and 21 percent of total tax revenue in 1980 and 1986, respectively. Earmarked social security taxes, equal to about 20 percent of gross wages, are not included.
Source: Ministry of Finance data.

sively eroded by inflation. Thus, despite accounting for about 20 percent of total value added in the economy, the agricultural sector contributed much less than 1 percent of total revenue in 1980. After a series of droughts and bad crop years in the early 1980s, agriculture was exempted from both direct and indirect taxes. This direct tax exemption extends to the year 2020, when agricultural income is once again supposed to be subject to taxation on a forfeit or presumptive basis for small farmers and on a net income basis for larger farmers. Other direct taxes, such as the urban property tax and the tax on dividends, also yielded small shares of total revenue. The contribution to the National Solidarity Fund consisted of a 10 percent surtax applied to personal and business incomes. The government introduced this fund to help defray some of the costs of the Saharan War.

Commodity and Sales Taxes

Before reform, indirect taxes contributed roughly 70 percent of total tax revenue. During 1972–86, taxes on international trade became more important than taxes assessed on domestic goods and services. This happened as a result of the imposition of higher nominal tariffs and the introduction of a special tax on imports, which was a surtax on the duty paid on the cost, insurance, and freight value of imports. Surtax rates rose steadily during the same period. The reduced role of excise taxes in the revenue system explains the diminished importance of domestic indirect taxes over time. Although in 1972 excise taxes were nearly twice as important as the sales tax, by 1986 excise taxes contributed less to total revenue than the sales tax. During this period chronic inflation undermined the real value of the per unit basis on which the excise taxes were collected.

Before its replacement in 1986, the sales tax had two parts, a tax on goods and a separate tax on services. The tax on goods was applied at the manufacturers-importers stage of production and exempted a number of basic necessities, such as sugar and bread, as well as all export sales. Borrowing from French experience, the tax on goods was a limited type of value added tax (VAT). While credit was given for purchases of capital goods, the credit was restricted in the case of material inputs to those that were embodied or used up in the process of production. In 1979 the basic tax rate for most commodities was 15 percent, while luxury items were subject to a 30 percent rate, and six other reduced rates were applied to a variety of processed foodstuffs, cooking oils, drugs, and basic utilities.

The other part of the sales tax, the tax on services, was a cascaded tax in that credit was not given for the tax on services against the tax collected on sales of goods, nor was credit for the tax on goods allowed against the tax on services. This tax was initially levied at a basic rate of 7.5 percent, but additional rates existed for a heterogeneous group of specific services. Small enterprises producing goods paid a rate of 7 percent, while similarly small enterprises furnishing services were subject to a 4 percent rate. As mentioned earlier, by 1983 the basic tax rate on goods had been hiked to 19 percent and that on services to 12 percent. Moreover, by then some items were taxed at differentially higher rates if they were imported.

Personal Income Tax Rates and Schedules

Morocco traditionally attempted to tax personal income on a schedular basis: different tax calculations applied to various kinds of income. A total of six distinct schedules existed: one for real estate rental values, both realized and imputed (the *taxe urbaine*); another for agricultural source income, now suspended; a tax on wages and salaries; a tax on portfolio income, defined to include both dividend and interest income; a tax on the capital gains resulting from the sale of urban real estate; and a complementary tax superimposed on these schedules in 1972 in an effort to include the total income of rich individuals in the personal tax net.

As 75 percent of imputed rental income (from homeownership) was excluded, the base of the *taxe urbaine* was essentially restricted to rents actually received. The tax was and is payable by both companies and individuals, and in the case of the latter, deductions are permitted for family status and rates are progressive, ranging from 10 to 30 percent. Ninety percent of the yield from this tax is distributed back to local governments.

The schedular tax on wages and salaries was withheld at source with a 10 percent deduction allowed for personal or professional expenses. Government employees paid about half of the tax. Pension contributions were deductible up to a limit of 6 percent of income and pension payments were taxable on receipt. Progressive tax rates applied within a range of 0 to 44 percent.[1] Steep progression in tax rates, combined with the lack of inflation adjustment in the early 1980s, augmented real tax burdens on labor incomes.

In the case of the schedule pertaining to portfolio income, a flat rate of 25 percent was used and the amount of tax payable was creditable against the complementary tax. Tax was withheld at the time of payment of the dividend or interest income, and no deductions were permissible for either family circumstances or the cost of earning investment income.

The tax on capital gains resulting from the sale of urban property was introduced in 1978. Its rate depends on the length of time that the owner has held the property before it was sold. For example, the rate is 25 percent if the owner has held the property for five years or less and only 5 percent if the property has been held for more than twenty-five years.[2]

The complementary tax was assessed against households that had multiple sources of income, and therefore functioned as a crude kind of global income tax. Introduced in 1972 at progressive rates between 0 and 45 percent, this tax affected only high-income individuals, because the initial zero rate bracket excluded most income recipients. The tax base consisted of the sum of the schedular bases, less the amount of schedular tax previously paid, plus the base of the business income tax if the household was taxable under the latter. With the addition of the complementary tax, dividends were likely to be triple-taxed, first at the company level, again under the schedular tax, and once more under the complementary tax! Under the schedular approach to taxing personal income, both nominal and effective tax rates varied widely according to the source of income. Nominal tax rates could reach as high as 76 percent in the highest income bracket.

Profits Tax

The business profits tax (*impôt sur les bénéfices professionnels*) applied to all enterprises, no matter what their organizational form, as well as to professionals' net earnings. It was the most important source of direct tax revenue for much of the period before the reforms. For companies, capital gains reaped from the sale of business assets were included in taxable income. Companies could deduct depreciation, although only on a historical cost basis using straight line methods, while individuals could deduct an allowance for dependents. Companies could also carry forward losses for four years. Individuals faced marginal tax rates ranging from 0 to 48 percent, while a three-tiered rate structure for companies was imposed in 1979. Under that system, small companies with earnings of less than DH500,000 in 1980 (US$89,000) paid tax at a 40 percent rate; companies earning more than that, but less than DH1.5 million (US$259,000) paid at a 44 percent rate; while even more profitable companies were subject to the highest rate of 48 percent. In the early 1980s, the government eliminated the intermediate rate and applied the smaller rate only to companies whose annual net earnings were less than DH250,000. However, for large and profitable corporations, the government levied a 10 percent National Solidarity Fund surtax, making the top nominal rate 52.8 percent.

Other Business Taxes

Smaller businesses were taxed on a forfeit or presumptive basis by applying a variable coefficient (depending on the type of business) to the enterprise's gross turnover. A minimum forfeit was also calculated by multiplying the estimated rental value of the business premises by a coefficient varying from one to five. Forfeit businesses too small to pay the minimum amount were subject to the local business license tax, the *patente*. Both small and large businesses were also subject to the *taxe urbaine* and a municipal property tax (*taxe d'édilité*), which used the same base as the *taxe urbaine*. Only the *taxe urbaine* was deductible from the business profits tax. In addition, businesses were affected by an implicit tax, the investment reserve requirement, which forced them to invest a portion of their profits in government bonds that paid a below-market interest rate. In the early 1980s, when government bonds paid 5 percent and market interest rates were around 15 percent, this implicit tax added around 1 percentage point to the effective business tax rate.

Excise and Trade Taxes

As already noted, Morocco levied a number of per unit excise taxes. These imposed a tax burden on a fairly narrow range of goods and services, including soft drinks, beer, wine and other alcoholic beverages, sugar and sugar products, milk concentrate, tires, matches, tea, coffee, gold, silver, and public entertainment. The excise tax system did not appear to have been designed with the objective of taxing either luxury consumption or consumption that was considered to be socially undesirable.[3] Nor were excise taxes closely coordinated with the tariff structure, as many high tariff rates were levied on luxury imports.

Until recently, Morocco operated with a highly differentiated tariff structure, and in 1978 imposed a special import surtax set at 12 percent. Moreover, a stamp duty acted as a 4 percent surtax on all the duties and taxes assessed against imports.[4] Thus the import tax structure consisted of a melange of measures, some intended to protect domestic producers and others designed to raise more tax revenue. Domestic producers also benefited from liberal exemptions on the payment of tariffs on intermediate inputs and capital goods. In addition to these protective devices, the government used import licenses to control the composition of imports. For some commodities, it simply banned importation.

Prior to the reforms, the tax system as a whole rested on a relatively narrow set of tax bases on which relatively high tax rates had been imposed. For example, Mateus (1988b) estimated marginal effective tax rates on new investment in Morocco, taking into account the taxes on capital income that would be applied at both the company and personal levels. He obtained an astonishingly high value: a marginal effective tax rate of 87 percent. The relatively high nominal rate on profits of nearly 53 percent and the nearly 20 percentage points attributable to the triad of capital taxes—the *taxe urbaine*, the business license tax, and the municipal property tax—explained most of the enormous marginal tax rate. Removing the personal taxes on interest and dividend income would have reduced the marginal effective tax rate by only about 13 percentage points. With such high marginal tax rates, it would have been surprising if much investment had occurred. What this calculation of marginal effective tax rates ignored, however, is two features of the fiscal system that diluted the impact of high nominal tax rates: the existence of a generous system of investment incentives and less-than-perfect tax administration.

Tax Compliance

As discussed later, the authorities have offered tax incentives to encourage a wide spectrum of economic activity. Whether Morocco's high nominal tax rates engendered the proliferation of tax incentives or whether the reverse was closer to the truth remains an interesting and open question.

Mateus (1988b) examined the tax files for ninety large firms and found that they used a blend of three techniques to minimize their exposure to high nominal tax rates. First, many of these companies manipulated their accounts to understate their taxable income. Many claimed generous deductions for anticipated bad debts, and a number of others used deductible shareholder loans to funnel income out of the enterprise. Second, most of these firms were highly leveraged. On average, their debt-to-equity ratio was 6.2 and their ratio of debt to net worth was 4.3. The deductibility of interest on this debt-dominated capital structure reduced the size of taxable profits considerably. Finally, many companies were affiliated into groups that included firms operating either in tax-exempt sectors or under favorable tax incentive conditions. This facilitated the transfer of profits to these low tax, or even no tax, operations. Through these diverse channels, Mateus estimated that companies successfully reduced their tax liabilities by DH3.8 billion in 1986, or approximately 14 percent of total revenue.

Tax Administration

The problems encountered in administering the profits tax highlight a general perception that overall tax administration was relatively weak before the reforms. The recurring impression is that tax officials were frequently overwhelmed by the paperwork the tax system generated and were incapable of enforcing the collection of arrears, let alone conducting effective audits. Mateus (1988b) suggested that tax officials performed about ten audits per year under the profits tax, which at the time had more than 200,000 annual filers. Moreover, the collection lags were extremely long in the case of income taxes, averaging nearly fifteen months and generating a loss in real tax revenue of 5 percent at a time when inflation was running at 12 percent. The large number of taxes and the proliferation of tax rates in each tax field also vastly complicated the task facing tax administrators. For example, one study (World Bank 1984) noted that customs duty rates varied from customs post to customs post, and a particular post might assess the same commodity at different rates at different times.

Several organizational weaknesses plagued Morocco's tax administration. For one thing, the schedular system expanded the amount of paperwork per taxpayer immensely, and also created the problem of each taxpayer having several identification numbers. For another, tax officials had no reliable way to detect nonfilers. Moreover, when taxpayers did file returns, officials typically did not verify the arithmetic, did not conduct consistency checks, and did not require receipts for any income and expense claims. Officials also carried out few audits: roughly 300 per year for a taxpaying population of more than 900,000 households and more than 200,000 filers under the profits tax. Those few audits did, however, reveal the existence of massive fraud: for the 129 audits carried out in 1984, the audited tax base was nearly three times as large as the declared base. Finally, enforcement was extremely weak in Morocco, as indicated by the growing volume of arrears.

In the light of these numerous deficiencies in tax administration, income tax coverage was patchy at best. Assessed wages and salaries under the schedular tax constituted around 20 percent of GDP in any given year. Given a paid labor share in GDP of about 30 percent, this suggests that the underreporting of labor incomes was about one-third. Observers have made even less favorable comparisons for the coverage of the profits tax. The census immediately prior to the tax reforms recorded about 225,000 individual enterprises, but only 50,000 of them regularly filed a return under the business profits tax. Mateus and Miltner (1987) suggested that while as much as 80 percent of total labor income was included in the income tax net, as much as 80 percent of total capital income had managed to elude that net. Large chunks of business-related capital income and the incomes earned by self-employed professionals effectively escaped income taxation because of weak tax administration.

An Assessment of the Prereform Tax System

Prior to the reform initiatives introduced in the 1980s, the tax system was extremely weak in almost every area. Morocco had simply too many taxes, too many rates, too many exemptions and fine distinctions in attempting to define tax bases, and too many special tax regimes. Neither taxpayers nor tax collectors could easily figure out what taxes were owing in some circumstances. Not only did this complexity breed tax evasion, it also prevented the achievement of a reasonable degree of horizontal and vertical equity. As a result of these uncertainties and complexities, there was a wide disparity in tax burdens across the various sectors and factors of production in the economy and a correspondingly large dispersion of effective tax rates on different kinds of economic activity. The tax system had steadily worked its way into a corner where relatively high tax rates led to inexorable pressures for a host of tax incentives and tax exemptions. These departures from a broad tax base in turn produced a tax system of bewildering complexity, encouraging extensive tax evasion and making it impossible for the system to have attractive efficiency and equity attributes.

As it stood on the brink of tax reform, Morocco provided an excellent example of how *not* to design a developing country's tax system.

- The direct and indirect components of the tax system were both riddled with exemptions and applied a highly differentiated rate structure to a relatively narrow base.
- The system was unindexed, and therefore vulnerable to distortions caused by inflation.
- The system imposed one tax on goods and another tax on services that operated on a turnover basis.
- The dependence on trade taxes was extremely high and numerous minor taxes generated extremely low yields.
- The complex investment and export codes offered tax incentives for nearly every type of economic activity, and a weak tax administration had limited ability to prevent widescale tax evasion.

Tax Reforms after 1986

The 1984 Framework Law (*Loi Cadre*) spelled out Morocco's intention to achieve comprehensive reform of all major taxes, diminish fraud and evasion, and impose a more equitable distribution of the tax burden across different socioeconomic groups and economic activities. The primary thrust of all these reform measures was to fashion broader tax bases and achieve lower marginal and average tax rates.

Under the Framework Law, Morocco introduced a new VAT and a new tax on business profits in 1986 and a new tax on personal incomes in 1989. The government also introduced a petroleum levy in 1986 to capture the windfall gains from falling international oil prices and started to reduce tax exemptions under the various investment codes in 1988. A 12.5 percent uniform import tax (the *prélèvement fiscal à l'importation*) was also introduced at the beginning of 1988 to replace the former special import tax and the relatively distortionary stamp duty.

These reforms led to a marked recovery of government revenues. As table 8.1 shows, tax revenues rose from 20.9 percent of GDP in 1987 to 26.4 percent in 1993. This was accomplished without increasing tax rates. Statutory rates for the most important taxes have actually been reduced, in some cases dramatically. Base broadening, improved administration, and certain other discretionary measures have made such tax rate reductions possible.

The goal of the VAT reform was to eliminate the cascading effects of the previous sales tax and to streamline the administrative machinery for taxing goods and services. The new VAT extended tax credits to the purchase of all types of business inputs; moved the tax point to the wholesale level; exempted small producers and traders; and initially applied four rates of 0, 7, 19, and 30 percent in place of the eleven previous rates. Food purchases were also exempted under the new VAT, as were both sales and input purchases made by the agricultural sector.

The new profits tax (*impôt sur les sociétés*) replaced the business profits tax. In addition, the government suspended the former investment reserves requirement . The new profits tax initially had a single and lower nominal rate of 45 percent, and was introduced alongside a minimum tax (a general rate of 0.5 percent of turnover) that was creditable against the profits tax. The business profits tax continued to apply to individual entrepreneurs (or unincorporated businesses) until the new global income tax was introduced. The payment period for the new profits tax was also shortened.[5]

The new global income tax on personal incomes (*impôt général sur le revenu*) was supposed to have been introduced alongside the new profits tax because of the significant interaction between the two pieces of legislation, but its passage was held up in parliament and it only came into effect in 1989.

In an independent initiative, in 1986 the World Bank persuaded Morocco to reduce its reliance on tariffs by cutting nominal tariffs. Morocco was unable to recoup the revenue loss from other revenue sources, however, and aborted this experiment in trade liberalization the following year. Sensing that tax reform would have to precede tariff reform as part of a rational sequence of policy reforms, the government postponed further tariff reductions until 1988.

An Evaluation of Reform Progress

The Value Added Tax

Initially the revenue record of the new VAT was disappointing. The sudden drop in the VAT yield from 1986 to 1987 of more than 1 percent of GDP compared to previous sales tax revenues seems to have been related to a number of disparate factors, including the elimination of tax cascading, the poor performance of the new tax at the wholesale level, a speed-up in settling claims for input credits, and the granting of a full VAT credit on inventories when the tax was introduced. The proliferation of tax rates and the administrative headaches that accompany any effort to make fine distinctions among taxable activities created another problem. For example, big automobiles were taxed at a 30 percent rate while smaller automobiles were taxed at the standard 19 percent rate. Furthermore, while soap was exempted from the tax, all beauty products faced a tax rate of 30 percent.

While the new VAT was originally designed to operate with four tax rates, the authorities quickly established a number of new rates and new rate categories, apparently in an attempt to make the new VAT rate structure conform as closely as possible to the old sales tax system. Thus there was a 30 percent luxury rate, the standard 19 percent rate, a 14 percent rate on construction contractors, a 12 percent rate on tourist establishments and on bank loan interest, a 7 percent rate on items such as school equipment and transportation services, and a zero rate for exports and a few other products. In addition, specific taxes were levied against wine and works of gold, platinum, and silver that were applied at a single stage and, in the case of wine, that gave rise to an element of tax cascading. Moreover, wine sold in a normal restaurant was subject to the standard rate of 19 percent while the same wine sold in a tourist establishment faced a lower rate of 12 percent.

Another problem with the new VAT was the presence of a buffer rule. Thus firms that acquired input tax credits in excess of the output taxes they owed were unable to use them, and were therefore subjected to an unintended higher effective tax rate. For example, the low 7 percent rate on transportation services led to an accumulation of unusable input credits, because most inputs were taxed at higher rates. The zero rate on agricultural inputs also invited the diversion of some of these inputs, such as fertilizers and tractors, to nonagricultural uses and provided an unwarranted subsidy to farm mechanization. In a number of other instances, the new VAT was not coordinated with previously existing excise levies and price controls. For example, VAT-exempt sugar was subject to both an excise tax and a price control subsidizing consumption.

The government also ran into trouble in trying to apply the tax at the wholesale level. Although the legislation was designed to control transfer pricing, wholesalers were overtaxed if they sold to a retailing subsidiary and undertaxed if their integrated activities were split up and reorganized to insulate their retail operations from the rest of the business.

Initial experience with the VAT offers a vivid illustration of Morocco's previous propensity to take a simple tax design and, by cluttering it up with numerous fine layers of distinction, compromise tax officials' ability to administer the tax properly. It also illustrates the difficulty a country may encounter in breaking old and bad tax habits and departing markedly from past traditions of taxation.

However, the authorities recently rationalized the rate structure and broadened the base. They reduced the number of rates to three by abolishing the 12 percent rate in 1992 and the 30 percent rate in 1993. They also reduced exemptions by extending the VAT to staples such as rice and flour (at the lowest rate of 7 percent).

Two major problems remain to be eliminated in the VAT: the buffer rule and the application of the VAT to interest payments. The VAT is intended to apply to added value in transactions, but in the case of financial transactions is applied to gross interest payments. Most countries do not apply a VAT to financial transactions at all. The authorities recently softened the distortionary effects of these problems, however, by reducing the VAT on interest from 12 to 7 percent in 1994 and simplifying the VAT's rate structure to reduce the distortions produced by the buffer rule. More reductions in the number of VAT rates would further ameliorate the problems created by the buffer rule. The problem with eliminating the VAT on interest entirely is the substantial amount of revenue it now raises.

Personal Income Tax

The base of the new personal income tax includes income from all sources except most dividends and interest payments, which are taxed on a schedular basis. The government is relying more on this tax and on the corporate income tax in collecting revenues: the share of total revenues from all income taxes (including the corporate profits tax) rose from 25.6 percent in 1989 to 29.1 percent in 1992. The growing importance of the personal income tax has increased the progressivity of the tax system as a whole.

Since 1990 the top marginal tax rate of the personal income tax was reduced in two stages from 52 percent to 46 percent in the 1994 Finance Law, and the taxable floor for annual income was successively raised from DH12,000 to DH18,000 (US$1,253 to US$1,879) in 1994. The result has been that the poor face lower income tax burdens, because the decreases in the amount of income taxed have been greater than the rate of inflation in the same period. The *Note Circulaire* (explanation bulletin) for the 1994 Finance Law asserts that some 240,000 people with modest incomes were removed from the tax rolls as a result of the hikes in the floor for taxable income.

Reductions in the top marginal rate have had a pronounced effect on revenue obtained from the personal income tax, because of the concentration of tax paid by those in the top income brackets. For instance, in 1992 those in the top two tax brackets of 46 percent and 48 percent paid just under half (49.9 percent) of all personal income tax.

A significant structural problem in the personal income tax is that wage and salary earners pay just under 90 percent (89.4 percent) of all personal income tax, while those having professional incomes pay only 9 percent. Thus, an obvious priority for the future is to increase compliance by taxpayers having professional incomes.

An important structural change in the taxation of components of personal income occurred in 1994, in terms of both raising revenues and leveling the playing field between private and public borrowers. Interest from government bonds became subject to taxation by automatic withholding at source, the same as for other forms of interest, which are taxed at either 30 or 20 percent. Interest paid to unidentified lenders, for example, holders of corporate bonds, is subject to a 30 percent tax that the borrower withholds. Where the lender is readily identifiable (as for bank deposits), interest is taxed at a 20 percent rate that the borrower withholds. This withholding tax is then credited against total personal income tax liabilities. If the credit is greater than the total personal tax liability, however, the government does not refund the difference.

Corporate Profits Tax

A separate corporate profits tax, with an initial rate of 45 percent, came into effect in 1986. During 1990–94, this marginal corporate tax rate was reduced by three separate finance laws to 36 percent. The contribution to the National Solidarity Fund adds a 10 percent surcharge to this rate. A significant liberalization in depreciation schedules, particularly for long-lived investments, accompanied the introduction of declining balance depreciation in the 1994 Finance Law. Other changes in corporate tax provisions are described later.

The introduction of quarterly installment payments for the corporate profits tax in 1993 was an administrative improvement that addressed the previous problem of long collection lags. Under the previous system tax payments were delayed by a year, which in inflationary conditions greatly reduced their real value.

Other recent structural changes in the corporate profits tax have important implications for private sector development. Thus state-owned enterprises are now required to pay the corporate profits tax like private corporations. This broadening of the base of the corporate profits tax has also leveled the playing field between private and public enterprises.

Tax Administration

Tax administrators have used minimum taxes extensively to shore up revenue bases. The authorities adopted minimum tax rates (with respect to turnover) to increase revenue from the corporate profits tax and have, in recent years, extended them to self-employed professionals. The general rate is 0.5 percent of turnover, with the self-employed professionals being subject to a rate of 6.0 percent. In 1994 a lower rate of 0.25 percent was introduced for businesses engaged in producing and distributing principal staples (sugar, flour, and so on) and energy and utilities (water, electricity, and hydrocarbon fuels). Some minimum taxes have also been imposed on personal incomes. For example, a minimum tax on real estate capital gains is automatically applied when sales are registered.

Tax administrators have also introduced measures to improve collection. In this respect, automatic withholding is a potent method of assuring tax collection, and has been extended from wage and salary earners to other sources of income, such as real estate income.

Administrative and compliance cost burdens have also been reduced. For instance, where wages and salaries are the sole source of income, individuals are no longer required to make an annual declaration of income—an important change given that wages and salaries account for just under 90 percent of personal income tax payments. Other taxes have also been simplified, as was the case, for instance, with stamp and registration taxes in 1989. Some administrative changes such as automatic withholding have also reduced administrative costs.

Computerization has also improved fiscal administration, as has reorganization of the tax branch of the Ministry of Finance. Computerization was increased with the introduction of the VAT in 1986. Reorganization of the tax branch has allowed increased coordination of collection of the different taxes and enhanced computerization has permitted more cross-checking of tax liabilities. Procedures for tax assessment, collection, and auditing have also been strengthened.

Assessing Tax Reform Using Effective Tax Rate Analysis

Sewell, Tsiopoulos, and Mintz (1995, 1996) measured the effects of Morocco's tax and incentive systems on the burden of taxation on investment and compared the resulting estimates of marginal effective tax rates on investment in Morocco with those for some other developing countries in the Mediterranean region. One of their central findings was that since tax reform commenced in Morocco, dramatic reductions have occurred in effective tax rates on investment. Table 8.4 shows that since 1986 the effective corporate tax rate has more than halved, dropping from 50.3 percent for domestic manufacturing firms and 44.2 percent for do-

Table 8.4 Historical Summary of Changes to Effective Corporate Tax Rates for Domestic Firms
(percent)

	Effective tax rates	
Change	*Manufacturing*	*Services*
1986 Finance Law (effective 1987)	50.3	44.2
1993 Finance Law	47.8	43.9
1994 Finance Law		
Reduction of corporate profit tax rate from 38% to 36%	47.3	43.7
Lowering of import tax from 12.5% to 10% on inputs of capital goods	45.7	43.0
Accelerated depreciation schedules	43.6	41.7
Impact of all changes	40.9	40.8
1995 Finance Law		
Elimination of import tax on inputs of capital goods	29.4	29.0
Tax credit of 10 percent on the percentage of equity used to finance investments	35.6	35.1
Impact of all changes	24.2	19.9

Source: Sewell, Tsiopoulos, and Mintz (1995, table 10).

mestic firms producing services in 1987 to 24.2 percent and 19.9 percent, respectively, in 1995. Table 8.4 indicates that most of this reduction has come about recently, as a result of Finance Laws for 1994 and 1995. Elimination of the special duty on capital goods imports in 1995 has been the most important source of change, followed by the introduction of accelerated depreciation in 1994 and the tax credit for equity finance granted in the 1995 Finance Law.

As concerns comparisons with other developing countries in the Mediterranean, the study contrasted effective tax rates on investment in Morocco with those in Greece and Portugal. The study found that effective tax rates for investments in Morocco that do not receive incentives in the investment codes were roughly comparable to, or even lower than, those for investments in Portugal. Effective tax rates on investment in Greece were generally lower than those in Morocco and Portugal, particularly if firms were able to take advantage of special accelerated depreciation provisions in Greece. These comparisons, nevertheless, provide some comfort for the Moroccan tax authorities in that they need not respond to the familiar pleas from special interests in all countries that overall tax rates need to be lowered in crash efforts to attract investment.

Estimated effective tax rates on investment carry important implications for growth. In a neutral policy environment effective tax rates would not differ for different investments, and the market would get to select and reward those investments that make the best use of economic resources. Sewell, Tsiopoulos, and Mintz (1996, pp. 584–85) found that substantial differences continue to exist in the effective tax rates on investments in Morocco for different types of assets, industries, methods of financing, and the investor's origins (foreign or domestic). Even without taking into account incentives in the investment codes, for instance, effective tax rates on investment in different types of assets varied from minus 132 percent in the case of debt finance to plus 67 percent without debt finance. Furthermore, the study found that even mild inflation of the type Morocco has recently experienced (around 5 percent per year) substantially penalizes the holding of inventories compared to investments in other types of assets because of the first in, first out (FIFO) accounting requirement. Removing or reducing these differentials in effective tax rates with a view to reducing the tax system's influence on investment choices is a suitable objective of future tax policy.

Remaining Problems: Tax Incentives and Exemptions

While acknowledging how far the tax system has moved toward simplification and more uniform treatment of all taxpayers since passage of the 1984 Framework Law is important, perhaps the tax system's most striking feature remains the number of instances in which it singles out particular sectors of the economy, industries, or other classifications of economic activity for special treatment. Such extensive intervention in markets reflects a belief in policy fine-tuning that is uncommon among contemporary fiscal authorities. It also denotes a past lack of faith in markets or market solutions that does not accord with the current thrust of government policy in Morocco. We believe that tax exemptions and incentives now constitute the principal remaining problem in the tax system, and we shall illustrate this by citing estimates of some of their effects.

That particular industries obtain advantages from incentive provisions in, say, the investment codes is not obvious. With eight sectors selected for encouragement in the codes, revenue still has to be raised from the tax system. In these circumstances, paradoxically, if everybody gets a tax break, nobody may be favored!

Apart from the income tax exemption for agriculture, separate investment codes provide tax and cash incentives for the industrial, tourism, mining, handicrafts, real estate, professional training, and maritime sectors. Additional incentives are in place for exporters, regardless of their sector of activity; for small business; and for investments that preserve energy, water, or the environment. Both the industry and tourism codes also attempt to promote regional balance by favoring activity outside the Casablanca area. The incentives provided under the industry code vary depending on whether the activity being undertaken is a new venture or an expansion of an existing activity. (See Berdai 1991 for a thorough description of the various codes, their history, and an analysis of their effects.)

Incentives are not limited to the formal investment codes. Conventions, or entirely ad hoc arrangements lowering fiscal burdens may be negotiated by the government with foreign firms offering substantial programs of investment. A number of such conventions have been negotiated over the years.

Finally, the goals outlined above of encouraging particular sectors, industries, or other classifications of economic activity have been promoted at one time or another by myriad incentive devices, including complete and partial corporate tax holidays for varying lengths of time, reserves for investment that can be accumulated tax free from business income, taxation of real rather than nominal capital gains in the case of real estate, accelerated depreciation, relief from import duties and VAT levied on imports of capital goods and inputs for production, subsidies varying inversely with the number of personnel employed by a firm, grants for small firms to acquire land, subsidies for borrowing costs, and relief from registration fees.

Attempts have been made to limit such tax concessions in the investment codes. For example, a 1989 reform reduced the scope and duration of benefits. Most exemptions were limited to 50 percent of income taxes that would otherwise have been owed, while the maximum period of exemption was reduced, in most cases to five years, but in a few cases to ten years. The authorities have also attempted to limit revenue losses from particular incentives. A side effect of these various measures to limit revenue losses arising from the incentives is that they may add to administration and compliance costs.

Prior to modeling the effects of the incentives in the investment codes on effective tax rates on investment, Sewell, Tsiopoulos, and Mintz (1996) made some preliminary inquiries to check on the application of these incentives. The primary finding was that incentives involving immediate expenditures have not been used in recent years because of budgetary stringency, thus leaving tax relief as the principal operational incentive in the codes. Berdai (1991, p. 166) reports, for instance, that cash incentives to small businesses, for example, to purchase land, have been dropped for these reasons. When reimbursement of the uniform tax on imported inputs was available to exporters, moreover, businesses told international financial institutions that the costs of red tape and compliance meant that they sometimes did not bother to apply for reimbursement. Moroccan civil servants often sum up these factors by stating that the tax exemptions are the only "automatic" incentives offered in the investment codes.

The investment codes entail other practical difficulties. First, the study by Sewell, Tsiopoulos, and Mintz (1996) reports that some of the incentive legislation cannot be implemented, because it is not able to deal with changes in the legal status of firms, such as mergers, takeovers, sale, or dissolution; with projects that fall under several codes; with changes in the location of a firm or its plants, which are important because of the regional incentives built into some of the codes; and with difficulties in enforcing other distinctions in the codes, such as that between newly created and expanding firms. In these and other respects, legislators and those drafting the legislation apparently did not anticipate the full complexity of what they were trying to do and the problems that would arise in implementing the legislation.

Measuring Tax Incentives

The nature of many of the interventions in the market through the tax system is that they are not transparent and their effects are far from self-evident. Two methods have been developed that make some of the hidden effects of taxes more visible. Each has it merits and disadvantages. Tax expenditure accounts provide estimates of the value or revenue cost of tax incentives. They involve making value judgments about "ideal" tax systems, however, and obtaining agreement on such normative goals is not easy (see Bruce 1988 for methodological issues in connection with tax expenditure analysis). Sewell (1995) describes an actual application of tax expenditure analysis in Morocco for housing that incorporates different views as to ideal tax bases. The calculation of effective tax rates on investment, by way of contrast, is a method of adding up the effects of tax and other incentives that is value free, although it does not provide an estimate of the revenue impact of these tax measures. In subsequent paragraphs we will outline some of Sewell, Tsiopoulos, and Mintz's (1996) findings concerning the effects of selected tax incentives in the investment codes.

Effective Tax Rates

According to estimates in Sewell, Tsiopoulous, and Mintz (1996), the provisions for tax holidays are clearly the most important individual incentives in the investment codes, and the study focused on establishing their effects. In an interesting outcome, the study found that the tax holidays given in these codes have virtually no incentive effect for some types of investments. Thus, table 8.5 indicates that the effective tax rates for typical investments by multinational firms in the service industries are 13.0 percent without a tax holiday, 12.8 percent with a five-year corporate tax holiday, and 11.2 percent with a ten-year corporate tax holiday. Such differences may not be material to investors, given the importance of other factors that affect investment decisions. The effects of tax holidays are somewhat larger for manufacturing, where typical investments by multinational firms face effective tax rates of 19.7 percent without tax holidays, 15.7 percent with a five-year tax holiday, and 10.5 percent with a ten-year tax holiday. These results reflect complex interactions between the tax holidays and the structure of different types of investments. They also reflect the effects of general tax reductions brought about by the 1994 and 1995 finance laws, however. Some of the provisions of the latter, such as the introduction of accelerated declining balance depreciation, in effect are substitutes for the incentives provided by tax holidays.

The study suggested that some of the objectives of the investment codes may have become obsolete or otherwise require reexamination. This is the case for the remaining tax relief for exports, as exchange market liberalization and tariff reduction reduce the need to compensate for dis-

Table 8.5 Aggregate Effective Corporate Tax Rates, 1995
(percent)

	No Incentives		Five-year 100% tax holiday (manufacturing)		Five-year 50% tax holiday (services)		Five-year 100% tax holiday followed by five-year 50% tax holiday	
Business	*Domestic firms*	*Multinational firms*	*Domestic firms*	*Multinational firms*	*Domestic firms*	*Multinational firms*	*Domestic firms*	*Multinational firms*
Manufacturing	24.2	19.7	17.9	15.7			13.2	10.5
Services	19.9	13.0			15.2	12.8	14.0	11.2

Source: Sewell, Tsiopoulos, and Mintz (1995, table 9).

tortions in these areas. The effectiveness of incentives for regional diversification also need to be examined, even if the objective itself is not in question. The study suggested that the authorities might wish to institute a formal evaluation process to examine the effects of these measures, as the investment codes themselves originally called for.

The study also raises questions about the merits of tax holidays as an incentive instrument. Tax holidays reward the creation of new firms rather than new investment, and are susceptible to use in several kinds of tax avoidance schemes. From the latter point of view, the continued existence of tax holidays could even be regarded as an accident looking for a place to happen. In addition, the study notes that a separate econometric inquiry for fourteen countries in the Mediterranean basin by Mintz and Tsiopoulos (1994) found no significant statistical relationship between the existence of tax holidays and levels of foreign investment. By way of comparison, the latter study did find a significant relationship between the size of effective tax rates and the level of foreign investment. For these and other reasons, Sewell, Tsiopoulos, and Mintz (1996) suggested that the Moroccan authorities might consider replacing tax holidays with simple reductions in statutory corporate tax rates.[6]

In recent years, by extending tax bases to include more taxpayers and improving tax administration, the Ministry of Finance has successfully reduced tax rates while preserving tax revenues. Replacing many existing tax and other incentives by means of simple reductions in statutory tax rates is an obvious extension of this strategy. Many countries of the Organisation for Economic Co-operation and Development have reduced tax rates while holding tax revenues constant by eliminating special tax incentives in the last decade or so, and to the public in these countries, such policy change has become almost synonymous with tax reform. One of the principal by-products of such reform to the Moroccan authorities might well be the reduced use of administrative resources that would thereby be made possible. The advantages for governance in terms of the transparency of policy instruments should also be self-evident. Above all, however, the adoption of such a reform would represent a natural progression toward allowing the market to determine the outcome of investments and abandoning detailed bureaucratic intervention in the economy. Such a reemphasis on the government's function in leveling the playing field for private investors, and thereby allowing the market to promote efficiency in the allocation of resources, would play a key role in Morocco's current policy thrust to promote development in the private sector.

Housing

The housing industry is now the second largest sector after agriculture to benefit from subsidies. This industry provides an excellent example of the extent to which the government has intervened in particular markets with the aid of a variety of fiscal instruments. A study by Sewell (1995) valued implicit tax subsidies to this industry and compared them with explicit subsidies.

Morocco has some highly visible housing subsidies. Mortgage interest rate subsidies may amount to as much as 5 percent on a 13 percent mortgage, and had a total value of DH190 million in 1992. That same year, transfers to housing developers accounted for a further DH383 million. Together, these two visible subsidies were equal to 3.6 percent of personal and corporate income tax revenues in 1992.

Turning to tax exemptions, the Building Investment Code provides real estate developers with a five-year exemption from the corporate income tax after construction is completed, and individuals constructing buildings for rent with a three-year exemption from income taxes derived from such buildings.[7]

Capital gains from the first sale of a building are exempted from capital gains taxation for five years after a permit for construction is obtained. The exemptions from corporate income taxes and capital gains taxes are full exemptions if the building is located outside the larger cities and half exemptions if the building is located in the larger cities.

Special treatment in the income tax is not the only case of special tax treatment for real estate in the Building Investment Code. The code also grants half or full exemptions from registration taxes. In the case of the *taxe urbaine,* without exemptions business users have to pay a flat rate of 13.5 percent of the rental value of their premises, although owner-occupied housing obtains a reduction of 75 percent of its equivalent rental value and then its owners pay according to a progressive rate schedule above a threshold level. The Building Investment Code gives new buildings a five-year exemption of payment from the *taxe urbaine* following construction, and low-income housing gets an additional exemption during the period when construction loans are being repaid.

As noted earlier, by foreign standards, real estate income is overtaxed with respect to the VAT on interest payments. The VAT is intended to apply to added value in transactions, but in the case of financial transactions is applied to gross interest payments. In 1992 the government reduced the

Table 8.6 Tax Expenditures in Capital Gains Tax on Building, 1992

Nominal tax rate (percent)	*Where tax applied*	*Annual rate of inflation 1992 (percent)*	*Total real tax rate*[a] *(percent)*
15.0 (real rate)	Capital gains on building	n.a.	15.0
52.0	Personal income, top rate bracket	4.9	56.9
52.4	Dividend income	4.9	57.3
20.0	Interest income, source undeclared	4.9	24.9
30.0	Interest income, source undeclared	4.9	34.9

n.a. Not applicable.
a. Columns 1 plus 3.
b. Column 4 times DH4,549 million, or estimated actual real estate profit in 1992.

VAT to 7 percent for housing and required it to be paid only for housing that is expensive by Moroccan standards; that is, housing valued at DH500,000 (US$59,000) or more in 1992.

The principal example of special tax treatment for housing other than owner-occupied housing is the fact that returns are taxed at a rate of 15 percent *after inflation*. Such taxation of real returns does not in itself constitute a tax incentive. Indeed, most proponents of an income tax would think it only appropriate to tax real and not inflationary income. The fact that the resulting tax rates for investments in real estate are so much lower than those for other assets does constitute special treatment, however. It is appropriate to compare the 15 percent tax on real estate income with the top tax rates on other types of income. On this basis, table 8.6 indicates that in 1992 the 15 percent tax on real estate income compared extremely favorably with the tax rates, taking into account inflation, of 56.9 percent for wages, 57.3 percent for dividend income, and 24.9 or 34.9 percent for interest income.[8]

Sewell's estimate of actual capital gains tax paid in housing in 1992 of DH600 million equals 4.3 percent of total revenues from personal and corporate income taxes in that year (Sewell 1995). The taxpayer's lowest conceptual opportunity cost, in terms of taxes paid on other sources of wage or investment income, would have been to have this income taxed at the rates of taxation for interest, which would have yielded revenues of DH997 million, or 7.2 percent of total corporate and personal income tax paid in 1992. Alternatively, taxation of real estate profits at the highest alternative rate applied to personal income—the rate for dividend income—would theoretically have yielded revenues equal to DH2,294 million, or 16.6 percent of total personal and corporate income tax paid in 1992. All other elements of tax expenditures and explicit subsidies for housing pale when compared to these estimates of taxes that would have to be paid if housing bore the same tax rates as other investments. Exemptions from capital gains taxation accounted for 24 percent of the to-

Taxable capital gains		*Tax expenditures*
Taxable capital gains: housing, industrial, and commercial real estate[b]	*Taxable capital gains in housing*[c]	*Opportunity cost in DH millions of present taxation of capital gains in housing*[d]
682	600	n.a.
2,588	2,278	1,678
2,607	2,294	1,694
1,133	997	397
1,587	1,397	797

c. Column 5 times 0.88.
d. Column 6 minus present taxable gains in housing of DH600 million.
Source: Sewell (1995, table 2).

tal value of tax incentives and exemptions, and the preferential real tax rates for housing for 71 percent of the total. The value of special tax treatment was more than four times that for visible subsidies in interest rates and transfers to developers.

Mortgage Deductibility

Some question arises as to whether one element in the tax treatment of housing—deductibility of mortgage interest costs in the calculation of personal income taxes—should be treated as a tax incentive or tax expenditure. The latter phrase is meant to indicate that expenditure programs can, in effect, be delivered through the tax system by means of departures from normal or benchmark taxes. Normative judgments enter into the choice of the benchmark system with which actual tax treatment is to be compared. Two well-known, but competing, normative benchmarks can be used to assess tax expenditures in Moroccan housing: the benchmarks of income and consumption taxes.

If housing were to be treated like any other investment in an income tax system, borrowing costs would be deductible and all income from the asset would be taxable. Under a consumption tax, those investing in owner-occupied housing would not be allowed to deduct the costs of such investments (such as mortgage interest) from taxable income as they would in a true income tax. However, under a consumption tax the yield from housing would not be taxed either in the form of the implicit rental value of the services from such housing or any capital gains when the asset is sold. Canada, for example, taxes housing according to such consumption tax principles. The principal tax expenditure for owner-occupied housing in Morocco under a consumption tax base would therefore be the deductibility of mortgage interest, because under a consumption tax the revenues from owner-occupied housing would not be, and indeed are not, taxed in Morocco.

Table 8.7 Distribution of Tax Expenditures for Mortgage Recipients by Income Class and Marginal Tax Rates, 1992
(dirhams)

	General program			
			Tax expenditures	
Annual income	*Distribution of income by income class*	*Marginal tax rate (percent)*	*General program*	*Foreign workers*
0–11,999	0.4	0	0	0
12,000–23,999	0.4[a]	14	155,239	18,120
24,000–35,999	7.2	22	4,391,049	512,528
36,000–59,999	30.0	36	29,938,696	3,494,507
60,000–89,999	20.5[a]	44	25,004,584	2,918,561
90,000–119,999	20.5[a]	46	26,141,156	3,051,223
120,000–200,000	10.5[a]	48	13,971,519	1,630,770
More than 200,000	10.5[a]	52	15,135,812	1,766,668
Total	100.0	n.a.	114,738,328	13,392,377

Note: Tax expenditures are calculated as a program's share of population in an income class times the marginal tax rate for that income class times the amount deductible for income tax for the program.

The distributional effects of mortgage interest deductibility in Morocco are also worth noting. Many housing programs have the specific redistribution objective of aiding low-income groups: in 1992 more than two-thirds—68 percent—of explicit mortgage interest subsidies were distributed for low-income housing. In the case of tax expenditures such as the deductibility of mortgage interest from taxable income, however, the distributional effects may be less evident, and may, indeed, be surprising to some. Sewell (1995) valued the distributional effects of mortgage deductibility at the marginal tax rate of those who could claim such deductibility. If one aggregates tax expenditures calculated in this way for all mortgage deductibility in the various mortgage programs in 1992, the total value was 81 percent of that for explicit interest subsidies for the same year. Clearly, the value of mortgage interest deductibility is an important element in the fiscal treatment of housing.

As table 8.7 demonstrates, it is also an element with perverse features, or at least a distributional outcome that is the opposite of that of the explicit interest subsidy program. A distributional feature of mortgage interest deductibility that is self-evident is that its benefits are confined to those who pay taxes. In Morocco's special program to provide mortgages to low-income people, 29 percent of the participants receive no benefits from mortgage interest deductibility because their incomes are so low that they are exempt from paying taxes. A related distributional effect of the deductibility of mortgage interest is that its benefits increase with income level. In 1992 the government, in effect, paid for 52 percent of the interest costs of mortgages held by those in the top personal income tax bracket. This is also an "open-ended" subsidy, in that the amount of mortgage interest that is deductible from income tax owed is not limited.

With respect to the overall distributional effects of mortgage interest deductibility, more than 7 percent of mortgage holders in Morocco's gen-

	Low-income program	
	Tax expenditures	
Distribution of population by income class	*General program*	*Foreign workers*
29	0	0
46[a]	13,353,736	726,160
25[a]	11,404,588	620,167
n.a.	n.a.	n.a.
n.a.	n.a.	n.a.
n.a.	n.a.	n.a.
n.a.	n.a.	n.a.
n.a.	n.a.	n.a.
n.a.	24,758,324	1,346,327

a. Totals interpolated between adjacent income brackets.
Source: Sewell (1995, table 4).

eral mortgage program and 25 percent of the mortgage holders in the special program for low-income people received assistance equal to 22 percent of their mortgage costs. Some 62 percent of those having mortgages in the general program received at least twice this level of assistance, because they could deduct mortgage interest costs at the marginal tax rate of 44 percent or more. Benefits to participants in the general mortgage program were estimated to be 4.6 times those for participants in the low-income mortgage programs.

Morocco is, of course, far from being the only country that permits mortgage interest to be deducted from taxable income. Advocates of tax expenditure accounts have labeled the regressive features of mortgage interest deductibility that we have noted as "upside-down equity" (see Surrey 1973, pp. 36–37, for the classic statement). They add that it is extremely unlikely that an expenditure program that offered similar substantial rates of subsidy to high-income groups and no subsidy at all to those too poor to pay any income taxes would be suggested, let alone defended and implemented.

Taxation and Energy Pricing

A final taxation topic that may warrant attention is taxation of energy. The revenue raised from petroleum products is such a substantial share of tax revenues—just under 17 percent in 1993—that an interesting question concerns the significance of these energy levies for Moroccan productivity, competitiveness, and macroeconomic goals.

Determining the exact contribution of taxes to the retail or "end" prices of particular petroleum products in Morocco is difficult because of complexities in the tax structure, cross-subsidization, compensation mechanisms for refiners, and other distortions in pricing. In its resulting complexity and lack of transparency, energy taxation is thus typical of much of Morocco's tax system before reform. In some respects, the industry is treated more generously than others: minimum income taxes are imposed on it at roughly half the rate of those on other industries. While end prices are not onerous for many uses, some are well out of line with foreign equivalents, and interfuel substitution is inefficient because of substantial differences in relative end prices. In transportation uses in 1993, for example, gasoline and diesel prices were, respectively, 20 and 30 percent below average European levels. Moroccan prices for heavy fuel oil were, however, 60 percent higher than the European average for heavy fuel oil in 1991, and this affects the troubled electricity industry for which heavy fuel oil is a principal input (UNDP and World Bank 1994, table 2.4). Among other things, a long-run solution to the fuel pricing problem would appear to involve imposing equal tax loads for alternative energy uses on final consumers.

Energy pricing arrangements may also pose problems for Morocco's recent hard-won achievement of relative budgetary equilibrium. Present pricing arrangements originated largely in a desire to capture windfall profits when world crude oil prices dropped. This source of revenue is now such an important source of total tax revenue that it would be threatened by a future rise in oil prices or fall in the exchange rate.

Lessons

This chapter was completed after passage of the 1995 Finance Law (or annual budget) of Morocco. A number of tax measures have been changed since then, one of the most important being the introduction of a uniform Investment Code in the interim budget for 1996. The new code has eliminated much of the variation in incentive measures between industries; a change that is strongly supported by the present analysis. The new code retains tax incentives such as tax holidays, whose value is questioned in the present study, however, and does not affect our recommendations about the merits of alternative incentive instruments and strategies.

Moroccan tax reform experience yields several generalizations.

1. *High tax rates give rise to small tax bases*. Before the tax reforms, high tax rates provoked demands for relief in the form of generous exemptions and tax incentives, which helped to reduce effective tax rates, but also increased the dispersion in effective rates across different activities and impaired the system's ability to avoid economic distortions and achieve equity in taxation. Widespread exemptions and incentives also facilitated tax evasion and caused further erosion of the tax base.

2. *Complex tax systems hamper administration and encourage evasion*. Combined with a finely layered multiple rate structure, a plethora of exemptions and incentives produces a tax system of considerable complexity.

Such a system not only provides numerous opportunities for tax evasion, but also handicaps tax administrators in their task of preventing it. In creating excessive paperwork that absorbed the energies of tax administrators, Morocco's complex tax system prior to reform had seriously compromised the ability of officials to enforce the system's regulations. Administrative improvements were compromised until some parts of the system were simplified to allow more efficient administration.

3. *Complex tax incentives can be unworkable and inequitable.* Unless or until tax administration can be improved and the government can place greater reliance on direct taxation, the objectives of obtaining more revenue and achieving greater vertical equity in taxation are likely to clash. Morocco's attempts at fine-tuning the tax system before tax reform more often gave the appearance of greater equity than the reality. Administering some of the distinctions legislators wanted was impossible, as in incentives in the investment codes pertaining to changes in firm ownership. Furthermore, some of the tax breaks introduced in the name of equity actually have perverse distributional effects, as in the case of housing. To be fair, however, Morocco is not the only country that has introduced such regressive tax expenditures for housing.

4. *Tax incentives often escape budgetary scrutiny and penalize goals that are attainable by other means.* Incentives in the investment codes involving actual expenditures—as opposed to tax incentives that do not involve visible outlays—have often not been implemented because of budgetary stringency. An example is interest rate subsidies for small business. Such suppression of incentives taking the form of outlays imparts a formidable bias in the use of policy instruments in favor of tax expenditures whose effects are hidden. This is not a desirable outcome for public policy, unless the effects of such measures can be made visible (as in calculating and publishing tax expenditure accounts) and the costs of these incentives are taken into account in policy formation.

5. *Old habits are hard to break.* Morocco's experience with tax reform also shows how the dead hand of the past intrudes upon the present. In trying to reform its system, the authorities frequently discovered the difficulties of deviating from previous practices and patterns. Political resistance to change seems to be stronger the longer a particular tax system has been entrenched, however undesirable many of its features.

6. *Multiple tax rates intended to increase equity render VAT performance unsatisfactory and lead to their own inequities.* Early experience with the VAT in Morocco demonstrates that the VAT is not necessarily a magical tax reform remedy. Like any other tax instrument, the VAT can be designed to work well or it can work poorly. In Morocco, numerous and complex exemptions and rates initially handicapped the operation of the VAT and condemned it to unsatisfactory performance. Indeed, the several rates still in effect for this tax complicate its administration and produce inequities in its application, as evidenced by application of the buffer rule.

7. *Structural means of deficit reduction should be distinguished from palliative measures.* Morocco's efforts to cope with large fiscal deficits also show that the size of fiscal deficits is not necessarily closely linked to the rate of inflation. Morocco managed to contain some of the inflationary impulses of its large fiscal deficits by imposing implicit taxes on the private sector in the form of significant arrears in payments. Such implicit or hidden taxes, however, cannot be sustained in the long run and are a poor substitute for transparent reform of the tax system.

8. *Successful tariff reform may be conditional on prior tax reform.* Morocco's unsuccessful early attempt at trade liberalization underlines the need to recognize the important revenue contributions of tariffs. Reducing tariffs may be impossible unless other parts of the revenue system can readily compensate for the loss of revenue. In this sense, tax reform may logically be a prior requirement of tariff reform.

9. *Tax holidays may be a poor vehicle for stimulating new investment.* These five- and ten-year holidays currently do not make a material difference to effective tax rates for investments in Morocco's service industries, although they do lower effective tax rates for investments in manufacturing. This reflects complex interactions between the holidays and the structure of different types of investment, and indicates that some of the general tax reductions introduced in the 1994 and 1995 Finance Laws, such as accelerated depreciation, are substitutes for the incentives provided by the tax holidays. Tax holidays also reward the creation of new firms rather than new investment and are susceptible to use in several types of tax avoidance schemes. An analysis of countries in the Mediterranean basin showed that reductions in overall effective tax rates have a significant and positive relationship with levels of foreign investment that is not evident for specific incentives, such as tax holidays.

10. *For the same revenue cost, general tax incentives lead to more efficient investment than specific sectoral incentives.* The finding that Morocco has cut effective tax rates for domestic investments by more than half since the government initiated tax reform in 1986 is significant. The inference is that the substitution of general tax advantages available to all taxpayers for specific sectoral tax incentives would further promote efficient investment in Morocco.

11. *Government "ownership" of the process of tax reform is critical to its success.* For tax reform to succeed, government "ownership" is important, in the sense that the government must be committed to the change and accept the social and economic impacts of the burden involved. The Moroccan government was determined both to undertake a long series of tax reforms and to stay the course over the extended period required to accomplish the task.

12. *The success of large-scale structural economic reforms depends on the existence of adequate government administrative capacity.* Another prerequisite for successful reform is that the government must invest in the admin-

istrative capacity needed to implement legislative tax reform. If, as is likely in a structural adjustment program such as Morocco's, other reforms such as trade liberalization, privatization and rationalization of public enterprises, formulation of medium-term expenditure programs, monetary policy to lower inflation, and liberalization of the financial sector are also to be undertaken, the World Bank's experience is that a country often may simply not have enough administrative capacity or competence to accomplish all these changes. One consequence is frequent shortfalls in revenues as a result of tax reform (Jayarajah, Baird, and Branson 1995, p. 136). With few exceptions (such as the introduction of the VAT early in the tax reform process), Morocco appears to have managed this process of wholesale change successfully. At least a partial explanation for this success lies in the improvement of tax administration, which, while it did not create as much of a splash as legislation introducing new forms of taxation, has been the focus of much reform effort in recent years.

Notes

The authors would like to thank Jamil Berdai and Oufaá Khallouk for their contributions to this work, Luc De Wulf for his comments, and John Underwood for his support.

1. In addition, social security taxes absorb nearly 20 percent of gross wages in Morocco. Social security funds are earmarked for the payment of family allowances, for labor training, and for workers' compensation.

2. Principal residences are exempt from the tax if they are held for eight years or longer.

3. Moreover, a complex rate schedule was elaborated for each excise tax. For example, in the case of the excise tax on mineral water, there were five classes of mineral water depending on the percentage of fruit juice contained and six bottle sizes, with each combination of class and bottle size having its own specific tax rate.

4. If CIF is the landed import price, SIT is the special import tax, CD is the customs duty, VAT is the value added tax rate, and SD is the stamp duty, the domestic price of an imported good can be expressed as CIF * [CD + SIT + (1 + CD + SIT) * VAT] * (1 + SD). The stamp duty was as high as 10 percent in 1979.

5. Also, for the first time, capital assets could be revalued for inflation, and any resulting capital gains to be included in taxable company income declined according to the length of the holding period. However, the business license tax remained in force and its historical cost basis continued to favor older enterprises over new ones.

6. Alternatively, they suggested that if circumstances warrant a general stimulus for investment, the authorities consider an investment tax credit.

7. In 1992 restrictions were imposed on the use of the income tax exemptions in the Building Investment Code. For instance, in the case of the five-year exemption from all or half of corporate and personal income taxes, a minimum payment equal to 25 percent of the amount that would otherwise have been payable is required. Furthermore, although the tax on capital gains in real estate is 15 percent, a minimum tax equal to 2 percent of the sale price is applied.

8. As noted already, interest paid to unidentified lenders (for example, holders of corporate bonds) is subject to a 30 percent tax that the borrower withholds. Where

the lender is readily identifiable (as for bank deposits), interest is taxed at a 20 percent rate that the borrower withholds. This withholding tax is then credited against total personal income tax liabilities.

References

Berdai, Jamil. 1991. "Les Codes d'Encouragement aux Investissements pour les Principaux Secteurs Productifs." Ministère chargé de l'incitation de l'économie. Rabat, Morocco.

Bruce, Neil, ed. 1988. "Tax Expenditures and Government Policy." Paper presented at the Seventh Roundtable on Economic Policy. Queen's University, John Deutsch Institute, Kingston, Ontario, Canada.

Jayarajah, Carl, Mark Baird, and William H. Branson. 1995. "Structural and Sectoral Adjustment: World Bank Experience, 1980–92." World Bank Report No. 14691, Operations Evaluation Study, Washington, D.C.

Mateus, Abel. 1988a. *Multisector Framework for Analysis of Stabilization and Adjustment Policies: The Case of Morocco*. World Bank Discussion Paper No. 29. Washington, D.C.

———. 1988b. "Profit Taxation, Corporate Policies and Investment Codes." World Bank, Europe, Middle East and Northern Africa Regional Office, Country Department II, Washington, D.C.

Mateus, Abel, and S. Miltner. 1987. "How Progressive Is the Moroccan Tax System?" World Bank, Europe, Middle East and Northern Africa Regional Office, Country Department II, Washington, D.C.

Mintz, Jack M., and Thomas Tsiopoulos. 1994. "Taxation of Foreign Capital in the Mediterranean Region." World Bank, Foreign Investment Advisory Service, Washington, D.C.

Sewell, David. 1994. "Morocco: Intergovernmental Fiscal Relations." World Bank, Europe and Central Asia, Middle East and North Africa Regional Office, Technical Department, Infrastructure Unit, Washington, D.C.

———. 1995. "Housing in Morocco: Visible and Less Visible Subsidies." Annex F in "Kingdom of Morocco: Housing Sector Strategy." World Bank Report No. 13930-MOR, Middle East and North Africa Region, Washington, D.C.

Sewell, David, Thomas Tsiopoulos, and Jack Mintz. 1995. *Tax Effects on Investment in Morocco*. World Bank Discussion Paper No. 15. Washington, D.C.

———. 1996. "Effects of Taxation on Investment in Morocco." *Tax Notes International* 12(February 19): 573–96.

Surrey, Stanley S. 1973. *Pathways to Tax Reform*. Cambridge, Mass.: Harvard University Press.

(UNDP) United Nations Development Programme and World Bank. 1994. *Morocco: Energy Sector Institutional Development Study*. Washington, D.C.: UNDP/World Bank Energy Sector Management Assistance Program.

World Bank. 1984. "Morocco: Industrial Incentives and Export Promotion." Report No. 4893-MOR. Washington, D.C.

9

Turkey's Struggle for a Better Tax System

Kenan Bulutoglu
Wayne Thirsk

Tax reform not only has a long tradition in Turkey, it has also engaged the energies of a wide variety of different forms of government in that country. In addition, reform efforts have exhibited a curious cyclical frequency. Beginning in 1949–50, a major attempt at tax reform, as well as a significant shift in political power, has characterized virtually every decade. Box 9.1 summarizes the major changes that have occurred in the Turkish tax system and distinguishes among the main episodes of tax reform.

This chapter has several aims. One is to describe the evolution of the Turkish tax system and set it in its historical context. Another is to discuss some of the important political forces that have motivated the successive waves of tax reform. Where possible, the chapter analyzes the impact of these reforms and evaluates their overall strengths and limitations. It pays particular attention to Turkey's effort to develop a modern income tax, to establish an efficient indirect tax system, and to create an effective tax administration apparatus. Although each reform episode is of some interest, the distance traveled along the road to reform is more interesting than the particular stops along the way. We focus on the broad lessons of tax reform that may be learned from the Turkish experience rather than on the narrower technical merits of alternative tax arrangements

Background

Turkey launched itself on the path to tax reform with a series of measures enacted in 1950. This initial tax reform package attempted to streamline a tax system that had become excessively complicated by wartime amendments designed to secure quick revenue increases. It also tried to lay the foundation for a modern income tax system. This reform thus

Box 9.1 Outline of Major Tax Reforms

1949–50

Global personal income tax (PIT) replaces schedular income taxes; agriculture and small business is exempt along with interest on government bonds and small deposits; rate brackets range from 15 to 45 percent.
Corporation income tax (CIT) introduced at a 10 percent rate plus 15 percent of after-tax profits (creditable against PIT if distribution occurs); government-owned monopolies exempt.
Small business tax levied on rent paid.
Agriculture made subject to a livestock tax.
Modest tax on urban real estate and inheritances introduced.

1956

Manufacturers' sales tax replaced by excise taxes on manufacturing inputs.
Small business tax abolished.
Upper rate of the PIT raised from 35 percent to 50 percent.

1960–63

Income tax extended to agriculture, with small farmers exempt and incomes of medium-size farmers taxed according to presumed expenses.
PIT rate brackets adjusted to a 10 to 60 percent range.
CIT rate raised to 20 percent; 20 percent withholding on after-tax profits.
Livestock tax abolished; land tax reintroduced.
Compulsory savings plan introduced.
Foreign exchange expenditure tax introduced.
Stamp duty on imports introduced.
A 30 percent investment deduction allowed in addition to normal depreciation.

1966

A cornucopia of investment incentives introduced exempting investment goods from import duties, stamp duties, real estate purchase fees, and bank and insurance transaction taxes.

1970–72

Seven new taxes introduced: 5 to 30 percent retail sales tax on services and luxury goods, capital gains tax on real property, motor vehicle purchase tax, sports lottery tax, building and construction acquisition tax, self-assessed land and building tax (replaced the old land and building tax collected by the provinces), and fiscal balance tax (a surtax on incomes replaced the compulsory savings program).
Earnings of those holding more than one job taxed as a single source of income.
Banking and insurance transactions tax (BITT) raised from 20 to 25 percent.
CIT rate raised to 25 percent plus an additional 20 percent on after-tax profits whether distributed or not.
Constitutional amendment permitted delegating powers to the government to change rates between the upper and lower limits stated in the tax law.

1980

New rate brackets and altered rate structure (40 to 60 percent range) for PIT.
Higher ceilings and rates for the lump-sum taxes paid by small businesses and the liberal professions.
Advance payments for income tax filers.

1980 (continued)

Narrower definition of tax exempt farmers.
Maximum deduction of 70 percent for presumed agricultural expenses of medium-size farmers.
Absentee landowners taxed under the provisions for rental income.
Five percent withholding tax on all agricultural sales.
Withholding on dividends and interest income raised from 20 to 25 percent and small deposit interest exemption eliminated.
Rental income from business real estate made subject to 25 percent withholding.
Corporate rate raised from 43 percent to 50 percent and government enterprises taxed at 35 percent.
Loss carry-forward reduced from five to three years.
Capital gains from corporate reorganization made nontaxable.
Higher motor vehicle and construction taxes.
New sales tax at a uniform 3 percent rate.

1982

PIT rates gradually altered to a 25 to 50 percent range.
CIT rate reduced to 40 percent and 20 percent of export-related revenues made deductible.
Introduction of living standards assessment system as a minimum tax on income tax filers.
Withholding on interest income reduced to 20 percent (30 percent on bearer certificates of deposit).
Real estate purchase tax and capital gains tax on real property replaced by a 4 percent real estate duty on both buyer and seller.
Extrabudgetary fund (EBF) for housing established; rapid expansion of other extrabudgetary funds in subsequent years.

1983

BITT reduced from 15 to 3 percent.

1984

Tax collection through commercial banks.
Withholding on interest income reduced to 10 percent.
State-owned enterprises made fully taxable.
Six percent consumption tax on petroleum products introduced.
More extrabudgetary funds created.

1985

Value added tax (VAT) introduced at a uniform 10 percent rate to replace seven other taxes: the production taxes on domestic production and imports, the business tax, the transportation tax, the sugar consumption tax, the services tax, the sports lottery tax, and the advertising services tax.
Lump-sum tax limits raised by a factor of four.
Fines and penalties increased tenfold.
Reduced withholding rate on interest income to 10 percent (3 percent on government bonds).
No withholding on dividends at the personal level.
Agricultural withholding revised to 3 percent on animal sales and 7 percent on all others.
Advance tax payments established at 50 percent of VAT obligations.Corporate tax rate raised to 46 percent plus a 3 percent surtax for the Fund to Support the Defense Industry.
More resources invested in tax administration.

(Box continues on following page.)

Box 9.1 *(continued)*

1986

Basic VAT rate raised to 12 percent and multiple rates introduced.

BITT reduced to 1 percent; 13.5 percent special VAT regime for small retailers canceled.

1987

Depreciation balances indexed to the rate of inflation.

Machinery purchases to enjoy 50 percent declining balance depreciation.

Expensing allowed for some sectoral investments.

1988

Increased rate differentiation in the VAT (seven rates including a zero rate) and basic rate reduced to 10 percent.

Agricultural withholding reduced to 2 percent on animal sales and 4 percent on all other sales.

Special VAT regime for small retailers reintroduced.

System of advance PIT payments based on VAT revised and geared to a fraction of the previous year's PIT.

Tax holidays given for investments in educational, health, and sports facilities.

Tax rebates on exports phased out during the year and subsidies from encouragement and stabilization fund paid to exports on a specific basis; subsidies removed in 1989.

A minimum tax, irrespective of profits or losses, introduced on corporations.

marked the transition from a schedular income tax to a global personal income tax (PIT) and the introduction of a separate tax on corporations' incomes. It also signaled the government's intent to shift a larger share of the tax burden toward higher-income groups in the economy. Moreover, the tax reform package included a new tax procedure law that set forth the general rules regarding taxpayers' obligations and the powers of the tax administration to assess taxes, obtain information, conduct audits, and collect taxes owed. The texture of the reforms was heavily influenced by the policy advice of a German Advisor, Fritz Neumark, and a number of the reforms carry many of the hallmarks of the former Federal Republic of Germany's tax system. However, by exercising their newly acquired political power, large farmers rapidly removed the stipulations of the new income tax bill that would have imposed a tax burden on their incomes.

Following democratic elections in 1950, a new conservative government gained power with a strong majority in parliament. During its first years in power, this government adopted a laissez-faire economic policy and implemented the earlier tax reform proposals without much alteration of their content. However, in the latter half of the 1950s tax revenues, particularly PIT collections, began to grow more slowly, while at the same time public expenditures grew rapidly. This era marked the beginning of a long series of problems with fiscal deficits, which have often been the impetus behind major tax reforms. Large fiscal deficits, in turn, have fueled a sharp rise in inflation. Annual price increases that had been running at about 5 percent in 1953 and 1954 had climbed to 23 percent by 1957.

Following a rescheduling of Turkey's external debt in 1958, the government introduced a stabilization program sponsored by the International Monetary Fund (IMF). The austerity measures included a freeze on bank credit, a major devaluation of the Turkish lira, and large contractions in the volume of public spending. Interestingly, the government did not adopt any major tax measures to try to raise more tax revenue. In fact, just the opposite occurred. The government continued to make a number of tax concessions, first to small traders and artisans in 1956 by abolishing the presumptive income tax, and then by abolishing the broadly based transactions tax on manufacturers' sales. It replaced the transactions tax with a narrowly based excise tax on manufacturing inputs such as cotton, yarn, fuel, electricity, plastics, and steel and other metals.

By 1960 Turkey's general political climate had deteriorated, marked by the government's continuing efforts to suppress political opposition, and a military coup took place. Along with its efforts to restore democratic procedures, the new military government tried to put the country's public finances on a solid footing. It adjusted tax brackets to account for the previous bracket creep induced by inflation and introduced a compulsory savings plan under which direct taxpayers were required to buy 6 percent government savings bonds in the amount of 3 percent of their tax bases. In addition, it once again extended the income tax to the agricultural sector and doubled the corporation income tax rate from 10 to 20 percent. Stronger administrative measures, such as requiring an annual declaration of wealth along with income tax returns, were introduced in an effort to curb tax evasion. Overall price stability was restored as the public sector deficit was eliminated through a combination of reduced public investment and several measures to increase revenues. The military government also created the State Planning Organization (SPO) to formulate and implement rational economic development policies.

A coalition government emerged from the free elections held in 1961 that introduced two kinds of tax changes. The first group of changes reduced revenues. They consisted of concessions made to assuage pressure groups that had been adversely affected by the reforms of the previous military government. The new minister of finance established a tax reform commission that included representatives from the business and agricultural communities in addition to some bureaucrats and university professors. On the strength of the commission's recommendations, the government once again softened the impact of income taxation on farmers and provided new tax incentives to the business community to encourage investment and exports. The government also introduced a new income tax rate schedule with more nominal tax progression.

The second group of tax changes, introduced in 1964, were intended to mobilize the public sector resources required to finance the first Five-

Year Development Plan. To the end the government increased customs duties on certain imports, introduced a general flat rate tax on all imports in the form of a stamp duty, created new taxes on foreign travel expenditures, transformed fees on real estate property transfers into a separate real estate purchase tax, and increased motor vehicle license taxes and fees. The most important feature of the development plan was that it mobilized resources by introducing new taxes and increasing the rates of existing ones rather than broadening the base of either the income or the sales tax. This tendency toward tax proliferation would also characterize future reforms.

Elections in 1965 brought a new conservative government to power. This government's first step was to make generous tax concessions to the business community by creating numerous exemptions to encourage private investment and assembly industries. The erosion of the income tax base was once more combined with greater public expenditures and higher state enterprise deficits. By 1970 the inflation rate had more than doubled to 11 percent per year. The government responded by bringing in a fiscal austerity program and allowing the lira to be devalued. As part of its efforts to reduce the budget deficit, the government introduced seven new taxes and increased existing tax rates across the board. Against a background of growing economic difficulties, the military intervened once again. As part of its perceived mandate to secure law and order, the military government acted to restrict constitutional rights and their safeguards. Preoccupied with the issue of political tranquility, the military-backed governments that followed made no changes in the country's tax laws.

Until recently, this pattern of unsustainable fiscal deficits followed by proliferating new taxes, and higher tax rates was the dominant theme in the evolution of public finances. Eventually, public finances began to display the strains of some severe internal contradictions. Seeking more rapid economic growth, Turkish policymakers sought on the one hand to stimulate private sector investment by granting generous investment incentives, and on the other hand to provide generous public sector funding for the investment activities of state-owned enterprises. Turkey failed, however, to impose the necessary sacrifices on consumption that a higher volume of investment required. Sometimes, foreign aid helped to cover up this inconsistency. At other times, the IMF insisted on revenue-raising measures to remove it. More often than not, however, these policy contradictions sparked bouts of inflation, and the authorities used revenues from the inflation tax to finance the deficit and curb consumption. For the tax system, however, the result was a revenue edifice of growing complexity; a tax structure totally lacking in coherence; and a system that was inexorably drifting away from the principles of neutrality, equity, and administrative feasibility.

After multiparty democracy was reinstated with free elections in 1973, a succession of weak coalition and minority governments appeared, none of which had sufficient support in parliament to pass sensible revenue measures. Even the tax adjustments needed to offset the effects of rising price levels could not be introduced. As will be seen later, the lack of indexation in the PIT acted to shift a major portion of the income tax burden from capital to labor incomes.

In the late 1970s the military once again wrested the reins of political power from the civilian authorities. After taking over in September 1980, the military promptly abolished parliament and all political parties and banned political leaders from founding any new parties. In the early 1980s, the military government introduced a plethora of tax changes that depended for the most part on the ability of government bureaucrats to gain access to the inner power circle and persuade the military authorities to adopt their ideas. The military rulers adopted these changes without any free-ranging discussion in the "consultative assembly" they had established in lieu of a parliament. Typically, mistakes in the design and implementation of these tax changes were discovered only after they had become law. A flurry of both minor and major tax changes took place during this period of military rule, which lasted until the return of elections in 1983. However, given the rapidity and volume of changes and the absence of an adequate communications channel between tax officials and taxpayers, tax compliance became increasingly difficult even for honest taxpayers.

With the elections of 1983, Turgut Ozal's party came to power with a clear parliamentary majority, a feat that no political party had achieved since 1969. The Ozal government introduced a host of new tax measures, discussed later. Its primary policy thrust was to give the economy more of an outward-looking orientation and to dismantle many of the import substitution policies of previous governments.

The major innovation in tax policy was the introduction and successful implementation of the value added tax (VAT) in 1985. A less promising innovation, however, was the government's growing tendency to resort to extrabudgetary funds, financed mainly by trade taxes, to promote specific expenditure plans. Coincidentally, a sizable public sector deficit remained as a continuing problem. Although the Ozal government was able to pass tax reform bills expeditiously, the legislation delegated broad discretionary powers to the cabinet to adjust normal tax rates and exemption limits, as though the legislature still operated under the threat of a political stalemate. This practice undermined the stability and transparency of tax rates and exemptions and increased taxpayers' uncertainty about their future tax burdens.

Pressures for Tax Reform

The sequence of tax reforms in Turkey is associated with, and in some cases is a by-product of, periodic economic crises. The latter are closely related to fluctuations in foreign aid inflows, as Turkey has been unable to reduce its dependence on foreign aid.

The foreign aid cycle has followed a regular pattern. A package of concessionary loans and grants have temporarily held the balance of payments problem at bay and increased nontax revenues. This has enabled governments to delay needed, but unpopular, adjustments in the foreign exchange rate, revenue yields, and state enterprise prices. These delays eventually culminate in slow economic growth and declining export levels as domestic prices rise faster than world prices. Finally, export stagnation and expanded imports combine to widen the foreign trade deficit beyond the limits that can be financed with foreign aid.

Trade deficits have gone hand in hand with public sector deficits. As noted earlier, successive governments have granted generous tax incentives to private investors while boosting public expenditures, often through the operations of state-owned enterprises. Such policies have typically elicited an increased inflow of foreign savings, either in the form of more foreign aid or of larger amounts of external borrowing. Eventually, a reduction in the inflow of foreign savings triggers both a foreign trade crisis and the related problem of adjusting to a burgeoning public sector deficit. After intense political discussion, Turkey has usually agreed to introduce an austerity package as requested by both international institutions and donors. This package typically contains a number of measures to improve the revenue performance of the public sector as well as a large devaluation of the exchange rate, higher interest rates, and upward adjustments in public enterprise prices. The life cycle of these crises has generally been around ten years.

The early 1950s witnessed a rapid growth in export revenues on the heels of a commodity price boom caused by the Korean War that coincided with three years of bumper crops. Flush with funds, the Menderes government used some of its newly obtained resources to grant tax relief to the poor by abolishing the transactions tax on food and financing a modest program of tax incentives. However, the government simultaneously initiated an ambitious program of public sector development projects. When the blush faded from the export boom, large fiscal deficits appeared, forcing the government to resort to inflationary financing and contributing to a foreign exchange crisis. A rescheduling of foreign debt payments and the infusion of fresh foreign aid in the Paris Moratorium of 1956 provided temporary relief from these problems.

When the military government assumed power in 1960, it dealt with the fiscal deficit directly by pruning public investments and increasing both tax rates and tax coverage. As part of the first Five-Year Development Plan (1963–67), the authorities designed a tax reform package to improve the tax system and mobilize the public sector resources required to finance annual expenditure programs. In 1963 the Organisation for Economic Co-operation and Development founded a consortium to help provide the external financing for the development plan. Under the aegis of the SPO, tax revenues increased steadily from 12.2 percent of gross national product (GNP) in 1962 to nearly 15.0 percent in 1967. The first development plan successfully reached its target of achieving a domestic marginal savings rate of 27 percent. Subsequent development plans relied to an even greater extent on improvements in public sector savings rates to increase domestic savings. However, the savings targets in these plans became increasingly unrealistic, and actual performance fell far short of them. The third and fourth plans, for example, called for tax efforts in the range of slightly more than 23 percent of gross domestic product (GDP). As expenditure programs were based on the assumption that planned tax efforts would be realized, the subsequent shortfalls in revenue created sizable fiscal deficits that were financed primarily through inflation.

After the 1970 stabilization program exports increased rapidly, and with higher workers' remittances the balance of payments crisis was resolved quickly. The first and second oil shocks during the 1970s enhanced the strain on the fiscal fabric and contributed to the large size of both the trade deficit and the fiscal deficit. The inability of weak coalition governments to pass tax measures to raise revenues and partisan attempts to gain political support by expanding employment in the public sector added to these fiscal woes. Thus inflation accelerated rapidly in the last half of the 1970s, rising from an annual rate of 26 percent in 1977 to 71 percent in 1979.

During the postwar period, the triumvirate of public sector deficits, inflation, and trade deficits has consistently signaled the need for tax reform. The availability of foreign aid has sometimes acted to delay tax reform, while foreign agencies such as the IMF and the World Bank have frequently fomented it. The 1970 tax reform was part of an IMF-supported stabilization program. The World Bank was more influential during the 1980s in encouraging a rationalization of trade taxes and the introduction of a VAT. Within the country, however, the combination of weak coalition governments, parliamentary gridlock, and the power of vested interests, both under democratic and military governments, has conspired to thwart needed tax reforms.

A World Bank (1987) study reinforces this overall impression of Turkey's fiscal dilemma. As table 9.1 indicates, although recent rates of economic growth have been respectable, the overall inflation rate has remained relatively high, around 30 percent in 1987 (and even higher, in the 70 to 80 percent range in 1988 and 1989). The presence of deficit-driven inflation is not evident from a quick reading of the data presented in table 9.2.[1] During the 1980s, total government revenues appear to have been more than adequate to finance the total spending of the public sector. Moreover, total spending itself has declined noticeably since the late 1970s, and more recently, the government's efforts to cut back on the levels of recurrent expenditure can be plainly seen. What cannot be seen, however, is that Turkey's obligation to pay interest on its public debt has required it to make substantial net transfers abroad, and that roughly four-fifths of the central government's fiscal deficit is attributable to the

Table 9.1 Macroeconomic Performance
(percent)

Year	GDP growth rate[a]	Inflation rate[b]	Gross domestic investment/GDP	Gross savings/GDP
1968	9.5	2.0	19.7	17.5
1969	6.1	5.0	19.8	17.8
1970	7.9	13.3	20.1	17.3
1971	12.0	16.0	19.0	15.1
1972	5.0	17.4	20.9	17.0
1973	5.3	23.5	19.1	16.2
1974	8.3	28.5	21.6	14.9
1975	10.4	14.8	23.2	14.8
1976	9.4	17.0	25.1	18.0
1977	4.4	24.3	25.2	17.1
1978	2.9	43.8	18.7	15.0
1979	1.3	71.1	18.6	14.5
1980	1.3	103.4	21.9	13.9
1981	4.4	41.8	21.9	16.2
1982	5.3	27.6	20.5	17.1
1983	4.4	27.9	19.9	16.1
1984	4.0	50.2	19.6	15.9
1985	5.1	43.8	21.1	18.4
1986	8.8	30.6	24.9	22.2
1987	6.0	29.4	23.8	22.0

a. Real growth rates were calculated as the difference between the change in nominal GDP and the GDP deflator, divided by one plus the change in the GDP deflator.
b. Annual changes in the GDP deflator. Other price changes, such as the annual variation in the consumer price index, track the GDP deflator closely.
Source: National accounts of Turkey.

operations of state economic enterprises and the need to raise funds for their investment activities.[2] Moreover, throughout the 1980s the budget made no provision for financing the tax rebates paid to exporters. Consequently, these rebates were financed by an extension of credit from the central bank, which contributed to the relatively high inflation during this period.

The need to borrow funds from the central bank and commercial banks has contributed to a relatively high level of real interest rates that has stimulated the level of private savings and threatened the investment plans of both the private and public sectors. To maintain the private incentive to invest, the Ozal majority government continued to rely on an extensive and expensive battery of tax incentives. It only narrowly avoided a vicious cycle in which high real interest rates contribute to larger fiscal deficits, greater borrowing needs on the part of the public sector, and even higher real rates of interest. A central conclusion, however, of the World Bank (1987) study was that to maintain its rate of economic growth and at the same time achieve a lower rate of inflation, Turkey must find ways to reduce the size of the fiscal deficit. Given the difficulty of further expenditure contraction, the implication of this conclusion is that more rounds of tax reform are in store for Turkey.

Table 9.2 Central Government Taxes, Total Government Revenues and Expenditures

Year	Recurrent spending/GDP	Total spending/GDP	Total revenues/GDP[a]	Central government taxes/GDP	Central government budget deficits/GDP
1967	12.6	22.0	19.7	14.7	—
1968	13.1	23.2	17.9	14.5	—
1969	13.7	23.4	18.4	15.4	—
1970	12.2	21.8	22.4	15.8	—
1971	14.3	23.4	21.3	16.3	—
1972	12.3	20.9	20.2	16.9	—
1973	12.5	19.8	19.8	17.6	—
1974	11.5	22.6	17.1	15.9	—
1975	12.3	24.9	20.3	18.3	—
1976	12.7	24.5	21.4	19.2	—
1977	13.5	26.7	21.3	19.5	—
1978	13.5	23.0	24.1	14.3	—
1979	13.6	23.3	23.9	18.8	—
1980	12.6	24.3	21.4	17.3	—
1981	10.9	24.4	22.2	18.6	5.9
1982	10.9	23.0	20.0	17.8	1.8
1983	10.2	20.3	21.8	16.8	2.2
1984	8.9	18.9	20.0	13.0	2.1
1985	8.5	20.1	21.7	14.0[b]	5.0
1986	8.8	22.8	22.2	15.3[b]	2.8
1987	9.0	22.5	—	15.6[b]	3.2

— Not available.

a. Total revenues include tax revenues, nontax revenues (surpluses of state-owned enterprises, revenues from state property, interest income, fines and penalties), extrabudgetary funds, and revenues shared with provinces and municipalities.

b. VAT revenues are recorded on a gross basis with VAT refunds appearing as an element of expenditure. If net VAT revenues only were recorded, the numbers for 1985–87 should be reduced by about 0.8 percent of GDP.

Source: National accounts of Turkey, 1967–87.

Reform of Direct Taxes

Personal Income Taxes

Turkey has for some time applied the personal income tax through three separate tax regimes: certain groups and forms of income are subject to withholding; those whose income is not subject to withholding or who receive relatively large incomes are required to file an annual income tax declaration; and lump-sum or presumptive methods of assessment are used for certain groups that are hard to tax, such as small traders and entrepreneurs, although more recently, the authorities have also adopted presumptive tax techniques for annual income tax filers in an effort to curb their evasion of personal income taxes. Withholding currently applies to all wage and salary earners, to the value of all agricultural sales at rates of 2 percent for animals and 4 percent for all other types of sales, to the receipt of nongovernmental interest income (currently at 10 percent), and to rental income and professional fees at a current rate of 15 percent.

In 1986, at a time when the nonagricultural labor force was approximately 5.5 million and virtually no one in the agricultural sector paid any significant amount of tax, 4.1 million people paid personal income tax. Of those who paid tax, workers and salary recipients accounted for 66 percent, annual filers accounted for 20 percent, and those assessed on a lump-sum basis accounted for 14 percent. The share of total personal income taxes borne by different groups was 65 percent for workers, 30 percent for annual filers, and a residual 5 percent for those subject to lump-sum presumptive taxes.[3]

Small businesses and certain wage earners who are taxed by the lump-sum method are currently divided into five presumed net earnings brackets according to the amount of gross income they report. They are subsequently taxed at a rate of 25 percent on this assessed income, subject to a minimum tax payment. An estimation commission in each region with representation from local chambers of commerce establishes the presumed values for net income.

As 87 percent of those filing a tax return report income that places them in only the initial tax bracket and nearly 97 percent of annual filers only admit to taxable income in the bottom two tax brackets, the government felt compelled to introduce a system of living standards assessment in 1983 for self-employed workers in agriculture, trade, and the liberal professions. Taxable income for presumptive tax purposes was subsequently based on the ownership of airplanes, automobiles, boats, houses, and race horses; the use of personal servants; and travel abroad. Presumptive income established by these criteria establishes a minimum value of taxable income. In recent years, about 84 percent of those filing annual declarations have been presumptively taxed under this system.

As table 9.3 shows, personal income taxes accounted for about one-quarter of total tax revenues during the late 1950s and the 1960s. This share rose dramatically to more than 50 percent during the 1970s and early 1980s, and by the mid-1980s had fallen back to a level of about one-third of total tax revenues. Subsequent sections focus on the political and technical constraints that together have determined the size of the PIT base and on the issues of rate progression and the kinds of income most effectively reached by this tax.

Agricultural Income Taxes

When agriculture was readmitted to the PIT base in 1960, administrative problems in taxing farmers' incomes shared the limelight with the issue of the then generous exemption limits. The original definition of a tax-exempt small farmer, determined according to the size of farm and the amount of gross income, excluded somewhere between 80 and 90 percent of all farmers from the scope of the PIT in 1960. Larger farmers were divided into two groups. Farmers with gross incomes in excess of LT150,000 were self-assessed, while those with incomes in excess of

Table 9.3 Tax Structure Changes
(percentage of tax revenues)

Tax	1958[a]	1963[a]	1969[a]	1980	1986	1987	1988
Personal income tax	25.3	26.6	25.9	51.2	35.2	34.2	38.5
Corporate income tax	3.0	4.9	6.4	4.9	15.9	14.7	14.1
Defense tax and monopoly revenues (excise taxes)	12.2	12.7	11.9	3.9	*	*	*
Import production tax	6.5	10.5	5.3	2.2	*	*	*
Domestic production tax	9.9	8.1	8.5	*	*	*	*
Customs duties	5.4	7.5	12.4	3.2	4.8	4.7	4.6
Sugar consumption tax	5.3	4.3	3.7	*	*	*	*
Banking and insurance transactions tax	3.5	4.1	4.2	*	1.6	1.7	1.2
Stamp tax	4.0	4.6	3.6	4.3	6.4	*	*
Petroleum import duty	5.9	5.6	5.3	*	0.1	0.1	*
Fuel oil customs duty	1.1	2.0	0.7	*	*	*	*
Fees	1.9	1.6	1.4	10.9[b]	3.2	*	*
Real estate purchase tax	1.6	1.4	1.9	*	*	*	*
VAT on domestic output	*	*	*	*	16.6[c]	19.2[c]	*
Supplementary VAT	*	*	*	*	3.0	2.9	26.0
VAT on imports	*	*	*	*	8.8	11.1	12.0
Miscellaneous taxes	4.2	15.1	12.4	19.4	4.4	11.4	4.0

* Tax was not in force.
a. Other taxes amounted to 4.2 percent of total tax revenues in 1958, 15.1 percent in 1963, and 12.4 percent in 1969. They include a foreign travel expenditure tax, a transportation tax, a motor vehicle tax, an inheritance and gifts tax, and a service tax on telephones and telegrams.
b. Taxes on fees in 1980 also include taxes on services, such as the banking and insurance transactions tax.
c. This figure includes VAT rebates, which amount to about 3 percent of total revenue.
Source: For 1958, 1963, and 1969: Kryzaniak and Özmucur (1973); for 1980: IMF data; for 1986, 1987, and 1989: Ministry of Finance data.

LT500,000 were required to calculate their taxable income from balance sheets. Medium-size farmers, those with an annual income of less than LT150,000, were allowed to report presumed expenses in the amount of a deduction equal to 70 to 90 percent of gross income along with other deductions for interest, other taxes, and rents.[4]

Not surprisingly, farmers reported little or even negative amounts of agricultural income, and the small farmer exemption encouraged the subdivision of many farms. The coalition government that succeeded the military government after the 1961 elections more than tripled these exemption limits, and as a result excluded more than 95 percent of all farmers from the income tax net. Under pressure from the financing needs of the first Five-Year Development Plan (1963–67), the government lowered these exemption limits so that about 10 percent of all farmers were subject to tax. In 1980 the military government also reduced the farm size exemption limit by 50 percent, and thus increased the number of farmers subject to tax considerably. After vigorous farmer protests, however, the same military government restored the exemption limit to its original values a year later. It also established a maximum deduction of 70 percent for presumed agricultural expenses. This step was intended to be a stopgap measure until studies could be carried out to determine more accurate estimates of average presumptive costs for different products. It remains in effect today.

In 1964 the government felt compelled to introduce a minimum tax according to types of crops to improve revenue collection from farmers. This was a temporary measure until regional committees could develop measurements for a presumptive farm produce tax. However, these committees were never set up as planned, and in 1980 minimum taxes fixed by type of product became a permanent practice. Minimum taxes by types of crops that were fixed in 1964 were not adjusted for later price changes and became totally eroded by inflation. In 1980 the military government increased the minimum taxes considerably, but in subsequent years inflation again eroded them as the government refused to link these minimum taxes to farm product prices.

Agriculture was and still is a tax-preferred sector. After almost three decades of trying to apply the income tax to the agricultural sector, the number of farmers filing an income tax return has increased from less than 1 percent to only about 2 percent. This small measure of progress was achieved during a period when the share of the agricultural sector in GDP had shrunk from 35 to 17 percent and its employment share declined from 77 to 55 percent. At present, farmers' income tax returns represent only 5 percent of all income tax returns and contribute probably no more than 2 percent of the total revenues obtained from the personal income tax. Most of the revenues from farmers is obtained from the withholding tax on farm product sales that was introduced in 1980. In 1984 the option of presumptive cost deduction was extended to all farmers who had paid withholding tax on their sales.

Despite the lenient treatment of the agricultural sector under the income tax, leaping to the conclusion that farmers have generally benefited from government policies would be a mistake. For one thing, growers of export crops, such as cotton, tobacco, dried fruits, and nuts, have faced high, but hidden, implicit taxes on their products during the many years of overvalued exchange rates. Producers of staple products have also faced low procurement prices in years when the government sought to supply cheap food to the urban population. During the 1970s the value of transfers delivered to farmers through the vehicle of subsidized fertilizer prices was probably far more significant than either the explicit or implicit taxes they paid.

Other Income Tax–Resistant Groups

Another group that is hard to tax consists of small traders, artisans, and street peddlers. The challenge of taming the shrewd and squeezing some revenue from this group has presented a continuing problem for tax administrators. Originally subjected to a lump-sum professional tax in 1950, small traders lobbied successfully to get rid of this levy in the mid-1950s.

In 1960 small traders and artisans operating in fixed premises were incorporated into the income tax and made subject to a lump-sum tax, assessed by the tax administration. In subsequent years an array of measures, including the use of presumptive profit margins, the taxation of current year income, and minimum taxation has applied in an effort to curb widespread tax evasion in this sector.

Presumptive average profit margins have always formed an integral part of the Turkish income tax. It has been used to obtain more accurate tax assessments on all retailers, not only small ones. Local committees consisting of representatives from chambers of commerce, members of locally elected bodies, and some tax administrators periodically establish presumptive average profit margins in each province for different trades and services. A burden of proof provision exists for those taxpayers who claim that their actual profit margin is lower than the one that has been presumptively set.

Taxes on Labor Income

Taxes withheld on labor income have traditionally been the mainstay of the income tax system. Moreover, a relatively large fraction of this revenue is derived from wages and salaries paid in the public sector. Given a highly progressive rate structure and the absence of any important collection lags, the PIT became the dominant source of revenue during the inflationary decade of the 1970s. As table 9.3 indicated, the share of the PIT in total revenues increased from about 26 percent in 1969 to 51 percent by 1980. Table 9.4 shows the annual increase in real revenue as a percentage of GDP, which is attributable to the phenomenon of bracket creep. By 1980, for instance, bracket creep by itself was responsible for revenue gains worth nearly 9 percentage points of GDP. Table 9.4 also shows Turkey's continued reliance on the inflation tax as an important revenue source. Since the early 1970s the inflation tax on holders of real money balances has contributed revenues ranging from 2.5 to 5.0 percent of GDP.[5] The size of the inflation tax, of course, simply mirrors the magnitude of Turkey's fiscal deficit.

Labor markets responded to the appearance of bracket creep in an interesting fashion. To ward off its effects, employers concocted new salary policies to avoid the cost-elevating impact of higher marginal tax rates. As fringe benefits and other perquisites were tax exempt, employers used them generously to top up workers' base salaries. They increasingly used free housing in the workplace, free transportation to and from the workplace, free meals while at work, free apparel for work, business trips, business meals, and gifts of food and other goods to provide income in kind that would not be taxable to the worker, but that would be deductible as a cost for the employer. Tax administrators challenged these practices in the tax courts, but failed to limit the interpretation of business-related costs.

Table 9.4 Inflation as a Tax Collector
(percentage of GDP)

Year	*Increase in real revenues due to bracket creep*	*Inflation tax revenues*
1963	0.6	1.0
1964	0.2	2.1
1965	1.4	1.1
1966	0.7	2.1
1967	1.4	2.0
1968	0.5	1.0
1969	1.2	1.5
1970	0.9	1.8
1971	1.9	4.0
1972	2.3	4.1
1973	2.5	3.1
1974	3.3	3.1
1975	1.4	3.5
1976	2.0	2.4
1977	3.1	5.1
1978	5.8	4.8
1979	6.2	4.8
1980	8.9	3.5
1981	4.3	4.0
1982	3.2	3.4

Note: After 1982, adjustments in PIT schedules may have reduced total revenues by some 0.4 to 0.8 percent.
Source: Erbas (1988).

Employers also resorted to paying a double salary, one that was taxable and another that was disguised as a deductible cost or paid under the table. The government itself resorted to paying tax-exempt fringe benefits to reduce the impact of inflation on its budget outlays. Allowances for heating, chauffeured cars, and free clothes are but a few examples of the fringe benefits that were provided in the public sector. Both the private and public sectors began to furnish their workers with lavish fringe benefits to control their labor costs as inflation shifted more and more workers into higher income tax brackets.

Taxes on Capital Income

Unlike the treatment of labor income, capital income has tended to be lightly taxed under the provisions of the personal income tax. Only the corporate income tax (CIT) has served to place an effective burden on capital incomes in Turkey, and it is one that is hardly onerous by international standards. Within the personal income tax, the authorities have relied on withholding mechanisms to collect a modest amount of revenue from capital incomes distributed to individuals. With the introduction of the corporate income tax in 1950, dividends were presumptively taxed through the 15 percent levy assessed against after-tax corporate profits. If dividends were actually distributed, taxpayers could claim a credit against their PIT liabilities for the amount withheld at the corporate level. In the early 1960s this withholding rate was raised to 20 percent.

In 1980 the presumptive dividend tax was converted into an imputation and withholding scheme. Dividends a corporation actually distributed became subject to withholding at a 25 percent rate. At the same time, shareholders were required to gross up their dividend payments by 50 percent, and in turn became entitled to a 50 percent dividend tax credit. This approach to integration, however, proved difficult to administer, and most likely resulted in overintegration in the case of many firms. Responding to these difficulties and the fact that bearer shares had successfully escaped the withholding tax, in 1985 the government eliminated the withholding tax on dividends and the accompanying imputation procedures. Thus dividends are currently not taxable at the personal level.

By contrast, interest income has always been subject to withholding. The only exceptions to this withholding rule were interest received from government bonds and, until the early 1980s, interest received on small deposits. Because small deposits were exempted individually at each bank branch, investors responded to this loophole by distributing their deposits among several branches. This encouraged the proliferation of bank branches, a process already stimulated by the negative real rates of interest paid on deposit accounts.

In more recent years, the effect of inflation on the taxation of interest income has become more significant for tax policy purposes than the presence of a few exemptions. Withholding taxes on interest income were set at 20 percent in the 1970s, were briefly raised to 25 percent in 1980, and were subsequently reduced to 20 percent in 1982 and 10 percent in 1985. The authorities viewed these withholding taxes as final taxes, thus individual taxpayers do not report interest income as a component of their personal income and are not entitled to claim any credit against personal income taxes for the amount of tax withheld. However, during the 1970s, and even into the 1980s, inflation imposed a large hidden tax on recipients of interest income. For example, when inflation is running at around 40 percent, a 20 percent withholding tax on interest income when nominal interest rates are 50 percent effectively represents a 100 percent income tax on real interest income.[6]

Not until the 1980s did the government recognize the importance of obtaining positive real rewards on bank deposits. From negative real yields in earlier decades, bank interest rates therefore rose during the 1980s to ensure that savers received positive real returns and the withholding tax rate was reduced to 10 percent. However, the authorities have not made any serious efforts to index the taxation of interest income and make real interest income the tax base.[7] In 1986 withholding was extended to interest received on government bonds held by companies, particularly banks.

Capital gains are effectively exempted from personal income tax. Capital gains realized on shares of companies with more than 100 shareholders are not subject to personal income tax. If companies with fewer than 100 shareholders generate a share-related capital gain, the initial LT100,000 is exempt. Amounts above that are only taxable if realization has occurred within one year. However, if these gains are realized beyond a year after they have accrued, they too are exempt from tax. Capital gains from real estate transactions are also considered to be nontaxable if the property has been held for longer than four years.

In 1971 the government introduced a new tax imposing a burden on capital gains from real estate gains realized after four years. This tax was collected at a progressive rate schedule ranging between 15 and 50 percent, and allowed an exemption for 10 percent of the sales value. Along with the 7 percent real estate purchase tax, this new capital gains tax produced significant lock-in effects in Turkey's real estate market that worsened as inflation grew during the 1970s. Underdeclaration of sales values subsequently became a serious problem for tax administrators. These administrative headaches help explain the elimination of the capital gains tax on real property in 1984.

Income Tax Returns

Turkey has always operated its personal income tax with a highly progressive nominal rate structure. At its inception in 1950, PIT brackets ranged between marginal rates of 15 and 35 percent. In 1955 and the early 1960s this range was elongated and encompassed marginal tax rates ranging from 10 to 60 percent. The severe inflationary pressures of the 1970s severely eroded and compromised this rate structure. By 1980 even relatively low income workers were being taxed at the top rate of 60 percent and the real value of their personal exemptions had virtually disappeared. All the weak coalition governments that ruled during the 1970s seemed powerless to adjust tax brackets and other nominal tax magnitudes for the effects of inflation.

By 1980 addressing the issue of inflation-induced bracket creep had become a top priority for tax reform. The authorities ruled out a swift return to the preinflationary status quo for revenue reasons. Instead, they implemented a gradual, phased approach to the reduction of nominal tax rates. The new income tax schedule, which was introduced in 1980, began with an unusually high first bracket rate of 40 percent and, partly to make up for a revenue loss, extended the top marginal rate from 60 to 66 percent. Despite its relatively high level, the 40 percent rate on the first income bracket actually represented a reduction in the average tax rate for most taxpayers, who had been pushed into the previous top marginal rate of 60 percent. Between 1981 and 1985 the authorities reduced both the top and bottom marginal rates, eventually arriving at a range of rate progression between 25 and 50 percent. Moreover, in 1981 they set the personal exemption level equal to the value of the minimum wage in that year; however, inflation reduced its real value until in 1986 it was only one-quarter of what it had been worth in 1982. Subsequently, personal exemptions were repealed.

Rather than restore the personal exemption level to its original real level, the Ozal government chose to introduce an expenditure rebate system to strengthen the complementarity between the existing PIT and the new VAT. Under this system, wage earners were entitled to tax rebates upon presenting their VAT receipts for the purchase of certain consumption items to their employers, who submitted refund claims to the local tax office. By giving wage earners a strong incentive to request and obtain receipts from retailers, the tax authorities had hoped to achieve the dual objectives of diminishing the tax burden on low-income workers and encouraging the issuance of sales receipts. In the 1970s the tax authorities had used fines and penalties to induce sellers to provide adequate documentation of their activities. This attempt was largely unsuccessful. In the 1980s Turkey abandoned the fiscal stick and replaced it with a fiscal carrot in the form of the expenditure rebate system. This system seems to have improved compliance with both the income tax and the VAT.

The rebates were initially confined to wage earners and essential consumption items. Over time, however, these benefits were generalized and liberalized to include all taxpayers and most consumption items. Rebate rates declined as the level of taxpayer income increased. Initially the average rebate represented about 6 percent of total expenditures for a typical minimum wage earner, but later increased to about 12 percent as the rebate was extended to a wider range of expenditures. In 1989, in an attempt to achieve better targeting of the expenditure rebates, the government trimmed the list of rebatable items by making consumer durables and food purchases in excess of LT100,000 ineligible for rebate.

The World Bank (1987) review of Turkey's tax system accepted the pivotal role of the expenditure rebate system in supporting the new VAT, but felt that it was a poor substitute for personal exemptions and should eventually be abandoned. Currently, minimum wage workers in the initial (25 percent) tax bracket can enjoy a 15 percent rebate, which after a few personal deductions gives them an average tax rate of 7 to 8 percent. The Bank criticized the rebate system on other grounds, namely: it included more items than the VAT covered, its ability to monitor receipts was weak, and it constituted a serious revenue drain. Moreover, the fact that the lowest rebate rate was less than most of the VAT rates created strong incentives for buyers and sellers to split the difference and bypass the VAT altogether.[8]

The Bank also had reservations about the rebate system based on its alleged disincentive to save. However, there is no inherent bias against saving and the Bank's concern on this score seems to have been misplaced. The point is that a dollar saved and invested at a normal return for the sake of tomorrow's consumption will obtain an even larger rebate than if it is spent now, but one whose present value should be about equal to the value of today's rebate. Contrary to appearances, the rebate system does not violate the requirements for intertemporal neutrality.

Table 9.5 Tax Incidence Estimates by Household Location and Income Bracket, 1968
(tax burden as percentage of disposable income in each income group)

A. Household location	Indirect tax burden	Direct tax burden	Total burden
Urban households	13.8	11.4	25.2
Rural households	12.2	0.3	12.4
All households	13.0	6.1	19.1

B. Annual income bracket (millions of LT)	Percentage of households	Average indirect tax rate	Average direct tax rate	Total tax rate
0–2,499	19.1	13.3	0.0	13.3
2,500–3,999	14.2	13.2	1.3	14.5
4,000–5,999	14.8	13.4	1.3	14.7
6,000–9,999	20.8	13.6	2.1	15.7
10,000–14,999	12.5	13.6	2.9	16.5
15,000–24,999	9.1	16.1	4.5	20.6
25,000–74,999	8.0	14.1	5.3	19.4
75,000 and over	1.5	14.5	20.4	34.9
All households	100.0	13.9	6.5	20.4

Note: The methodology for panel B assumes complete backward shifting of direct taxes and complete forward shifting of indirect taxes.
Source: Panel A: Krzyzaniak and Özmucur (1974); panel B: Krzyzaniak and Özmucur (1973).

Attempts to measure the incidence of taxes in Turkey have been few and far between. Those that do exist are now out of date. For what it is worth, table 9.5 presents an estimate of tax incidence in 1968. At that time, the burden of direct taxes represented a little more than 11 percent of urban household income and a negligible fraction of the income of rural households. Estimates of average PIT rates in relation to the size of income, shown in panel B of table 9.5, suggest that as income rises, the direct tax burden increases moderately until in the highest income bracket, which represents 1.5 percent of all households, the average tax rate reaches a value of slightly more than 20 percent. Since 1968, however, the PIT has unquestionably become less progressive as more and more households were catapulted into the highest rate bracket during the 1970s. Inflation transformed the PIT into something more akin to a flat rate tax with no real tax exemptions. As mentioned earlier, however, the reforms of the early to mid-1980s helped to restore a greater degree of progressivity to the distribution of PIT burdens despite the reduction in rates applied to interest income.

However, as inflation accelerated in the late 1980s back toward the levels of 70 percent reached in the late 1970s, the familiar problem of bracket creep and the resulting inequities have resurfaced as an important policy concern. For instance, unpublished estimates of effective burdens by the World Bank's resident mission in Ankara suggest that the effective tax rate on nonagricultural paid labor incomes rose from 19 to 32 percent between 1987 and 1989, while the effective tax rate on other nonagriculture incomes fell from 11 to 9 percent during the same period.

Tax Administration

In recent years the authorities have made concerted efforts to reduce the length of the collection lag in the personal income tax and to prevent any inflationary erosion of revenue yields that might result from relatively high inflation rates. In 1980, for example, the authorities introduced the principle of paying taxes on the basis of the current year's accrued income for traders and self-employed professionals. However, this turned out not to be administratively feasible, and the government abandoned the attempt in 1981. Nonetheless, the government came back in 1986 with a renewed effort to tie this year's tax payments to this year's income. Under the new approach, an amount equal to half of the monthly VAT tax collections was required to be withheld for income tax purposes and could be credited toward payment of the annual income tax. After widespread complaints, this approach toward withholding was canceled in 1988 and in its place the government introduced a new system of advance payments. Currently, tax payments for the current year are made in monthly installments and are equal to half of the income tax paid in the previous year subject to a certain minimum.[9] For annual tax filers, the minimum tax payment is set in accordance with their particular living standard indicators. In addition to being an effort to shorten the length of collection lags, the minimum tax principle is also an indication of the tax administration's frustrations about tax evasion.

In general, the Ozal government has recognized the vital importance of tax administration in assessing liabilities and collecting revenues. On taking office, it immediately set aside resources for 25,000 new inspection posts and began the critical task of computerizing many of the tax administration's functions.[10] Turkey has an adage that no tax reform is needed; implementing existing taxes effectively would suffice. In the early 1970s, Shoup (1972), after a short mission to Turkey on the introduction of a VAT, concluded that the tax administration's inability to collect all the taxes was actually a blessing for the Turkish economy, for if adequately implemented the tax system could capture most of the national output. Nevertheless, successive finance ministers have followed the time-honored habit of introducing new taxes to increase revenues as the administration fails to collect the existing ones.[11] Aware that revenue buoyancy now greatly depends on administrative efficiency in tax implementation, governments in the 1980s introduced several measures to strengthen tax administration and revamp tax litigation procedures. According to Ministry of Finance officials, the ratio of collections to assessments rose from 80 to 90 percent during 1984–88.

One of the major factors behind the tax administration's inefficiency has been the fragmentation of tax collection powers into departments that do not communicate with each other: tax administration is nearly totally cut off from the tax audit board, which operates directly under the minister of finance; litigation procedures are handled by the administrative judiciary, a body lacking in expertise in tax matters; and from the early 1930s to 1984, a separate ministry (Ministry of Customs and Monopolies) collected foreign trade taxes.

In their efforts to increase tax revenues, most governments neglected the option of improving the efficiency of the tax administration. Despite their critical role in the public sector's financial equilibrium, until recently tax administration personnel have remained inadequately recruited, poorly trained, and blatantly underpaid and undermotivated. Although tax offices created in 1950 to accompany the Tax Procedure Law stand as a landmark in the modernization of the tax administration, the new tax offices failed to develop their capability to respond to the challenge of an increasingly complex and overburdened tax system. Many petty tax officers still engage in selling bookkeeping services to taxpayers or taking graft in exchange for tax hints to top up their lamentably low salaries. In recent years, tax administration staff have been offered a benefit program, of which the main component is free housing.

A school of finance set up and run by the Ministry of Finance was the major source of recruitment for the ministry's mid-level civil servants, but educating the students was a secondary job of the senior staff of the ministry's senior staff, the curriculum often did not keep pace with new developments, and materials were frequently not relevant to the practice of modern tax administration. As universities expanded around the country, the school eventually lost its raison d'être as an institution for supplying skilled labor tailored to the needs of the Ministry of Finance.

Starting in 1971, the audit capacity of the tax offices was strengthened by creating positions for tax auditors, who were recruited from among university graduates, often after passing a special examination. Until recently, the tax administration did not conduct any formal in-house training programs. Since the introduction of the VAT in 1985, it has been endowed with a development fund and has started an in-house training program at two levels: a short program for provincial finance and tax office directors and another program for rank-and-file functionaries. By the end of 1988 more than a third of all tax administration functionaries had gone through a formal training program.

Computerization of the tax administration started in the early 1970s, but made little progress until recent years. A German technical assistance program provided hardware and software to the tax administration and trained some of its staff members in German tax administration. However, low government salaries relative to those of the private sector, as well as the administration's reluctance to initiate and accelerate computerization of its operations, have been major factors in the loss of computer-trained tax officers to the private sector. Not until 1987 could tax transactions be carried out with computers in certain tax offices. Now that the computers of all provincial tax administration offices are linked to both the regional and central systems, the bottleneck lies in generalizing the transfer of data into the network. The use of computers to detect tax dodgers and discover hidden tax bases through cross-checks is still at an early stage.

The customs administration is responsible for clearing goods through the customs gates and assessing their cost, insurance, and freight (CIF) values. The first task depends on the clearance capacity at customs gates and surveillance capacity near borders. Smuggling has always existed, but distortive foreign trade and tax regimes have sometimes contributed both to smuggling and to corrupt practices by customs officers. Overvalued exchange rates, export taxes, and bans have stimulated illegal exports that sometimes bypass custom gates altogether and sometimes pass through custom gates with fraudulent documents. Import shortages, prohibitions, and high import duties have increased the attractiveness of smuggling with or without the cooperation of customs officers. In the 1980s more realistic exchange rates, the economy's increased openness, and some moderation in custom duty rates attenuated the pressure for customs fraud. Nevertheless, the government policy of allowing export subsidies in the form of tax rebates generated fictitious exports, with fraudulent papers being arranged at customs gates with the cooperation of tax officers. Products of little or no value were declared as high-value exports to get proportionally high export refunds. Export tax rebates and extrabudgetary lump-sum levies on imports and exports dramatically increased the falsification of papers.

When ad valorem tariffs became dominant with the reform of 1954, the assessment of CIF values at competitive border prices became the major task of the customs administration, but the procedure is difficult when international prices and exchange rates fluctuate continuously. Only gross departures from world prices can be easily investigated given the difficulty of identifying small variations and the expense of investigations. Computerization of customs clearance procedures made no progress until recent years. The lack of control after merchandise has been cleared through customs is a major factor in a tax officer's cooperation with fraudulent exporters and importers.

In 1960 the authorities introduced several procedure and penalty measures to improve compliance with the tax laws. In 1960 they made wealth declaration part of annual income tax returns, and in 1964 added disclosure of personal and family expenditure. When the statements of accumulated personal wealth and family expenditures were inconsistent with a taxpayer's declared and tax-exempt incomes, the difference was taxable. Finally, all taxpayers were obliged to hang tax certificates at their premises showing their tax bases and income taxes paid in the last five years. The authorities also sought third-party cooperation in discovering hidden tax bases by extending to all taxes a time-honored practice used in collecting custom duties: rewarding the informant with a fraction of the resulting tax collected. Right-wing parties promised to abolish the compulsory annual wealth declaration because of its negative impact on entrepreneurs' motivation. The Demirel government did so in 1967, but the president vetoed the bill. The declaration remained in effect until 1984, when the Ozal government abolished it. As a compensatory measure, the Ozal government introduced a minimum tax in the form of the living standards assessment system for income tax payers filing annual returns.

The Tax Audit Board was founded in 1945 under the direct auspices of the minister of finance. It recruits able university graduates through a highly competitive examination. It is a prestigious and elitist board, with a strong tradition of in-house, master-to-apprentice training of its recruits. It sets up its own audit program all over the country, but given its small number of staff, cannot audit more than a tiny fraction of annual returns. As the Tax Audit Board is completely outside the tax administration system, the files it picks up do not necessarily correspond to the priorities of tax offices. The board's elistist tradition has restricted the number of board members; accordingly, they earn prime salaries that reflect their scarcity value to the private sector. The Tax Audit Board, together with its rival, the Board of Inspectors, which focuses more on internal auditing, has been a nursery for the recruitment of directors in the public sector. Because of this drain, the percentage of active senior members of the board has declined over the years, as has its in-house training capacity.

In 1971 another corps of auditors was created directly in tax offices under the General Directorate of Revenue. Their number increased rapidly and they brought professional auditing and expertise to local tax offices. In the 1980s more power was delegated to controllers at the level of tax offices to increase the capacity for routine reviews of assessments. Despite the increased capacity, even companies that benefited from generous tax exemptions were not regularly audited. As table 9.6 shows, inspection officers of tax offices carry out the bulk of the inspection of returns (more than 90 percent).

Despite efforts to increase the percentage of returns reviewed, the present coverage represents only 3 to 4 percent of all income and corporate tax returns. The percentage is much lower when one takes other tax returns into account. The administration is currently trying to enhance the coverage and selection of returns by more extensive computerization of files.

Additional assessments made as a result of audits used to represent, on the average, more than 100 percent of the declared tax base. In recent years, however, the percentage of discovered additional tax bases has declined to somewhere between one-fourth and one-third of declarations.

Table 9.6 Tax Audits and Additional Assessments, 1987

Responsible party	*Number of audits*	*Tax base covered (LT millions)*	*Additional assessment (LT millions)*
Board of Inspectors	457	7,988	9,244
Board of Auditors	3,224	2,250,983	293,179
Revenue comptrollers	2,585	414,934	286,476
Tax review officers	73,998	275,716	175,491
Total	80,264	2,949,622	764,390

Source: Ministry of Finance, General Directorate of Revenue data.

More investigation would be needed to reach strong conclusions about any voluntary improvement in compliance in the submission of self-assessed returns.

The authorities revamped tax litigation institutions and procedures in 1982. Before this reform, an administrative complaint commission and a two-layer judiciary system comprised a three-level approach to tax conflicts. At the first level, that of the administrative or tax complaint commission, two officers from the tax administration and two from locally elected bodies (the municipality and the chamber of commerce) would review the case at a taxpayer's request. The decision could be reported to the tax appellate commission. The final court of appeal was the administrative tribunal.

A taxpayer's objection used to stop the recovery of the tax automatically, and it could not be collected until the commission rejected the taxpayer's demand. This safeguard, meant to protect honest taxpayers from administrative abuses, was often abused by taxpayers to defer their payments. The tendency to contest assessments increased the backlog of files in the tax complaint commissions, prolonged average settlement time, and thus encouraged taxpayers to file even more complaints.

The authorities have introduced several measures to speed up the tax litigation process and to dilute the incentives to file complaints. First, in 1982 permanent judiciary courts replaced the administrative complaint commissions. Second, to reduce the burden on the Supreme Administrative Court, regional administrative courts were established that had the final say on complaints involving small amounts. Third, in 1985 tax presettlement commissions were set up in local tax offices, where the officer who had reviewed a particular case attempted to reach agreement on the amount that the taxpayer would agree to pay. Finally, the penalty on deferred payments has been regularly increased above the market interest rate to deter taxpayers from using taxes owed as a cheap line of credit. The increase in the capacity for complaint resolution has substantially reduced the backlog of complaints and the average settlement time.

Despite these improvements, the government did not abandon the old practice of allowing total or partial tax amnesty on the condition that taxpayers pay the principal of their debt before a deadline. This practice has been reinstated several times in the postwar period to increase immediate revenue and to reduce the burden of complaints in the courts. The administration thus condoned the behavior of tax dodgers relative to honest citizens who complied with their obligations on time. After the dramatic reduction in the volume of complaints, the government introduced its last amnesty in 1988 to obtain a quick revenue gain.

Corporate Income Tax

The corporate income tax was introduced at the same time as the personal income tax, but the authorities have made only limited and sporadic efforts to integrate these two tax bases. As table 9.3 shows, the yield of the CIT was relatively stable from the 1950s to the mid-1960s at around 4 percent of total tax revenues. Since 1980, however, the yield has risen substantially, reaching a level of nearly 15 percent by 1987, although by 1989 it had declined to 14.1 percent. To a large extent this trend toward higher yields reflects a growing tax base that resulted from the increasing attractiveness of the corporate form of business and the inclusion of state-owned enterprises in the tax base. To a certain extent, however, the growth in revenues is also attributable to a steady increase in nominal corporate tax rates.

The CIT is assessed against the profits of limited liability partnerships and also the business activities of legal entities such as associations, foundations, and state bodies that do not distribute profits, but engage in some form of business enterprise.[12] Partnership profit distributions, whether in the form of dividends or profit shares, are also taxed at the personal level. However, the profits of personal partnerships having unlimited liability are taxed only once as the personal income of the partners. Originally, state-owned enterprises were excluded from the scope of the corporate income tax. During the 1970s they were brought into the CIT base, but were differentially taxed at concessionary rates. In 1980, however, the concessionary rate was repealed, and now both private and public enterprises face the same nominal tax rate. From the beginning, cooperatives have been exempted from the corporate income tax. This exemption has contributed to the proliferation of cooperative forms of business activity, many of them weak and badly managed, especially in the housing sector.

For most of the postwar period, the relatively low corporate tax rates have placed incorporated businesses at an advantage in accumulating capital compared to personal businesses. Nominal tax rates rose from 10 percent in the 1950s to 20 percent in the 1960s, 25 percent in the 1970s, and subsequently wobbled around much higher levels in the 1980s. In 1980 the government doubled the nominal rate to 50 percent, but reduced it to 40 percent a year later, and it is currently set at 46 percent. Since 1987 the government has tried to encourage wider ownership of corporations by offering lower rates to more widely held businesses. Thus a special 40 percent rate applies if 25 percent of the company's paid-up capital is held by more than 200 individuals, and even lower rates of 35 percent and 30 percent are applicable if the paid-up capital shares increase to 51 percent and 80 percent, respectively.

During the 1950s the corporate form of enterprise was more advantageous than a personal business despite the relatively low 35 percent bracket rate that was applicable to personal incomes. In addition to the 10 percent corporate rate, a 20 percent PIT was withheld on profits in the following year, whether retained or distributed as dividends to shareholders. Shareholders could claim the amount withheld at the corporate level as a tax credit against their PIT liabilities. On undistributed profits the total tax rate was thus equal to 28 percent.[13] In 1960, however, authorities doubled the CIT rate and did not allow any credit at the personal level for the amount of tax withheld on dividends at the corporate level. At the same time they increased the top marginal PIT rate from 50 to 60 percent.

In the 1970s inflation-driven bracket creep in the personal income tax increasingly enhanced the attractiveness of the corporate form of business. A deadlock in the legislature prevented any adjustment in the income tax schedule to correct for the escalation of the tax burden on real personal incomes. Toward the end of this period, even small firms opted for the incorporated form of business to reduce their tax burden, which became a maximum of 40 percent when all earnings were retained.[14] This rate was substantially lower than the nearly 60 percent average income tax rate that had become applicable to even modest profits generated by unincorporated businesses.

In 1980 the authorities doubled the CIT rate to 50 percent and provided a 50 percent tax credit for the dividend recipient. A year later, however, they lowered the corporate tax rate to 40 percent, and as a result incorporation became advantageous at all income levels. Finally, in 1986 they elevated the CIT rate to 46 percent to compensate for the elimination of dividend taxes at the personal level. Dividends had become increasingly hard to reach because of the growing use of bearer shares. The dividend tax credit also disappeared at this time and was replaced by a 10 percent withholding tax on all of a corporation's after-tax profits.

These changes amounted to a partial return to the schedular income tax system that had existed before 1950. Interest income was now taxed only at the personal level at a relatively low flat rate of 10 percent. Equity income, by contrast, was now taxed only at the corporate level, and the top marginal income tax rate became closely aligned with the nominal corporate rate at around 50 percent. Shareholders in lower tax brackets became worse off relative to those who were subject to the highest personal tax rate. As a result, small shareholders' incentive to purchase shares was weakened. Moreover, the growth of generous tax incentives during the 1980s increased the attractiveness of the corporate form of business.

Until 1984, intercompany dividend distributions were exempt from the corporate income tax if the parent company's share was not less than 10 percent of the participating firm's equity and had been held for at least the previous two years. Because of its inducement to corporate concentration, the authorities removed the minimum share requirement for enjoying exemption in 1984 and all transfers of income between corporations became tax exempt. In the same year all forms of corporate reorganization became exempt from capital gains tax.

In recent decades several factors have increasingly encouraged corporate firms to splinter from their parent companies, and even to restructure some of their departments into new companies. These factors include the deferral of PIT withholding, the absence of corporate tax cascading on intercorporate profit transfers, the tax savings realized through transfer pricing, the evasion of official price markups, the enhanced eligibility for bank credits and foreign exchange allocations, and the prestige that individuals may garner from owning many companies. Dormant or invoice-producing shadow companies have also proliferated as instruments for transfer pricing and rent seeking.

More recently, two factors have restrained the rapid growth of corporations. First, the liberalization of imports and of the pricing of public sector outputs have reduced the scope of rent seeking activities, although opportunities for tax evasion through the manipulation of invoices by parent firms have continued. Second, and more important, in 1988 the government applied a minimum tax to all corporate firms. Now, regardless of their profit or loss situation, all companies must contribute a monthly provisional tax that is creditable toward their annual income tax. The introduction of this minimum tax has prompted a wave of liquidation of shadow companies.

Also working in the direction of corporate consolidation was a long-standing tax provision that allowed the losses of acquired companies to be offset against the taxable income of acquiring companies. This provision stimulated a rash of mergers in the 1970s and 1980s, when a number of companies that had been financed by the Turkish Investment Bank, and which were in large part underwritten by the foreign remittances of Turkish workers, became unprofitable.

Despite a two-digit inflation rate that has lasted for more than two decades, Turkey has been slow in introducing a form of inflation accounting or indexation for determining of the appropriate size of the CIT base.[15] In 1983, however, firms were allowed to revalue their fixed assets and to add the extra value to their equity capital without incurring any tax obligation. This asset revaluation also reduced the tax base because of the higher depreciation charges it triggered. In addition, it reduced the impact of capital gains tax whenever revalued assets were sold. It thereby temporarily eliminated the lock-in effect on firms' physical assets. This revaluation measure therefore increased the mobility of physical assets and facilitated their allocation to more profitable uses. In 1987 complete inflation adjustment of depreciation allowances became law. Nonetheless, inventory valuation is still done on a historical cost basis that magnifies the size of the corporate tax base under inflation. Last in, first out inventory accounting is not permitted as a way of eliminating the inflationary overstatement of profits.

Turkey has allowed inflationary adjustments for assets, but after some consideration rejected a corresponding adjustment for liabilities. Indexation, therefore, remains only partial in scope. Consequently, the deductibility of nominal, rather than real, interest expenses reduces the size of the CIT base in proportion to the rate of inflation.[16] Although Turkish companies are not highly leveraged by international standards, this feature has only added to the incentive to rely upon debt rather than equity finance.[17]

Prior to the 1980s, when government-regulated interest rates became increasingly negative under inflationary conditions, enterprises increased the debt component of their capital base. In closely held corporations, shareholders displayed a strong preference to lend their own money to firms rather than increase the amount of equity capital. Larger company groups set up their own banks or borrowed their capital by issuing bonds rather than increasing their equity base. Not only did these financial practices erode the CIT base, they frequently led to the payment of interest on bearer bonds that went undeclared on PIT returns. Moreover, since 1978 firms have been able to deduct any losses they incurred on their foreign currency debts. However, these same firms can still revalue assets that have been bought with foreign currency. This serves to make the foreign currency loss deduction redundant. Prior to 1978 firms were also insulated from the risk of devaluation under the fixed exchange rate system, which provided a strong inducement for external borrowing.

Whereas inflation has historically been an important determinant of effective corporate tax rates, mainly through its influence on corporate debt and depreciation deductions, a panoply of tax incentives has been at least as significant. The SPO gives out investment incentive certificates to eligible firms, and also monitors, to some extent, the performance of tax-preferred investments. A rich variety of incentives exists to promote the growth of special sectors, less developed regions, exports, and investment in general. The abundance of investment incentives is perhaps matched only by the rich variety of forms that they assume. Currently, accelerated depreciation is available to all firms in the form of a general investment allowance of 30 percent, and other generous allowances are permitted for specific investments. The allowance rate is boosted to 40 percent for investments in agriculture and in areas of general regional development; to 60 percent for investment in priority regions; and to 100 percent (complete expensing) for investments in scientific research and development, tourism, communications, and energy. On top of this, all firms are permitted to use 50 percent declining balance depreciation on the indexed cost base of new equipment.

Moreover, 18 percent of sales revenue may be deducted from taxable income for exports of industrial goods provided that annual total sales revenue is greater than $250,000. Exporters who do not produce the goods but market them, receive one-quarter of the value of this deduction for themselves. Similarly, 20 percent of the revenues from tourism activities that are obtained as foreign currency may also be deducted from corporate taxable income. Through a combination of these deductions, most firms should be able to "zero out." If not, two other incentives may help them to do so. Under the terms of the financing fund, one-quarter of this year's taxable income may be deducted to finance current investment expenditure and is subsequently added back to next year's taxable income. In addition, in 1988 the authorities established a five-year tax holiday (ten years in the poorest regions) from both corporate and personal income taxes for taxpayer investments in educational, health, and sports facilities.

Tax Incentives

The economic history of tax incentives in Turkey is replete with examples of how the authorities have viewed the tax system as a potent instrument for shaping economic development. The delicate issue of striking a balance between the goals of resource mobilization and incentive generation was a central theme in tax policy discussions during the financing of the first development plan. The first team of planners recruited by the military government stressed the resource mobilization goal, as did Kaldor (1962) in his report on the reform of the Turkish tax system. Later, however, the coalition government and the Tax Reform Commission set up in Istanbul stressed the importance of incentive generation by granting selective tax incentives. In an internal report on tax reform commissioned by the minister of finance in 1962, Professors Bernd Senf and Gunter Schmolders from Germany also advised a policy of growth promotion through tax incentives for savings and investments. In their reform efforts, successive governments have inconsistently tried to satisfy both goals by riddling a broadly based corporation income tax with numerous exemptions and deductions and attempting to recoup the revenue tax from several narrowly based new taxes.

The first tax incentive package for economic development was introduced in 1964. It included an investment allowance, the postponement of import duty on machinery and equipment, an allowance for accelerated depreciation, a longer loss carry-forward period, and a lump-sum tax rebate on exports. The investment allowance was deductible in addition to normal depreciation. Except for depreciation and the carry-over of losses, the encouragement concessions were not automatically available and were granted only on a discretionary basis by the SPO, which issued an encouragement certificate upon approval of an investment project submitted by the beneficiary.

The SPO's discretionary power to decide whether encouragement measures would be granted to an investment project was severely criticized by investors. Refusal of an incentive certificate to certain investments, particularly on the grounds of the existence of excess capacity, prompted strong and justified protests from investors. Past practice amounted to the blocking of new entry and competition, but later the SPO implemented a more liberal policy in issuing encouragement certificates, although entrepreneurs continued to complain about the SPO's bureaucratic delays and its abuses, inconsistencies, and even corrupt practices.

Tax incentives are targeted regionally, sectorally, and according to project size by an annual list that specifies the exact conditions for eligibility. The SPO generally favors large projects and only grants incentives to investments above a certain size. This attitude stands in sharp contrast to certain World Bank loans recently granted to Turkey that are reserved for small and medium-size enterprises.

Tax incentives for economic development are, for the most part, designed to reduce the cost of physical capital relative to labor. Only the regionally differentiated personal exemption, which is eight times higher in the poorest regions than in the richest regions, works in the opposite direction. However, reducing capital costs by means of such measures as the investment allowance, accelerated depreciation, foreign currency allocations of overvalued lira, and the treasury's assumption of the lira's depreciation risk have encouraged investors to overspend on machinery and equipment. This policy bias in favor of capital use was further accentuated by high taxes on labor income, foreign currency allocation on the basis of physical capacity, gradual strengthening of labor unions in the formal sector, and tax-induced labor migration from the formal to the informal sector. Only in 1985 was a limited income tax exemption introduced for industrial employment created in certain less developed regions.

In the early 1980s the government introduced other tax incentives to boost exports. The cornucopia of export incentives included export subsidies in the form of input tax rebates, a tax exemption on export credits, the exemption of corporate profits from construction work carried out by Turkish contractors abroad, and a 20 percent deduction from corporate earnings on exports of manufactured and agricultural products and from the foreign currency proceeds of freight and tourism services.

In the 1980s the central bank granted generous lump-sum tax rebates to exports on the provision of evidence that export earnings were transferred to Turkish banks. This export subsidy induced fraudulent and overstated export claims. To cash in on high tax rebates, exporters stepped up the declared value of their exports or made fictitious export claims by getting fraudulent shipping documents from customs. Lump-sum tax refunds on exports became even more attractive when the VAT was introduced with its own mechanism of tax refunds under zero rating. Foreign currency corresponding to either fictitious or overpriced exports was purchased from the remittances made by workers abroad or from tourists in Turkey. As exports expanded, so did the overpriced and fictitious exports. The central bank kept on paying tax refunds through monetary expansion as no budget appropriation existed for them. As the abuses became too evident and widespread, as central bank financing created unsustainable monetary expansion, and as the pressure of dumping accusations by foreign countries accumulated, the government was compelled to phase out lump-sum tax refunds on exports starting in 1989. Currently, the only tax refund linked to exports is provided by the zero rating under the VAT on the submission of invoices, and these payments are made out of the budget.

As an upper bound, the tax administration has calculated the cost of the incentive legislation in terms of revenue loss as 1.2 percent of GDP. With the rich assortment of investment and export incentives, the average effective tax rate on enterprises is much lower than the present statutory CIT rate of about 50 percent would suggest. In a sample of twenty profitable high-income corporations taken in 1988, nine paid no taxes, seven paid taxes of less than 10 percent of profits, and four paid taxes of just about 10 percent of profits. In total, they paid LT6 billion out of a total of LT150 billion in profits, which represents an average tax rate of 4 percent. Given that CIT revenues as a percentage of GDP are currently about 2.5 percent, if corporate profits represent, on average, about 12 percent of GDP, the average corporate tax rate for the economy as a whole is roughly 20 percent. In many sectors of the economy, average tax rates are, of course, much lower than this figure. In 1988, for example, the top twenty banks in Turkey paid an average effective corporate tax rate of only 7.6 percent.

Of greater relevance to investment behavior than the average rate is the pattern of marginal effective tax rates (METRS) that reflects the joint influence of corporate tax provisions and tax incentive legislation. Table 9.7 presents some recent estimates of METRS. A negative number in the table indicates that the investment is on balance subsidized rather than taxed through the revenue system. Table 9.7 leaves little doubt that METRS vary widely according to the type of asset purchased, the region in which the investment activity occurs, the mode of finance a firm selects, the underlying rate of inflation in the economy, and the firm's ability to enjoy the benefits of various tax incentives.

One can draw a number of implications from the estimates in table 9.7. First, investments in machinery are highly favored over investments in buildings, a bias in the composition of investment that favors greater capital intensity at the expense of more employment. In a number of instances machinery investments received a net subsidy from the tax system. Second, debt is highly favored over equity as a source of finance for investment spending. Third, investments in priority regions face a substantially lower tax burden than those in developed regions. Fourth, inflation in the absence of any indexing raises the METR on investment, but does so less as the debt-equity ratio increases. Fifth, by themselves tax incentives have increased the dispersion in METRS and may therefore have en-

Table 9.7 Estimates of Marginal Effective Tax Rates on Machinery and Buildings by Type of Region and Rate of Inflation

		Industry type					
		Negative[a]		Normal[a]		Incentive[a]	
Type of region	Type of asset	Inflation 0%	Inflation 50%	Inflation 0%	Inflation 50%	Inflation 0%	Inflation 50%
Total equity financing							
Developed region	Machinery	2.3	41.4	–39.1	–7.2	–57.2	–39.1
	Buildings	36.6	43.3	17.7	41.1	8.6	22.5
Priority region	Machinery	–24.0	9.1	–78.2	–41.2	–112.8	–58.4
	Buildings	27.6	41.4	11.3	32.3	–1.9	20.0
One-half equity financing							
Developed region	Machinery	–24.0	9.1	–78.2	–41.2	–112.8	–58.4
	Buildings	27.6	41.4	11.3	32.3	–1.9	20.0
Priority region	Machinery	–347.5	–203.5	–347.5	–203.5	–347.5	–203.5
	Buildings	–73.3	–14.2	–73.3	–14.2	–73.3	–14.2

a. The State Planning Organization classifies industries as negative (investment to be discouraged), normal, or incentive (to be encouraged).

Source: World Bank (1987).

couraged a significant misallocation of investment resources.[18] Inflation, by contrast, has reduced the dispersion in METRS. To some extent, the declining real value of depreciation deductions has offset or diluted the impact of higher nominal interest expenses at higher rates of inflation.

In the area of corporate tax administration, the tax authorities have progressively moved toward shortening collection lags and using the corporation as an effective withholding agent for foreign remittances. Corporations are currently required to withhold 20 percent of foreign-paid royalties, 25 percent of all management fees paid abroad, and 5 percent of all foreign lease payments (but note that dividends paid to foreigners are not subject to any withholding tax). Until recently, corporations were required to make advance payments in the amount of half of the VAT collections they made. After protests from businesses and some criticism from the World Bank about the crudeness of this approach, the system was revised so that now corporations pay one-twelfth of one-half of the previous year's tax payments each month, although the Council of Ministers has the discretionary authority to raise this payment to 100 percent of last year's taxes. Finally, although Turkey does have legal restrictions on "above normal expenses" for interest and labor payments that are intended to prevent profit shifting, these restrictions have seldom been invoked.

Land and Property Taxes

During the debates on the financing of the initial development plan in the early 1960s, Kaldor (1962) proposed an income tax on the potential, rather than the actual, value of farm products as the best way of taxing farmers without harming their production incentives. The net product of land in every region was to have been determined from the agricultural output estimates of the National Institute of Statistics. This proposal boils down to a land value tax, because land's net product (that is, its rental value) determines its market value. The government rejected the proposal after vehement objections from farmers' groups.

A land value tax introduced after the abolition of the Islamic tithe in the 1920s was still collected in addition to the income tax, but its base had been seriously eroded after several decades of inflation. In 1960 the military government multiplied the tax base by ten to approximate the increase in farm product prices. However, the coalition government that came to power following the 1961 elections reduced this coefficient to three to address the double taxation complaint of farmers also subject to income tax. The differences in land values that had developed over the decades had magnified the horizontal inequities among assessed land values to the extent that the government preferred a correction that lagged far behind market values.

The reform package of 1970 merged the land tax and the buildings tax into a real estate tax. The major improvement lay in the principle of land value declaration by owners, subject to review by tax administrators. To make the tax "modern," the authorities adopted the principle of aggregation of the values of immovable properties as well as a progressive rate schedule, but later abandoned these features. Typically, land owners declared low values. During the military intermezzo of 1971–72, an amendment to the constitution stipulated that indemnification to owners when land was expropriated could not exceed the value owners declared for tax purposes. This measure was introduced as a self-enforcement device to induce owners to declare values closer to market values. The government and municipalities availed themselves of this stipulation to expropriate private land at low declared values for public service needs. As the rule was part of the constitution, the weak coalitions of the seventies could not amend it. However, interest groups subsequently succeeded in repealing this stipulation by means of a unique ruling of the Court of the Constitution. Finally, in the 1980s the real estate tax was abolished altogether on farm land.[19]

Until 1970 local bodies (special provincial administrations) collected the urban building and land tax. Without changing their local character, the reform package of 1960 tried to restore the eroded base of these taxes by multiplying their assessed values by inflation coefficients that roughly reflected the price changes since the last assessment. Moreover, the building tax was no longer to be creditable against the income tax collected from the rents of buildings; instead it could only be deducted as a cost. Responding to taxpayers' protests, the subsequent coalition government restored the rule for tax crediting in 1961.

Gradually, the provincial administrations in charge of collecting of land and building taxes weakened and surrendered most of their public service functions to the central government. They were held responsible for the lack of review of assessments and the poor collection rate for property taxes. In 1970 the central government took over the administration of these two taxes, hoping that it could squeeze out more revenue by revamping them with the rule of self-assessment, aggregation of property values, and progressive rates. However, the bulk of revenue (80 percent) was still earmarked for local bodies (municipalities and provincial governments).

This valuation reform at first increased revenues, but later met with growing resistance from interest groups engaged in land speculation around the cities. Speculative land purchases mushroomed as a hedge against inflation and as a convenient and lucrative investment at a time when the urban population was increasing by more than 5 percent per year, bank deposits were yielding negative interest rates, and the capital market was at best embryonic. In addition, they provided collateral for bank credits that were often extended at negative real interest rates. Banks also bought speculative land or ended up owning collateral land parcels as they liquidated their bad loans. A coalition of these vested-interest groups finally obtained the abolition of the constitutional rule of expropriation at self-assessed value. After this change, taxpayers started to declare lower-than-market values without any fear. The declared values were further eroded by inflation until the next self-assessment in five years. The central tax administration soon became unable to keep pace with the review of taxpayers' returns as the practice of underdeclaration became generalized. Gradually it lost interest in this exercise as its share of tax proceeds was only 20 percent, and the costs of collection could be as high as 30 percent of the gross revenue. Ultimately, in 1984 the administration of the urban property tax was again returned to local authorities, this time to the municipalities.

The tax reforms in the 1980s generally increased both the revenues and powers of local bodies. The return of the administration of the buildings and urban land tax to municipalities was part of this trend. With their built-in interest in an increase in urban property tax revenues, the municipalities were willing and able to increase the assessed values and collections quickly.

Apart from the building tax, the construction costs of certain types of urban infrastructure were also distributed among urban property owners according to legally determined benefit criteria. The penalty threat on underassessment was restored with the reintroduction (as part of the law, but not of the constitution) of the rule of expropriation of buildings and developed land at taxpayers' declared values.

To garner more revenue, governments have also converted certain fees into independent taxes. The first example of this practice was the transformation of fees on unrequited property acquisition into a gift and inheritance tax in 1927. The reform package introduced to finance the first development plan in 1963 also converted the fees on registering immovables acquired through purchase or construction into an independent real estate acquisition tax with much higher rates (7 percent on the sale value of immovables). This tax still exists at a high rate (8 percent) and causes a lock-in of real estate values. This is a serious drawback in a country where the financial sector is repressed and the stock market is undeveloped. The reform package of 1970 transformed the motor vehicle registration fee into a motor vehicle acquisition tax collected on the registration of a vehicle by the owner. The same package also transformed construction permit fees into an independent tax. In 1972 the financial balance tax replaced the compulsory savings program and was imposed on the same income bases at the same rate of 3 percent.

Reform of Indirect Taxes

Domestic Indirect Taxes

Turkey has a long history of some form of general sales taxation. In 1926, a few years after several European countries introduced a multiple-stage sales tax, Turkey introduced a similar tax to make up for the revenue lost from abolition of the Islamic tithe. As in France, the comprehensive multiple-stage sales tax was quickly converted into a cascaded manufacturers' sales tax. This tax was in place until 1957, but underwent several changes: exports and the wholesale stage were first included and then excluded; presumptive and then actual deduction of taxed inputs was permitted; and small producers were initially included, but later excluded, from the tax base. The conservative opposition party that came to power in 1950 had promised to get rid of the general sales tax, a major source of complaints by smaller producers despite the small-size exemption that was determined by such external indicators as the number of employees, amount of workshop space, and installed machine power. The authorities also narrowed the manufacturers' transaction tax base in 1950 by excluding food items, then abolished it altogether in 1956 and replaced it with a set of selective excise taxes on manufactured primary inputs.

The base of the so-called outlay tax (referred to as the production tax in table 9.3) was much narrower than that of the sales tax and cascading became inevitable. Manufactured primary inputs such as cotton yarns; steel, copper, and other metals; rubber; plastics; and energy constituted the bulk of the tax base. Further processing of these primary inputs was exempt. Rebates for input tax were provided on a product-by-product basis according to specified schedules that imputed input tax content. These rebates neglected to offset taxes on imported inputs and were paid only after long delays. The import of primary inputs and the finished products made out of them were also taxable, along with a restricted number of finished goods, foods, and drinks. To yield revenue comparable to that of the abolished sales tax on a narrower base, the authorities had increased rates considerably, and they often reached as high as 30 to 35 percent.

The impact on the private sector of the tax on primary inputs was substantial only in cotton yarn. Other affected products were largely the output of the public sector. Most public enterprises that produced products subject to the outlay tax were already operating in the red. The impact of the tax was regressive, as the effective rate declined on consumer goods with more processing and value added. Higher tax rates had also increased considerably the attractiveness and practice of tax evasion in the private sector.

After the abolition of the outlay tax on primary manufactures in 1985, its counterpart on financial services remained intact. The bank and insurance transactions tax (BITT) was a popular vehicle for revenue generation with various governments until the World Bank cited it as the main culprit behind the poor development of Turkey's financial sector. Its rate had risen from 10 to 15 percent in 1957, then to 20 percent with the reform package of 1963, and finally to 25 percent with the austerity measures introduced in 1970. Thus, in addition to the government's intervention in controlling interest rates, the escalation of the BITT further repressed the development of the financial sector. After the adoption of the VAT in 1985, the BITT continued to exist independently of the VAT, but its rate was gradually reduced to 3 percent by 1988.

The tax on transport services did not yield much revenue because private sector transport (passenger and cargo transports and intercity buses) was exempt and the taxation of public sector transport (seaborne, urban transit, and railways) made little sense because these services were provided at subsidized prices.

Over the years, all efforts to increase the buoyancy of the outlay tax failed. In the 1960s, as the manufacture of finished products progressed, an increasing number of items was added to the list of taxed products, such as electrical household appliances and car assembly. These changes, however, stopped short of transforming the outlay tax into a manufacturers' sales tax, and as the major growth sectors, such as light industries (textiles and garments, footwear, furniture, cosmetics) and construction materials, remained exempt, these measures failed to generate substantial revenues.

In 1970, as part of its austerity package, the government created a new tax on the retail sale of certain goods and services under the uninspiring title of the business tax. Thus the government jumped onto the retail sales stage after its retreat, fifteen years earlier, from the manufacturers' tax to a primary inputs tax. The implementation of the tax was generally poor because taxable items were selective, rates were variable, and retail outlets in small towns were exempt. Exemptions granted according to the size of agglomeration where the retail shop was located increased the possibilities for evasion; many firms set up branches in small towns to make their sales from these shops tax free. The business tax was the most poorly conceived of the numerous new taxes introduced in 1970.

The government recognized the desirability of adopting a VAT as early as 1963, when Turkey became an associate member of the European Community, although the agreement imposed no obligation for tax harmonization during the twenty-two-year transition period. However, serious discussion about the introduction of a VAT did not start until the 1970s, when debate about such a tax was widespread in universities and chambers of commerce. The government even sent a team to the Federal Republic of Germany to find out how to go about implementing a VAT. The government prepared several draft laws, but they were not even submitted to parliament because of the lack of agreement among coalition partners. It was not until Ozal's civilian government took power that the authorities prepared a VAT bill in 1984, passed it into law, and put it into effect expeditiously in early 1985, only four months after its passage.

The commencement of the VAT signaled the demise of eight other indirect taxes: the production tax on domestic production (the outlay tax), the production tax on imports (another part of the outlay tax), the business tax, the transportation tax, the sugar consumption tax, the services tax, the advertising services tax, and the sports lottery tax. Because these taxes had applied at only one level (production, wholesale or retail) and because most products were subject to one or more of these taxes, calculating the tax burden on any specific good or service was virtually impossible. As a result, effective tax rates across different consumption items varied widely and tax burdens across consumers were distributed haphazardly. The VAT promised to be much more neutral in both its coverage and incidence.

Table 9.5 attempts to measure the incidence of this earlier cluster of indirect taxes. Not surprisingly, in light of the wide range of consumer goods subject to one tax or another, urban and rural households shared a similar tax burden, although urban households were slightly worse off. On the basis of the size of income, these taxes exhibited a mainly proportional distribution of tax burdens because of the taxation of numerous services and the exclusion of most basic foodstuffs from the tax base. Nonetheless, the indirect tax burden borne by low-income households was relatively high at 13 percent of total income.

Because the government had collected substantial revenue from the production taxes on alcoholic beverages and tobacco products, a supplementary VAT, which was in reality a new set of excise taxes, was ushered onto the tax stage at the same time as the VAT appeared. Most of these excise taxes, whose rates range from 50 percent for tobacco and alcohol to 15 percent for beer, wine, and soft drinks, are collected from the state monopoly TEKEL. In 1985 they generated revenues that were a little more than one-fifth of the amount the VAT yielded from the domestic base.

Implementation of the VAT has been a success on the grounds of both revenue generation and incentives. Accumulated experience with the implementation of the manufacturers' sales tax, the outlay tax, and the business tax were decisive factors in the tax administration's quick and successful adoption of the VAT. Also, the abolition of eight old and somewhat cumbersome taxes and their replacement by the VAT released administrative resources for implementing the VAT. Moreover, the transfer of the administration of real estate taxes to municipalities in 1984 released additional manpower to the central tax administration. The share of the eight taxes the VAT replaced represented 13.3 percent of all tax revenues in the fiscal year before their abolition. The VAT almost doubled that share in its first year of implementation and reached 28.4 percent of all tax revenues in its third year. The World Bank (1987) has estimated that the domestic VAT base encompasses about 60 percent of GDP, a relatively high percentage by international standards.

Initially, the adoption of a single VAT rate of 10 percent on all sales greatly simplified both tax collectors' and consumers' understanding of the tax. Subsequently, the government gave concessions to pressure groups that destroyed the rate uniformity. In 1986 the government raised the standard VAT rate to 12 percent and applied lower rates of 5 percent to medicines, books, and newspapers, and 1 percent to agriculture. Basic foodstuffs were zero rated, as were exports. Turkey also uses a buffer rule when input tax credits exceed gross VAT obligations to deter firms from underreporting their sales and claiming bogus refunds. Under this rule, excess input credits can only be carried forward. By 1988 the basic rate remained intact, but basic foods were now taxed at 3 percent; natural gas deliveries at 5 percent; the special rate for medicines, books, and newspapers was reduced to zero; a new luxury rate of 15 percent was applied to purchases of cars, jewelry, furs, and the like; and the supplementary VAT on tobacco products and alcoholic beverages was raised to 100 percent. Greater rate dispersion and a larger number of rates is the political price that Turkey has paid for increasing the basic rate from 10 to 12 percent.

Basic foodstuffs are taxed at a low or zero rate with a refund of the VAT paid on inputs. The practice of providing tax refunds to income tax payers on the presentation of receipts for purchases of consumer goods reduces the need to lower the VAT rates on essential consumption items. The government, moreover, does not seem to have drawn a link between the policy of giving tax refunds to low-income taxpayers and the level of the VAT on basic necessities. Instead, the government has viewed the rebates as a substitute for personal exemptions under the income tax.

VAT administrators are confident that taxpayers have not used fraudulent invoices to reduce their income tax payments under the expenditure rebate system. Government printers supply the sales documents that vendors must use and the number assigned to each vendor should correspond to that on receipts submitted by a purchaser. Random checks are made on the degree of correspondence, and thus far no problem has apparently arisen. A more serious concern is that too much bargaining over receipts may be going on between seller and buyer. For example, a seller may offer to reduce a price from LT10,000 to LT8,000 if the buyer forgoes a receipt, in which case the buyer realizes a net gain of LT1,000 from the price reduction of LT2,000 less the LT1,000 of expenditure rebate foregone. Tax inspectors currently pose as customers, and in cases where vendors do not issue receipts, the firm is either fined or closed down. The recent exclusion of consumer durables from the expenditure rebate system has, however, diluted the incentive for collusion between buyers and sellers.

Farmers who are exempt from income tax are also exempted from the VAT. This implies that exempt farmers absorb the VAT paid on agricultural inputs as a cost item. The exemption of most farmers has proven to be a source of weakness in the effective coverage of the food processing industries. The tax administration's logistical constraints prevent more intensive checks to detect untaxed shipments.

The extension of the VAT to the retail stage has posed a huge challenge for the tax administration because of the presence of numerous small retailers. However, the administration has capitalized on its experience with small traders under the income tax. Retailers that are exempt from income tax have also been exempted from the ordinary VAT, but are burdened on a presumptive basis through a compensatory VAT rate mark-up on their purchases of merchandise. Currently, small retailers, who are also lump-sum income tax payers, apply a 20 percent uplift to the value of their input purchases and pay that amount less any input tax credit as their VAT obligation.[20] The amount remitted to the VAT authorities also cannot be less than 10 percent of small retailers' lump-sum income. Retailers that are subject to the VAT are compelled to register their sales by compulsory use of tamper-proof electronic cash registers and by the expenditure rebate practice of refunding tax to income tax payers on the presentation of retail invoices.

A major flaw of the Turkish VAT is the partial inclusion of capital goods into the tax base by the denial of an immediate tax credit for investments that are not encouraged. This procedure converts the tax into a GNP type of VAT instead of a consumption type of VAT. The taxation of capital equipment is somewhat attenuated by the rule of rebating the tax in installments spread over three years (originally five years). The World Bank (1987) estimated that at current interest rates, this deferral of credit was equivalent to an investment tax (a negative tax credit) of 3.6 percent. Construction activity, whether commercial or for housing, is also subject to the VAT. However, an exemption for residences smaller than 150 square meters was introduced that was to remain in effect until 1988, but was later extended to 1992.

The coverage of services by the VAT is quite extensive. This is a considerable achievement considering the long delays involved in extending the VAT to services in the European Community countries. As a standard practice in the European Community, bank and insurance services are excluded from the VAT base, whereas in Turkey the old BITT is still applicable to financial services. As this tax is not creditable under the VAT, interest firms pay on bank loans is a cost item and constitutes part of their VAT base. Thus both the VAT and the BITT strike the same base. By itself the BITT increases the cost of financial services and induces firms to gain direct recourse to household savings by issuing bonds and shares. However, in the 1980s the rate of the BITT was substantially reduced (to 3 percent) in an effort to lower the cost of credit and expand interbank transactions and financial intermediation.

Taxation of Foreign Trade

During the first three decades of the postwar period, Turkey's revenues from foreign trade taxes depended more on the foreign trade regime and the value of the foreign exchange rate than on customs duty rates. During periods of overvalued exchange rates, revenue from customs stagnated, or even declined, as shrinking export earnings reduced import capacity and import values failed to keep pace with inflation. As imports shrank because of foreign exchange shortages, the government raised additional taxes under different names or requested advance deposits for granting import licenses at the official foreign exchange rate to soak up part of the importers' rents. However, these practices failed to restore the buoyancy of import taxes because they were poor substitutes for correcting the overvaluation of the lira.

Overvalued exchange rates depressed export prices and diverted export goods to the domestic market. Some goods such as livestock became illegal exports. However, certain export goods could not be easily diverted to the domestic market in the short run (for instance, crops with long gestation periods or minerals) and the domestic market could not absorb them without a price collapse. Thus overvalued exchange rates operated like a hidden tax on producers of exportables without yielding any revenue to the government and transferred rents to holders of import licenses. The government shared these rents by raising additional taxes on imports as the foreign exchange shortage became acute. The stamp duty on imports, first introduced in 1964, was subsequently maintained as a rent absorber on imports. Illegal export profits and the foreign currency earnings of Turkish workers abroad financed illegal imports, which grew in importance during periods of foreign exchange shortages and further eroded the tax revenue from imports.

The first foreign exchange crisis of the postwar period occurred in the 1950s. At the end of the Korean War export boom, the ensuing exchange crisis was resolved through a stand-by agreement with the IMF in 1958. During the following two decades, Turkey continued to adhere to a fixed exchange rate regime, one with frequently overvalued exchange rates. The government financed the resulting foreign trade deficit first with foreign aid, and later, in the 1960s, with workers' remittances. Foreign aid helped to close both the public sector deficit and the foreign trade deficit, but workers' remittances only closed the foreign trade deficit without generating any revenue for the government. The foreign exchange crisis that developed toward the end of the 1960s was again resolved through a stand-by agreement with the IMF plus a tax reform package to increase government revenue that has already been described.

In the 1970s the foreign exchange crisis took less time than before to pressure the government into action because of the concomitant oil price shock. The servicing of short-term external debts owed to commercial banks at high interest rates, incurred to finance oil imports, severely burdened the country's import capacity. Before the financing of the first oil shock had been completed, the second oil shock set in and ravaged the economy. To overcome the foreign trade crises caused by the oil shocks, during 1978–81 three different governments introduced a total of four austerity policy packages and concluded three stand-by agreements with the IMF and three debt rescheduling agreements with donor countries and banks. After this prolonged crisis, the authorities introduced and maintained a policy of more realistic (if not flexible) exchange rates. The oil shocks taught Turkish governments a vital lesson: the importance of avoiding overvalued exchange rates in the pursuit of economic stability.

Since the second half of the 1960s, along with the policies that repressed foreign trade, import duty exemptions granted under the investment encouragement policy help explain the low buoyancy of import duties. Equipment imports were duty free and assembly industries enjoyed low or zero duty rates on their imports of inputs. As Turkey's industrial base developed, domestic producers of machinery and equipment started to exert pressure on the government for higher protection of their value added. Despite adjustments in customs tariffs to obtain more protection for the domestic equipment industry, the SPO continues to grant encouragement certificates that secure duty-free equipment imports for new investment projects.[21]

Turkey's association with the European Community prompted a gradual reduction of customs tariff on goods imported from member countries. After an initial 10 percent tariff reduction, however, Turkey could not make any further reductions after the onset of the first oil crisis. Since then, Turkey has not entered into a tariff reduction agreement with the European Community, although it applied for full membership in 1988.

After 1980, exports were promoted not only by continuously adjusting foreign exchange rates, but also by the extending low interest credit and other subsidies paid in the guise of a tax rebate. When the first development plan was put into effect, a lump-sum tax rebate was granted to exports. Before this measure, domestic taxes that went into the cost of an export product were refunded on submission of invoices showing the paid taxes. This procedure involved delays and was unable to guarantee the refund of taxes the suppliers of the exporter paid. Refunds of exporters' tax payments could not be verified without long delays. To expedite the refund, the authorities abandoned the rule of allowing tax refunds on actual payments (invoices) and adopted a presumptive lump-sum tax refund instead. A government committee estimated the average tax costs on the sale prices of seven commodity groups. Rebate rates were based on assumptions about the cost incidence of direct and indirect taxes. They were also arbitrary and high. Tax rebates were used later on as a substitute for devaluation when overvaluation of exchange rates eroded the competitiveness of Turkish exports.

After 1980 the government increased export rebates and granted additional tax rebates to exporters above a given turnover ceiling. The latter stipulation induced small producers to use the umbrella of big exporters to share the subsidy. Attractive rebates triggered the growth of fictitious exports, particularly in high-value items such as silk carpets, fashionable leather garments, embroidery, gold wire, crustaceans, and expensive mushrooms. The central bank paid tax rebates without any evidence from the tax administration about tax payment and based only on documents that showed that the goods had passed through the customs line and that foreign currency had been transferred to a Turkish bank.

Fictitious or not, export rebates represented about 20 percent of the value of commodity exports. This was a major factor in inflationary monetary expansion, because the central bank did not receive any budget allocation for its export rebate payments. Lump-sum rebates paid by the central bank amounted to a multiple exchange rate practice, but unlike the common multiple exchange rate practice, the central bank could not recoup tax rebate payments by selling foreign currency to importers for more than the basic rate. With the introduction of the VAT, a separate and additional lump-sum tax rebate can no longer be justified, because the VAT has its own built-in rebate system with zero rating of exports. However, the refunds under the VAT are granted after an audit of exports, which implies delays, and occasionally induces corrupt practices intended to accelerate the procedure. After scandalous and widespread abuses, the government was compelled to scale down lump-sum tax rebates by the central bank, and finally began to phase them out in 1989.

In the 1980s the government made major progress in liberalizing foreign trade. It removed quantitative restrictions, and in 1984 replaced the positive list of freely importable goods with a negative list that only included prohibited imports. After 1984 statutory tariff rates were also reformed and reduced. Currently, import duties on basic materials are subject to the lowest rates of 0 to 15 percent, intermediate goods have duties ranging from 10 to 30 percent, while those on finished products range between 30 and 40 percent. Furthermore, average import duties declined in the 1980s, but contrary to the liberalization policy, the widespread use of levies tied to extrabudgetary funds (EBFS) increased the overall average tariff rate on imports. With EBFS the tariff rates on some commodities exceeded 100 percent and the dispersion in tariff rates rose noticeably. During the 1980s governments increasingly resorted to discretionary trade taxes to finance the EBFS. While revenues from import duties declined, EBF tax revenues on foreign trade increased and approached the level of nonoil import taxes.

In contrast to the policy of general encouragement of exports, the government has taxed certain agricultural exports heavily. Among the products that have been subject to export taxes, such as cotton and hazelnuts, only the tax on hazelnuts could be shifted, to some extent, in world markets because of Turkey's near monopoly position. Even this advantage should not be overestimated, because a prolonged monopolistic exploitation could eventually invite substitutes into the market.

Extrabudgetary Funds

Originally, the government created some EBFS to lend money at low interest rates to activities it wished to encourage. It had left the management of funds allocated to these activities to banks that, after examining the candidates' eligibility, made these preferred loans. The funds were created because loans could not be efficiently handled under existing budgetary rules for spending. In the 1980s EBFS were increasingly used to bypass normal budgetary procedures in scrutinizing current public expenditures. By 1990 some 125 funds promoted particular sectors, investments, and exports, as well as transfers to the poor and cheap housing credit.

Revenues earmarked for the EBFS are now substantial, and represent a major departure from efficient resource mobilization, that is, the EBFS seriously comprise neutrality with regard to relative prices and equity in the distribution of the tax burden. Half of the funding for EBFS is raised through taxation; the other half comes from budget allocations, external and internal borrowing, deposits, penalties, and the operating income of infrastructure facilities.

Taxes earmarked for EBFS are surtaxes on existing taxes and most are based on foreign trade. They have taken in twice the revenue of customs duties in recent years and represent additional distortions to the patterns of effective protection arising from normal customs duties. Import duties that are earmarked for EBFS are not proportional to the value of imports, but tend to be lump-sum dollar amounts on physical units of imports. Effective protection secured by the lump-sum EBF taxes is higher for low-quality (and low-value) products and lower for high-value products, because the same levy applies to a given quantity of product irrespective of its value. Because the government can change the EBF taxes by decree at any time, they are a source of uncertainty regarding the effective protection given to domestic producers. Entrepreneurs frequently complain about the unpredictable policy changes that offer golden opportunities to possessors of inside information.

Extrabudgetary funds have been poorly coordinated and have showed a propensity to work at cross-purposes with each other and with other government policies. For example, the government raised taxes on trade while simultaneously pursing a broad strategy for export promotion. Similarly, one fund taxed agricultural exports at the same rate as another fund subsidized those exports. In addition to the taxes on foreign trade, some EBFS rely on an additional 3 percent defense surtax on personal and corporate income taxes. Excise taxes on certain luxury products (alcoholic beverages and cigarettes) and on petroleum products are major contributors to EBFS. EBFS appear to represent a triumph of politics over economics.

An Assessment

Measured against the traditional public finance trinity of efficiency, equity, and ease of administration, tax reforms in Turkey during the past forty years play to mixed reviews.

In comparison with 1950–80, when the tax system degenerated amidst inflationary chaos and severe tax and revenue imbalances occurred, the 1980s witnessed modest progress toward a broader-based and more neutral tax system as minimum taxes were imposed on both individuals and corporations, top bracket personal rates declined, state enterprises were included in the corporate tax base, export subsidies were introduced, discriminatory taxation of financial institutions was drastically reduced, and withholding rates on lightly taxed agriculture were raised. Nevertheless, the relatively much lighter combined corporate and personal taxation of capital income in comparison to labor income meant that Turkey's direct tax system was actually much closer in its operation to an expenditure tax than to a true income tax. Turkey's liberalization efforts and the associated relaxation of foreign exchange controls have contributed to this trend by making it much more difficult administratively to tax the components of capital income that are received at the personal level.

On the income distribution side, Turkey's current tax system is unlikely to make more than a modest contribution toward a progressive pattern of tax burdens. What is more likely is that substantial horizontal tax inequities characterize the system as a whole. For direct taxes, while poor rural workers are kept outside the tax net and there is probably a modest degree of progression in the tax burdens facing urban workers, higher-income, self-employed taxpayers, who are not subject to withholding pay, still average no more than the first bracket tax rate of 25 percent, and sometimes substantially less than that. Nevertheless, the direct tax system is undoubtedly much better today than it was during the inflationary turbulence of the late 1970s and early 1980s. During that era, bracket creep shifted more and more of the income tax burden to lower-income workers, and vertical and horizontal inequities became increasingly rampant because of the proliferation of fringe benefits and evasion among those earning business and professional incomes.[22] It would also be more than mildly surprising if the current indirect tax system had a noticeable redistributive impact. The best that one could hope for is that the zero rating of food under the VAT removes the main regressive feature of the tax and that the supplementary VAT on luxury consumption imposes a somewhat greater tax burden on higher-income households.

If the distributive effects of these reforms are difficult to ascertain, any improvements in economic efficiency resulting from them are even harder to pin down, because so many significant nontax distortions abound within the Turkish economy. The VAT is no doubt a significant improvement over the motley array of taxes that it replaced, but it, too, is inherently flawed as long as investment purchases remain effectively taxed. While the reduced reliance on trade taxes is also to be applauded, the rapid growth of EBFS has been diligently working in the opposite direction. Similarly, no easy judgment can be offered on the allocative impact of the incentive-based corporate and personal income tax. No doubt, the incentive structure has effectively compensated for some of the tariff and spatial distortions that would otherwise produce an inefficient pattern of investment. However, these taxes have seriously eroded the tax base, and the enormous variation in METRS that these incentives generate creates considerable concern about the quality and overall productivity of investment in Turkey. Moreover, as the CIT system moves closer to expensing, the continued allowance of interest deductibility increases the likelihood that the tax system may inadvertently be drifting toward subsidizing of investment at the margin.

What seems clearer, however, is that in recent years Turkey has taken important strides to upgrade and enhance the capacity of its tax administration. Some of the major accomplishments in this area are more resources; better training; improved flows of information; and greater emphasis on control, collection, and auditing. Presumptive taxation on agriculture and high-income groups has also been effectively applied. By curbing tax evasion, mainly on the part of the rich, and by reducing the degree of differential tax evasion across activities, these measures may have contributed to making the tax system both more equitable and more efficient.

Since the perilous decline in tax effort in 1984, the revenue system has rebounded and the new VAT has more than replaced the revenues that the earlier indirect, antiquated tax system had collected. Nonetheless, a sizable fiscal deficit remains and will likely continue to fuel inflation unless the government can squeeze more revenue from the current system. In this connection, while inflation had dipped to a relatively low rate of about 30 percent in the mid-1980s, toward the end of the decade the rate had climbed back to 65 to 70 percent and the perennial problem of bracket creep had begun to reappear as a serious flaw in the tax system. Although the tax effort has rebounded from 15.6 to 16.8 percent of GDP during 1987–89, the deficits of state-owned enterprises grew even faster because of the failure to allow the regulated prices of these enterprises to keep pace with the rate of inflation. Monetization of these deficits has recently produced an even higher rate of inflation. This appears to be an instance where the lack of indexation, only this time on the expenditure rather than the tax side of the budget, has fostered macroeconomic instability.

All in all, perhaps a fair judgment of Turkey's tax reform progress to date is that at a minimum, Turkey has stopped making matters worse. It is perhaps too early to judge whether matters have become significantly better. While there have been some unambiguous improvements in the tax system, particularly the adoption of the VAT, so many profound problems still exist that serious doubts remain about whether Turkey has progressed very far along the road to substantive tax reform. The road not yet traveled encompasses indexing the PIT; restoring personal exemptions; imposing more demanding levels of presumptive taxation; extending corporate indexation to the liability side of the balance sheet and to inventories; streamlining a tangled thicket of investment-related tax incentives, eliminating EBFS, or at least greatly diminishing their role; and reducing reliance on agricultural export taxes.

Lessons

While an overall assessment of Turkey's tax reforms remains problematic, a number of significant lessons can be drawn from Turkey's tax reform experience:

1. *In stimulating a useful flow of information to tax administrators, the fiscal carrot may produce more satisfactory results than the fiscal stick.* Turkey

had relatively little success in trying to compel the use of adequate documentation by levying fines. The expenditure rebate system, which rewards the solicitation of documents, has worked much better.

2. *The failure to index the personal income tax and the resultant bracket creep* encourages widespread tax evasion and horizontal inequity through the inducement it provides to pay workers in the form of untaxed fringe benefits.

3. *Inflation is an overall sign of fiscal weakness and produces regressive effects on income distribution* by promoting bracket creep and larger tax burdens on the poor, and also by reducing the value of wealth taxes and various presumptive taxes that are paid predominantly by the rich.

4. *Governments without a solid political base* cannot carry out fundamental tax reforms involving base broadening and lowering rates.

5. *Unconditional foreign aid may relax any latent pressures for tax reform,* while conditional lending may, by contrast, be an important stimulus to tax reform.

6. *Politically weak governments are likely to produce internally inconsistent revenue systems* in which different parts of the tax system work severely at cross purposes with each other. In Turkey export incentives and import taxes have coexisted in harmony for many years. EBFS have sprung up in recent years to reimpose the tariff protection that had been systematically dismantled in previous years.

7. *A subsidy granted to one group or sector in the economy can become generalized* through the political process. In Turkey, a reasonably well targeted expenditure rebate system was quickly stretched out of shape to include virtually all taxpayers and all components of consumption.

8. *The political price for imposing a higher VAT rate* may be the acceptance of a multiple rate VAT system.

Notes

1. Part of the difficulty in interpreting the available data is that such data neglect some components of Turkey's four-sector public sector: the central government, provincial and municipal governments, state-owned enterprises, and extrabudgetary funds. Frequently, information is readily available only on the activities of the central government.

2. The World Bank (1987) study reveals the tricky nature of measuring fiscal deficits. While the public sector as a whole is currently running a noninterest current account surplus, it barely runs a noninterest (primary) surplus and is in substantial deficit when interest payments on public debt are considered. Turkey has been financing its net external resource transfer through greater monetization of public debt. The Bank study also favors a stock approach to measuring the fiscal deficit, looking at annual changes in the stock of net public sector liabilities, because of the statistical problems inherent in the use of flow-related income and expenditure accounts.

3. A World Bank (1987) report on tax reform indicated that in 1986 annual filers accounted for 25.0 percent of PIT revenue while those assessed under the lump-sum system contributed only 3.5 percent. Moreover, that same year, 18.0 percent of all personal income taxpayers filed an annual return and 15.0 percent were subject to a lump-sum assessment.

4. Presumptive costs are estimated by committees in each province that represent the tax administration and farmers' associations, subject to approval by a central committee at the national level. As presumptive costs include the value of farmers' own labor, this labor market activity is automatically excluded from the tax base.

5. The inflation tax is calculated as the value of the real change in the base money, or required reserves, of the financial system. This procedure nets out the redistribution occurring between borrower and lender and captures the extent to which households, on balance, have to depress their spending to maintain the real value of their monetary balances.

6. Using the numbers in the text, the real after-tax return would become 0.5 (1.0 – 0.2) – 0.4 = 0 .

7. Using the numbers in the text, indexing the real after-tax return would become (1.0 – 0.2) (0.5 – 0.4) = 0.08.

8. For example, if a high-income purchaser forgoes a receipt on a item worth $1.00 that is taxable at 12 percent this represents a sacrifice of a $0.05 rebate. Without a receipt he would gain if the vendor agreed to sell to him at a price of $1.06, and the seller pocketed the extra $0.06.

9. Note that the Council of Ministers has the discretionary authority to raise the withholding rate to 100 percent.

10. During 1985–88 the number of local tax offices swelled from 25,000 to 40,000 and nearly three-quarters of these offices were fully computerized.

11. Hirsch and Hirsch (1966) emphasized that Turkey has historically levied taxes on the basis of the ability to collect. They also expressed some alarm that during 1955–60 average taxable income declared by taxpayers was unchanged, while nominal nonagricultural income doubled.

12. However, in 1964 the government obtained discretionary authority to exempt businesses owned by foundations from all taxes, including the CIT. A number of foundations have availed themselves of this privilege.

13. On a dollar of corporate income the rate is calculated as 0.1 + 0.2 (1.0 – 0.1) = 0.28.

14. On a dollar of corporate income the rate is calculated as 0.25 + 0.2 (1.0 – 0.25) = 0.40.

15. Given the numerical example in the text, with indexing the real after-tax return would become (1.0 – 0.2) (0.5 – 0.4) = 0.8.

16. In this case, the absence of indexation may be destabilizing because it has contributed to larger fiscal deficits and higher rates of inflation. The World Bank (1987) study estimated that allowing only real interest deductibility would have generated additional corporate tax collections in 1985 to the tune of 3.6 percent of GNP, roughly the same amount as the tax actually collected.

17. Note, however, that the effective absence of a capital gains tax and the personal dividend exemption achieve financial neutrality in the choice between retained earnings and new issues of equity.

18. However, many of the incentives are intended to combat the congestion externalities associated with investments in large cities such as Istanbul, so that a wider social accounting framework is required to make convincing statements about the efficiency of resource allocation.

19. In 1960 the land tax was deductible in the determination of farmers' taxable income. From 1963–80 the land tax could be credited against the income tax paid on income from the same land.

20. Turkey operated a similar compensatory system in 1985, but shelved it in 1986 in an effort to apply the tax uniformly to all sellers and prevent retailers from collecting the tax and keeping it for themselves. The administrative burden of this decision forced a later reversal of policy.

21. To restore incentives for domestic capital goods, a subsidy of 15 percent of their purchase price was introduced in 1988 and raised to 20 percent in 1988 and 25 percent in 1989. However, the subsidy was eliminated at the end of 1989.

22. As inflation ratcheted upward in the late 1980s, this pattern of troublesome bracket creep appeared to be repeating itself.

References

Erbas, S. Nuri. 1988. "Effects of Inflationary Finance on Tax Revenue under Progressive Tax Regimes: An Application to Turkey, 1963–82." *Socioeconomic Planning Sciences* 22(6): 277–85.

Hirsch, Eva, and Abraham Hirsch. 1966. "Tax Reform and the Burden of Taxation in Turkey." *Public Finance* 31(3).

Kaldor, Nicholas. 1962. "Report on the Turkish Tax System." Confidential report prepared for the State Planning Organization. Prime Ministry of the Republic of Turkey, Ankara.

Krzyzaniak, Marian, and Sleyman Özmucur. 1973. "The Distribution of Income and the Short-Run Burden of Taxes in Turkey, 1968." *Finanzarchiv* 32(1).

———. 1974. "The Short-Run Burden of Taxes on Turkish Agriculture in the Sixties." Paper No. 52. Program of Development Studies, William Marsh Rice University, Houston.

Shoup, Carl S. 1972. "Tax Reform Planning in Turkey: A Report to the UNDP." Preliminary draft (translated from Turkish). United Nations Development Programme, New York.

World Bank. 1987. "Turkey: Fiscal Policy and Tax Reform: Issues from the Past and Options for the Future." Report No. 6374 (restricted access). World Bank, Europe and Central Asia Region, Washington, D.C.